Oxidative Stress in the Newborn

Oxidative Stress in the Newborn

Guest Editor

Ru-Jeng Teng

MDPI

Basel • Beijing • Wuhan • Barcelona • Belgrade • Novi Sad • Cluj • Manchester

Guest Editor
Ru-Jeng Teng
Department of Pediatrics
Medical College of Wisconsin
Milwaukee
USA

Editorial Office
MDPI AG
Grosspeteranlage 5
4052 Basel, Switzerland

This is a reprint of the Special Issue, published open access by the journal *Antioxidants* (ISSN 2076-3921), freely accessible at: https://www.mdpi.com/journal/antioxidants/special_issues/1UFD0PA1MT.

For citation purposes, cite each article independently as indicated on the article page online and as indicated below:

Lastname, A.A.; Lastname, B.B. Article Title. *Journal Name* **Year**, *Volume Number*, Page Range.

ISBN 978-3-7258-7450-7 (Hbk)
ISBN 978-3-7258-7451-4 (PDF)
https://doi.org/10.3390/books978-3-7258-7451-4

Contents

About the Editor

Ru-Jeng Teng

Ru-Jeng Teng, M.D., is a Professor of Pediatrics in the Division of Neonatology within the Department of Pediatrics at the Medical College of Wisconsin. He graduated from the National Taiwan University Medical College and trained at the University of Alabama at Birmingham. He is passionate about teaching students and conducting bench research. He has won Best Teacher Awards at the National Taiwan University Medical College and the University of Florida. His research subjects include innate inflammation, free radical biology, vascular biology, bronchopulmonary dysplasia, programmed cell death and senescence, mitochondrial biology, autophagy, unfolded protein response, AMP kinase signaling, LKB1 signaling pathway, and nephrogenesis.

Preface

Antioxidant treatment has long been considered beneficial because oxidative stress (OS) is thought to damage tissues and organs. Recent research has clarified its essential role in organ patterning and growth in utero. This may explain why antioxidant treatments have inconsistent results in preventing prematurity-related health problems. Exposure to harmful factors like placental insufficiency, nicotine, air pollution, and cocaine, among others, during pregnancy can easily disturb the natural balance, leading to fetal organ damage. The sudden surge in OS during the perinatal transition is especially challenging for neonates, particularly those born prematurely with underdeveloped antioxidative defenses. Respiratory support, such as supplemental oxygen and positive-pressure airway support, can further increase OS. Unstable blood flow in sick neonates worsens OS through ischemic-reperfusion injury. Proper nutrition and careful management of tissue oxygenation are vital for helping these vulnerable infants. Since breast milk is the best nutrition for neonates, the mother's diet is particularly important because it affects breast milk composition. Neonates develop their gut microbiomes from their mothers, and these microbiomes significantly influence the neonate's OS status; we can expect maternal gut microbiomes to affect health outcomes, making them valuable for extensive research. OS and innate inflammation activate each other, which explains why inflammation often occurs in many OS-related neonatal disorders. This is one reason systemic corticosteroids were once routinely used to treat premature neonates with chronic lung issues. Addressing OS alone may not be sufficient to treat prematurity-related conditions. Therefore, a new systems pharmacotherapy approach is proposed. Melatonin, which has both antioxidant and anti-inflammatory properties, has been studied for its potential effects on multiple therapeutic targets. Despite many successes in treating OS-related neonatal problems, researchers face the challenge that most common animal models are born with mature antioxidative systems. This makes it hard to accurately mimic prematurity-related health issues. Advances like organoid systems and bioinformatics may assist in researching these problems, but only time will reveal their true value.

Ru-Jeng Teng
Guest Editor

antioxidants MDPI

Editorial

Oxidative Stress Is a Double-Edged Sword for the Neonate

Ru-Jeng Teng [1,2]

[1] Department of Pediatrics, Medical College of Wisconsin, Children Corporate Center, Suite C410, 999N 92nd Street, Milwaukee, WI 53226, USA; rteng@mcw.edu

[2] Children's Research Institute, Medical College of Wisconsin, 8701 W Watertown Plank Rd., Wauwatosa, WI 53226, USA

Fetal oxygen tension is low, with an umbilical venous pO_2 of 30–37 mm Hg in the third trimester, while arterial pO_2 is around 20 mm Hg. Placental PaO_2 can be as low as 18 mm Hg at 8–10 weeks of gestation. These values are relatively lower than those in ambient air, which is approximately 150 mm Hg. This "physiological hypoxia" is necessary for placental growth and tissue differentiation because it stabilizes hypoxia-inducible factors (HIF-1 and HIF-2) [1]. Acting as master regulators, HIFs promote vascular endothelial cell growth by upregulating glycolysis and inducing vascular endothelial growth factor (VEGF) expression [2]. This process is coordinated by reactive oxygen species (ROS) produced during "physiological hypoxia." The right amount of ROS is vital for developmental senescence, which is essential for organ patterning during fetal development [3]. Therefore, an optimal level of ROS is indispensable during fetal growth.

The sudden increase in oxygen during the perinatal transition, with pO_2 rising sharply from below 40 mm Hg to 150 mm Hg, can effectively turn off this HIF pathway. However, if neonates require supplemental oxygen for extended periods due to respiratory distress, oxidative stress (OS) can sequentially activate HIF1α and HIF2α. Activation of HIF1α during the initial phase is proinflammatory, while subsequent HIF2α activation supports angiogenesis and organ repair [4]. Persistent OS may cause organ damage by impairing mitochondrial function [5], triggering the unfolded protein response [6], autophagy [7], oxidation-induced senescence [8], and cell death pathways such as necrosis, necroptosis, pyroptosis, ferroptosis, and others [9], along with fibrosis [10]. The damage worsens if prematurity is involved, as premature neonates are born with surfactant deficiency and an underdeveloped antioxidative system. The oxygen supplementation and positive airway pressure required to maintain tissue oxygenation can further increase ROS production in premature infants, potentially overwhelming their immature antioxidant defenses.

Excessive ROS in sick premature neonates can damage cellular DNA, especially in rapidly dividing stem and progenitor cells. OS-induced DNA damage can lead to cell death or senescence, depriving immature organs of growth potential by depleting these progenitor cells [11]. At the same time, dead or dying cells release high-mobility group box 1 (HMGB1), which triggers innate inflammation [12]. This inflammation, mediated by cells containing inducible nitric oxide synthase and myeloperoxidase (MPO), releases MPO and NO, creating highly reactive oxidants such as hypochlorous acid (HOCl) and peroxynitrite ($ONOO^-$) that damage immature organs [13]. Thus, inflammation amplifies OS. The interaction between OS and inflammation [14] creates a self-sustaining, destructive cycle in neonatal organs under respiratory support.

HMGB1 is a non-histone nuclear DNA-binding protein that is released from the cell upon injury. Once outside, HMGB1 acts as a danger signal, activating damage-associated molecular pattern (DAMP) signaling with proinflammatory effects [15]. Its receptors include the Receptor for Advanced Glycation End Products (RAGE) and Toll-like receptors (TLRs). When HMGB1 binds to these receptors, it activates NF-kB signaling, leading to macrophage and neutrophil activation. HMGB1 contains three sulfhydryl groups, and its effects depend on its redox state. Fully reduced HMGB1 acts as a chemoattractant, the disulfide form stimulates inflammatory cells, and the fully oxidized form is anti-inflammatory. Thus, HMGB1 is a key link between OS and inflammation.

Prematurity and intrauterine growth restriction predispose to OS and inflammation even before birth [14]. Paradoxically, the heightened OS upregulates the HIF pathway through ROS production. As a result, VEGF expression increases to maintain tissue oxygenation, but it can also lead to disorganized pathological angiogenesis (dysangiogenesis). This abnormal vessel growth is associated with several prematurity-related conditions, such as bronchopulmonary dysplasia (BPD) and retinopathy of prematurity (ROP), including the pulmonary hypertension seen in BPD [16]. Persistent activation of HIF signaling also promotes lung fibrosis via endothelial-to-mesenchymal transition [17]. Elevated HIF1α is also observed during the second ischemic phase of ROP. HIF1 can further boost insulin-like growth factor-1 (IGF-1) signaling by increasing IGF1 binding protein levels [18]. Interestingly, hyperoxia correlates with decreased IGF1, which strongly links to ROP development in meta-analyses [19]. These conflicting findings highlight the complexity of OS's influence on HIF signaling in neonatal organs.

For a long time, OS has been considered the main cause of all prematurity-related complications. Over the past thirty years, many antioxidants have been studied, including N-acetylcysteine, vitamin C, vitamin E, superoxide dismutase, melatonin, selenium, and lactoferrin. Unfortunately, consistent clinical benefits have not been established. Although intramuscular vitamin A has been shown to reduce BPD [20], it is not an antioxidant itself. Several reasons may explain why broad-spectrum antioxidants do not work effectively: (1) basal OS is necessary for developmental senescence, which is vital for organ patterning and growth; (2) the timing of HIF1α and HIF2α changes is critical for organ recovery after injury; (3) OS plays complex roles across different organs simultaneously. Determining the precise level of OS that promotes premature organ growth while minimizing damage remains very challenging. Additionally, the interaction between OS and inflammation must be considered. A systems pharmacological approach targeting multiple signaling pathways might be the most effective strategy for managing OS-related neonatal disorders. Decades of research have led to a better understanding of OS in neonatal disorders. However, we are still far from finding the silver bullet. The animal models we have been using do not recapitulate what happens in premature neonates. Other animal models and new approaches will be needed to identify appropriate OS-oriented therapies.

In this Special Issue, entitled "Oxidative Stress in the Newborn", we are excited to present five research articles and six reviews for our readers. In contribution #1, an ex vivo organotypic slice of a neonatal brain was challenged with oxygen–glucose deprivation and reperfusion (OGDR). Cell death, redox state, and proteomic profile of inflammatory mediators were studied in the corticostriatal and hippocampal regions. Melatonin was used as the therapeutic agent. Not surprisingly, OGDR increased cell death in both regions, but melatonin markedly attenuated it. Although melatonin increased the GSH/GSSG ratio in both regions, only the hippocampal region had a GSH/GSSG ratio below 1 under OGDR, indicating a regional difference in response to OGDR. In contrast, the OGDR-induced inflammatory response was more pronounced in the corticostriatal region, and melatonin also ameliorated it. In contribution #4, prenatal melatonin injection was shown to facilitate the postnatal lung maturation process in an antenatal inflammation-induced preterm birth.

https://doi.org/10.3390/antiox15030400

Together, these two contributions suggest that melatonin has both anti-inflammatory and antioxidative properties that can protect neonatal organs.

In contribution #2, Endesfelder and Bührer used a rat hyperoxia exposure model (80% O2) for 3 or 5 days to show that caffeine can restore alveolar development and limit the fibrosis. Although caffeine transiently enhances the fibrosis in room air-exposed fibrosis, in the end, there was no morphometric difference. Their study provided another mechanism by which caffeine attenuates BPD severity. In contribution #3, Shvetsova et al. demonstrated the maturational change in superoxide dismutase expression and activity, and their effect on the vasomotor activity of the systemic artery. Apparently, SOD and catalase activities are significantly higher in young rats (10–15 days) than in adult rats (3–4 months). The endothelial-independent vasomotor response in the systemic artery is more pronounced in young rats than in adult rats. The possible age-related difference in response should always be kept in mind when we treat patients.

In contribution #5, the milk composition was compared between vegan and omnivorous women. A significantly higher cortisol concentration and lower iron, vitamin B6, and antioxidant status were seen in the milk from vegan women than from omnivorous women. With the popularity of the vegan diet, health professionals need to be aware of the low iron and vitamin B6 content. Whether antioxidant depletion increases susceptibility to OS-induced organ damage warrants further exploration.

The cross-talk between the OS and inflammation has been clearly described in the literature. In contribution #6, Kaltsogianni et al. reiterated this relationship in several neonatal pulmonary disorders. With no obvious benefit from antioxidant therapy, they recommended judicious use of oxygen to avoid both hypoxia and hyperoxia, thereby reducing OS. The gut microbiome has attracted our attention over the past several years and has been linked to almost all diseases. In contribution #7, Morozan et al. suggested that maternal dysbiosis is associated with increased risk of perinatal asphyxia. The link is mainly through its association with maternal metabolic disorders, immune response, and impaired placental function, which might render an increased susceptibility to perinatal asphyxia in their offspring.

As inflammation results in direct organ damage and increased OS, corticosteroids were once almost routinely given to premature neonates due to the belief that their use could decrease the incidence of BPD. Later, a randomized controlled study showed a significant increase in neurodevelopmental deficit with early postnatal systemic steroid use, which led the American Academy of Pediatrics to recommend not prescribing systemic steroids to premature neonates [21]. Further studies suggested that the timing of steroid treatment might be critical for this association. In contribution #8, Galletta et al. carefully reviewed the topic and offered suggestions for addressing the issue.

In contribution #9, Teng et al. developed a new hypothetical destructive cycle based on their previous studies to explain the development and progression of BPD. The cycle is initiated by OS from supplemental oxygen treatment, followed by HMGB1-mediated DAMP signaling and the unfolded protein response, and then subsequent OS amplification, forming a self-perpetuating cycle. In this contribution, they suggested a systems pharmacotherapeutic approach to prevent or attenuate BPD.

With E-cigarettes gaining more popularity nowadays, their impact on fetuses cannot be ignored. In contribution #10, Gambadauro et al. provided evidence that E-cigarettes are not as safe as users believe. Although caffeine has been shown to reduce BPD in premature infants through the Caffeine for Apnea of Prematurity (CAP) trial [22], its original hypothesis was that it reduces apnea of prematurity, thereby improving neurodevelopmental outcomes. Hypoxia, because of apnea and subsequent hyperoxia after recovery,

 https://doi.org/10.3390/antiox15030400

is similar to the ischemic–reperfusion episode that will generate OS. Although the study did not show early-age neurodevelopmental benefits, its subsequent long-term follow-up did demonstrate efficacy. Because most enrollees were intubated early in the CAP trial, there was no way to demonstrate that caffeine reduced episodes of apnea of prematurity (AOP); it is impossible to show that AOP is associated with increased OS. This is described in contribution #11.

Unfortunately, we were unable to include all OS-related, prematurity-associated morbidities in our Special Issue due to time constraints. OS is believed to play a critical role in retinopathy of prematurity, acute kidney injury, necrotizing enterocolitis, periventricular–intraventricular hemorrhage, periventricular leukomalacia, and birth asphyxia. We hope researchers will offer their expertise and research to readers so that we can better understand pathophysiology and formulate the best ways to prevent or attenuate these morbidities.

Funding: The author is supported by the 2025 Mellowes Endowed Chair funds, the 2025 Department of Pediatrics Internal Support funds, the 2026 Mellowes Center CRI-Focused Award, and NHLBI R44HL166018, R01HL128371, and R01HL174635.

Conflicts of Interest: The author declares that the research was conducted in the absence of any commercial or financial relationships that could be construed as a potential conflict of interest.

List of Contributions

1. Beldarrain, G.; Montejo, U.; Chillida, M.; Alart, J.A.; Álvarez, A.; Alonso-Alconada, D. Region-Dependent Responses to Oxygen–Glucose Deprivation and Melatonin in Neonatal Brain Organotypic Slices. *Antioxidants* **2026**, *15*, 13. https://doi.org/10.3390/antiox15010013.
2. Endesfelder, S.; Bührer, C. Caffeine Protects Against Hyperoxia-Induced Structural Lung Injury and Restores Alveolar Development in Neonatal Rats. *Antioxidants* **2025**, *14*, 1497. https://doi.org/10.3390/antiox14121497.
3. Shvetsova, A.A.; Shateeva, V.S.; Semenovich, D.S.; Schubert, R.; Gaynullina, D.K. The Vasomotor Impact of Cu/ZnSODs Is Higher in Arterial Smooth Muscle of Early Postnatal Rats Compared to Adult Animals. *Antioxidants* **2025**, *14*, 1231. https://doi.org/10.3390/antiox14101231.
4. Park, S.H.; Cho, H.Y.; Jun, J.H.; Song, H.; Lee, J.Y. Prenatal Melatonin Therapy Enhances Postnatal Lung Development in a Mouse Model of Inflammation-Induced Preterm Birth. *Antioxidants* **2025**, *14*, 1094. https://doi.org/10.3390/antiox14091094.
5. Chrustek, A.; Dombrowska-Pali, A.; Olszewska-Słonina, D. Composition and Antioxidant Status of Vegan Milk—Pilot Study. *Antioxidants* **2025**, *14*, 505. https://doi.org/10.3390/antiox14050505.
6. Kaltsogianni, O.; Dassios, T.; Greenough, A. Oxidant Stress, Hyperoxia/Hypoxia and Neonatal Respiratory Disorders. *Antioxidants* **2025**, *14*, 1389. https://doi.org/10.3390/antiox14121389.
7. Morozan, V.-P.; Ionescu, M.I.; Zahiu, C.M.D.; Catrina, A.M.; Racoviță, A.; Chirilă, A.-T.; Dogaru, I.-A.; Ciotei, C.; Pircalabioru, G.G.; Zăgrean, A.-M. Does the Maternal Gut Microbiome Influence the Outcome of Perinatal Asphyxia? *Antioxidants* **2025**, *14*, 1134. https://doi.org/10.3390/antiox14091134.
8. Galletta, F.; Li Pomi, A.; Manti, S.; Gitto, E. Inhaled and Systemic Steroids for Bronchopulmonary Dysplasia: Targeting Inflammation and Oxidative Stress. *Antioxidants* **2025**, *14*, 869. https://doi.org/10.3390/antiox14070869.
9. Teng, M.; Wu, T.-J.; Pritchard, K.A., Jr.; Day, B.W.; Naylor, S.; Teng, R.-J. The Destructive Cycle in Bronchopulmonary Dysplasia: The Rationale for Systems Pharmacology Therapeutics. *Antioxidants* **2025**, *14*, 844. https://doi.org/10.3390/antiox14070844.
10. Gambadauro, A.; Galletta, F.; Andrenacci, B.; Foti Randazzese, S.; Patria, M.F.; Manti, S. Impact of E-Cigarettes on Fetal and Neonatal Lung Development: The Influence of Oxidative Stress and Inflammation. *Antioxidants* **2025**, *14*, 262. https://doi.org/10.3390/antiox14030262.
11. Thompson, L.; Werthammer, J.W.; Gozal, D. Apnea of Prematurity and Oxidative Stress: Potential Implications. *Antioxidants* **2024**, *13*, 1304. https://doi.org/10.3390/antiox13111304.

References

1. Park, A.M.; Sanders, T.A.; Maltepe, E. Hypoxia-inducible factor (HIF) and HIF-stabilizing agents in neonatal care. *Semin. Fetal Neonatal Med.* **2010**, *15*, 196–202. [CrossRef] [PubMed] [PubMed Central]
2. Liu, Y.; Cox, S.R.; Morita, T.; Kourembanas, S. Hypoxia regulates vascular endothelial growth factor gene expression in endothelial cells. Identification of a 5′ enhancer. *Circ. Res.* **1995**, *77*, 638–643. [CrossRef] [PubMed]
3. Sunuwar, S.; Heo, Y.S. Reactive Oxygen Species in Embryo Development: Sources, Impacts, and Implications for In Vitro Culture Systems. *Life* **2026**, *16*, 136. [CrossRef]
4. Zhang, Y.; Cao, L.; Zhang, R.; Pi, W.; Wee Yong, V.; Xue, M. Oxidative stress as the orchestrator of the HIF axis switch in intracerebral hemorrhage. *Free Radic. Biol. Med.* **2026**, *244*, 312–322. [CrossRef] [PubMed]
5. Xu, X.; Pang, Y.; Fan, X. Mitochondria in oxidative stress, inflammation and aging: From mechanisms to therapeutic advances. *Signal Transduct. Target. Ther.* **2025**, *10*, 190. [CrossRef] [PubMed] [PubMed Central]
6. Bhattarai, K.R.; Riaz, T.A.; Kim, H.R.; Chae, H.J. The aftermath of the interplay between the endoplasmic reticulum stress response and redox signaling. *Exp. Mol. Med.* **2021**, *53*, 151–167. [CrossRef] [PubMed] [PubMed Central]
7. Yun, H.R.; Jo, Y.H.; Kim, J.; Shin, Y.; Kim, S.S.; Choi, T.G. Roles of Autophagy in Oxidative Stress. *Int. J. Mol. Sci.* **2020**, *21*, 3289. [CrossRef] [PubMed] [PubMed Central]
8. von Zglinicki, T. Oxidative stress and cell senescence as drivers of ageing: Chicken and egg. *Ageing Res. Rev.* **2024**, *102*, 102558. [CrossRef] [PubMed]
9. Shen, S.; Shao, Y.; Li, C. Different types of cell death and their shift in shaping disease. *Cell Death Discov.* **2023**, *9*, 284. [CrossRef] [PubMed] [PubMed Central]
10. Otoupalova, E.; Smith, S.; Cheng, G.; Thannickal, V.J. Oxidative Stress in Pulmonary Fibrosis. *Compr. Physiol.* **2020**, *10*, 509–547. [CrossRef] [PubMed]
11. Jing, X.; Jia, S.; Teng, M.; Day, B.W.; Afolayan, A.J.; Jarzembowski, J.A.; Lin, C.W.; Hessner, M.J.; Pritchard, K.A., Jr.; Naylor, S.; et al. Cellular Senescence Contributes to the Progression of Hyperoxic Bronchopulmonary Dysplasia. *Am. J. Respir. Cell. Mol. Biol.* **2024**, *70*, 94–109. [CrossRef] [PubMed] [PubMed Central]
12. Pritchard, K.A., Jr.; Jing, X.; Teng, M.; Wells, C.; Jia, S.; Afolayan, A.J.; Jarzembowski, J.; Day, B.W.; Naylor, S.; Hessner, M.J.; et al. Role of endoplasmic reticulum stress in impaired neonatal lung growth and bronchopulmonary dysplasia. *PLoS ONE* **2022**, *17*, e0269564. [CrossRef] [PubMed] [PubMed Central]
13. Beckman, J.S.; Viera, L.; Estévez, A.G.; Teng, R. Nitric oxide and peroxynitrite in the perinatal period. *Semin. Perinatol.* **2000**, *24*, 37–41. [CrossRef] [PubMed]
14. Teng, M.; Wu, T.J.; Jing, X.; Day, B.W.; Pritchard, K.A., Jr.; Naylor, S.; Teng, R.J. Temporal Dynamics of Oxidative Stress and Inflammation in Bronchopulmonary Dysplasia. *Int. J. Mol. Sci.* **2024**, *25*, 10145. [CrossRef] [PubMed] [PubMed Central]
15. Tang, D.; Kang, R.; Zeh, H.J.; Lotze, M.T. The multifunctional protein HMGB1: 50 years of discovery. *Nat. Rev. Immunol.* **2023**, *23*, 824–841. [CrossRef] [PubMed]
16. Bui, C.B.; Pang, M.A.; Sehgal, A.; Theda, C.; Lao, J.C.; Berger, P.J.; Nold, M.F.; Nold-Petry, C.A. Pulmonary hypertension associated with bronchopulmonary dysplasia in preterm infants. *J. Reprod. Immunol.* **2017**, *124*, 21–29. [CrossRef] [PubMed]
17. Nam, J.K.; Choi, S.H.; Kim, J.H.; Choi, K.J.; Lee, H.J.; Lee, J.; Seo, H.R.; Lee, Y.J. Combined HIF-1α blockade and CHIR99021 treatment reverses pulmonary fibrosis via modulation endothelial-to-mesenchymal transition. *iScience* **2025**, *28*, 114028. Erratum in *iScience* **2026**, *29*, 114722. https://doi.org/10.1016/j.isci.2026.114722. [CrossRef] [PubMed] [PubMed Central]
18. Kajimura, S.; Aida, K.; Duan, C. Insulin-like growth factor-binding protein-1 (IGFBP-1) mediates hypoxia-induced embryonic growth and developmental retardation. *Proc. Natl. Acad. Sci. USA* **2005**, *102*, 1240–1245. [CrossRef] [PubMed] [PubMed Central]
19. Modrzejewska, M.; Zdanowska, O.; Połubiński, P. The Role of HIF-1α in Retinopathy of Prematurity: A Review of Current Literature. *J. Clin. Med.* **2024**, *13*, 4034. [CrossRef] [PubMed] [PubMed Central]
20. Rakshasbhuvankar, A.; Pillow, J.J. Vitamin A and bronchopulmonary dysplasia: The next steps. *Lancet Respir. Med.* **2024**, *12*, 503–505. [CrossRef] [PubMed]
21. Yeh, T.F.; Lin, Y.J.; Lin, H.C.; Huang, C.C.; Hsieh, W.S.; Lin, C.H.; Tsai, C.H. Outcomes at school age after postnatal dexamethasone therapy for lung disease of prematurity. *N. Engl. J. Med.* **2004**, *350*, 1304–1313. [CrossRef] [PubMed]
22. Schmidt, B.; Roberts, R.S.; Davis, P.; Doyle, L.W.; Barrington, K.J.; Ohlsson, A.; Solimano, A.; Tin, W. Caffeine for Apnea of Prematurity Trial Group. Caffeine therapy for apnea of prematurity. *N. Engl. J. Med.* **2006**, *354*, 2112–2121. [CrossRef] [PubMed]

https://doi.org/10.3390/antiox15030400

antioxidants

MDPI

Article

Region-Dependent Responses to Oxygen–Glucose Deprivation and Melatonin in Neonatal Brain Organotypic Slices

Gorane Beldarrain, Unai Montejo, Marc Chillida, Jon Ander Alart, Antonia Álvarez and Daniel Alonso-Alconada *

Department of Cell Biology and Histology, School of Medicine and Nursing, University of the Basque Country (UPV/EHU), 48940 Leioa, Spain
* Correspondence: daniel.alonsoa@ehu.eus

Abstract

Hypoxic–ischemic encephalopathy remains a major cause of neonatal mortality and long-term neurological disability. Therapeutic hypothermia is currently the only available treatment in hospitals, but its efficacy is limited, making the search for alternative neuroprotective strategies essential. Melatonin has shown promising results in other models of hypoxia–ischemia, acting as a potent antioxidant and anti-inflammatory molecule. Here, we studied the effects of hypoxia–ischemia and melatonin treatment in two brain regions that are particularly vulnerable to hypoxic–ischemic injury. Neonatal rat organotypic slice cultures from the corticostriatal and hippocampal regions were subjected to oxygen–glucose deprivation and reperfusion (OGDR) and treated with melatonin (50 μM). Cell death (propidium iodide staining), redox state (GSH/GSSG ratio) and the inflammatory profile (Proteome Profiler) were analyzed. OGDR markedly increased cell death in both regions and melatonin treatment significantly reduced it. The GSH/GSSG ratio decreased only in the hippocampus after OGDR, but melatonin treatment elevated this ratio in both regions. In contrast, the inflammatory profile was more pronounced in the corticostriatal region, where the treatment strongly reduced proinflammatory mediators. These findings reveal region-specific mechanisms involved in the response to hypoxic–ischemic damage and support the potential of melatonin as a promising therapy for neonatal brain injury.

Keywords: neonatal hypoxia–ischemia; melatonin; oxygen–glucose deprivation; organotypic slice cultures

1. Introduction

Currently, 2–3 out of every 1000 term newborns experience some form of neurological dysfunction referred to as hypoxic–ischemic encephalopathy [1,2]. Approximately 24% of these infants die in neonatal intensive care units, and a significant number of survivors face serious long-term consequences [3]. These include cerebral palsy in 10–20% of cases, visual and hearing deficits in 40%, as well as motor and behavioral disorders such as epilepsy, global developmental delay, or autism [3].

Since therapeutic hypothermia was implemented in clinical practice, the risk of dying or developing such consequences has significantly reduced. However, several studies indicate that the benefits of hypothermia in neonates after hypoxia–ischemia (HI) might be limited [4–6], urging for new neuroprotective strategies. One of the reasons of a lack of an effective treatment is the complex physiopathology of hypoxic–ischemic encephalopathy, with concomitant processes encompassing oxidative stress and neuroinflammation, leading to cell death and brain tissue damage [7–9].

Melatonin (N-acetyl-5-methoxytryptamine) is a hormone synthesized in the pineal gland in response to light/dark cycles, so it is commonly used as a treatment for sleep disorders in children with few or no side effects [10]. It is also a potent antioxidant that scavenges highly toxic free radicals, increases antioxidant enzyme expression and inhibits nitric oxide synthase activity, the source of many free radicals that accumulate in cells after hypoxic–ischemic damage [11]. Several studies have demonstrated its neuroprotective potential, reporting reduced brain injury in rats in the short and long term [12,13], augmented neuroprotection when combined with hypothermia in piglets [14] or decreased cell death and inflammation in sheep [15], among others. However, much less is known about its region-specific impact within the brain. Most available studies addressing differences after melatonin treatment in different regions rely on either neuron cultures (which lack the architectural complexity of the brain), or in vivo studies (where the analyses are often restricted to histological outcomes). Organotypic brain cultures preserve the structural organization of the brain, including neurons, glia and their interactions. This provides an intermediate level of complexity that allows the study of molecular mechanisms such as oxidative stress and inflammation under controlled conditions across different brain regions. As reviewed by Humpel [16], organotypic slice cultures offer a more in vivo-like microenvironment than cell lines or dissociated primary cultures, while still allowing precise experimental control, reduced variability and a decrease in the number of animals required. Despite some limitations (they do not preserve the full neurovascular unit, systemic circulation or peripheral immune components), organotypic slice cultures have been widely used to efficiently study the effects of HI in the brain [17–23].

In neonates surviving intra-partum HI, the regional pattern of brain injury may change depending on various clinical factors, such as the gestational age or the severity of the insult [24,25]. Thus, different cerebral areas might not respond in the same way to the same insult, and a single treatment may not be able to offer neuroprotection in all brain regions

Using organotypic brain slice cultures from neonatal rats, the aims of this work were (i) to compare the extent of the damage in corticostriatal and hippocampal regions by evaluating cell death, oxidative stress and inflammation and (ii) to determine whether melatonin differentially modulates these mechanisms across the two regions.

2. Materials and Methods

2.1. Organotypic Slice Cultures

All animal procedures were performed following the European Union regulations for animal research (Directive 86/609/EEC) and approved by the Animal Welfare Committee of the University of the Basque Country.

Organotypic slice cultures of the corticostriatal and hippocampal areas of the brain were prepared as previously described [22,26]. Briefly, brains were extracted from 6-day-old Sprague Dawley rats and the two hemispheres were separated. Coronal sections (350 µm thick) were obtained using a McIlwain Tissue Chopper (Ted Pella, Inc., Redding, CA, USA; Cat. No. 10180-220, 220V) and separated individually under a stereomicroscope. From these sections, those containing the corticostriatal region were selected and from adjacent sections, hippocampi were carefully dissected out. Both corticostriatal and hippocampal slices were then placed in Milicell culture inserts (Merck, Darmstadt, Germany; Cat. No. PICM0RG50) immersed in 1 mL of complete culture medium consisting of 50% MEM, 25% HBSS, 25% horse serum, supplemented with glucose (25 mM), B27 1X (Thermo Fisher Scientific, Waltham, MA, USA; Cat. No. 17504044), GlutaMAX 1X (Thermo Fisher Scien-

 https://doi.org/10.3390/antiox15010013

tific, Waltham, MA, USA; Cat. No. 35050061), fungizone (Amphotericin B, 1.25 μg/mL; Thermo Fisher Scientific, Waltham, MA, USA; Cat. No. 15290018), penicillin–streptomycin (50 U/mL; Thermo Fisher Scientific, Waltham, MA, USA; Cat. No. 15070063), and gentamicin (50 μg/mL; Sigma-Aldrich, St. Louis, MO, USA; Cat. No. G1397). Slices of the same experimental group always originated from different rats to prevent intra-animal redundancy. Cultures were maintained in a 37 °C humidified incubator gasified with 5% CO_2/95% O_2 and medium was changed three times weekly for 14 days. Slices that showed altered morphology after 14 days in culture were discarded from the analysis.

2.2. Oxygen-Glucose Deprivation and Reperfusion (OGDR)

After 14 days in vitro, oxygen–glucose deprivation (OGD) was performed to simulate hypoxia–ischemia. Slices were placed in a hypoxic chamber (MIC-101, Billups-Rothenberg, San Diego, CA, USA) infused with a mixture of 95%N_2/5% CO_2 gas and the complete medium was changed to a glucose-free medium. Corticostriatal slices were maintained in such conditions for 1 h whereas hippocampal sections for 45 min. For the next 24 h, reperfusion was simulated by placing the cultures under normoxic conditions and replacing the medium with reperfusion medium, which had the same composition as the complete medium except for the horse serum. A shorter OGD duration was used for hippocampal slices to avoid extensive and irreversible hippocampal cell death due to its regional vulnerability, thus obtaining a similar level of injury in both regions that would still allow the detection of potential neuroprotective effects of melatonin. These different OGD times were also in line with previous works [18,23].

2.3. Preparation of Melatonin Treatment

Melatonin (Sigma-Aldrich, St. Louis, MO, USA; Cat. No. M5250), was dissolved in 0.05% DMSO and diluted in the reperfusion medium to a final concentration of 50 μM. The treatment was added immediately after the OGD process and maintained in the medium for 24 h. The final concentration of Melatonin was selected in accordance with previous studies demonstrating neuroprotective effects in organotypic and in vitro models of hypoxia–ischemia [13] and after dose–response experiments carried out by our group.

2.4. Experimental Groups

(1) Control: Slices maintained under normal culture conditions with complete medium and no treatment.
(2) OGDR: Slices subjected to oxygen–glucose deprivation followed by 24 h of reperfusion, to mimic HI and reperfusion-induced injury.
(3) OGDR + MEL: Slices subjected to oxygen–glucose deprivation and treated with 50 μM melatonin followed by 24 h of reperfusion to assess the potential protective effects of melatonin against HI and reperfusion-induced injury.

Experimental procedures, image analysis, and quantification were conducted in a blinded manner to ensure unbiased evaluation.

2.5. Cell Death Quantification

Dead cells were counted by staining the slices with propidium iodide (PI). PI is uptaken by dead cells as it is a polar molecule that can only enter the cell when the membrane is damaged. Once inside the cell, PI binds to DNA generating red fluorescence. PI (5 μg/mL) was added to the cultures after OGDR and maintained in the medium for 30 min. Slices were then observed using an inverted fluorescent microscope and the total number of PI positive cells (dead cells) were counted in situ at 10× (high power field, HPF) within

https://doi.org/10.3390/antiox15010013

the target region (hippocampus or corticostriatal area). Results are expressed as the ratio between the mean number of PI-positive/dead cells in the OGDR or OGDR+MEL groups and the control group (fold change).

2.6. Glutathione (GSH) and Oxidized Glutathione (GSSG) Quantification

To assess the redox state of the cells after OGDR and melatonin treatment, GSH and GSSG levels were measured using the Glutathione Fluorescent Detection Kit (Invitrogen, Waltham, MA, USA, Cat. No. EIAGSHF). For each experimental group, slices from a single culture were pooled to obtain sufficient tissue for analysis. The pooled samples were homogenized and processed according to the manufacturer's instructions.

Briefly, we first obtained free GSH levels by measuring the fluorescence at 510 nm. The same samples were then incubated with a mix containing glutathione reductase and fluorescence was measured again to obtain total GSH levels. The GSSG levels were calculated dividing the difference between total GSH and free GSH by two:

$$GSSG = (totalGSH - freeGSH) \div 2$$

The free GSH/GSSG ratio was calculated as an overall indicator of the redox state of the cell. Each measurement was performed in triplicate to ensure technical reproducibility. Data therefore represent technical replicates from a single pooled biological sample per condition and are presented as representative values.

2.7. Proteome Profile

The proteome profile of supernatants was assessed using a membrane-based sandwich immunoassay (Proteome Profiler Rat Cytokine Array Kit, R&D Systems, Minneapolis, MN, USA; Cat. No. ARY008). Briefly, supernatants of the same experimental groups were pooled, total protein was calculated using bicinchoninic acid assay and 400 µg of total protein from each pool were used to perform the assay. Each sample was mixed with a cocktail of biotinylated antibodies and incubated overnight in contact with a nitrocellulose membrane containing 29 different antibodies printed in duplicate. The next day, several washes were performed and streptavidin-HRP and the chemiluminescent reagent mix were added. The chemiluminescent signal was captured using a Syngene G:BOX Chemi HR16 (Syngene, Cambridge, UK) and pixel densitometry was obtained using Protein Array Tool (v2.0.0.1) [27] on MATLAB(R2024b, 24.2.0; The MathWorks, Inc., Natick, MA, USA), which turns the captured images into a list of relative protein levels.

2.8. Statistical Analysis

The normality of the figures was analyzed using the D'Agostino–Pearson test. For parametric data, a two-tailed, unpaired Student's *t*-test was performed. In the same way, non-aparametric data were analyzed using the Mann–Whitney test. Statistical analysis was performed using the GraphPad Prism 10 software package (GraphPad Prism 10, GraphPad Software, San Diego, CA, USA) and data were considered significantly different if $p < 0.05$.

3. Results

3.1. Cell Death

In all experiments, PI uptake of the control group slices was considered as the basal level of cell death, so values of the other experimental groups were expressed as fold change vs. control.

https://doi.org/10.3390/antiox15010013

In the corticostriatal region, slices that underwent OGDR showed significantly higher values of cell death compared to control, with the number of PI+ cells being 12 times higher in this group ($p < 0.0001$). Treating cells with melatonin after OGDR significantly reduced the number of PI+ cells observed in the OGDR group, as the fold change reduced from 12 to 1.9 ($p < 0.001$). Further, the treatment reduced the number of dead cells to control/basal levels, with no significant differences when comparing OGDR+MEL and control groups. Results are shown in Figure 1.

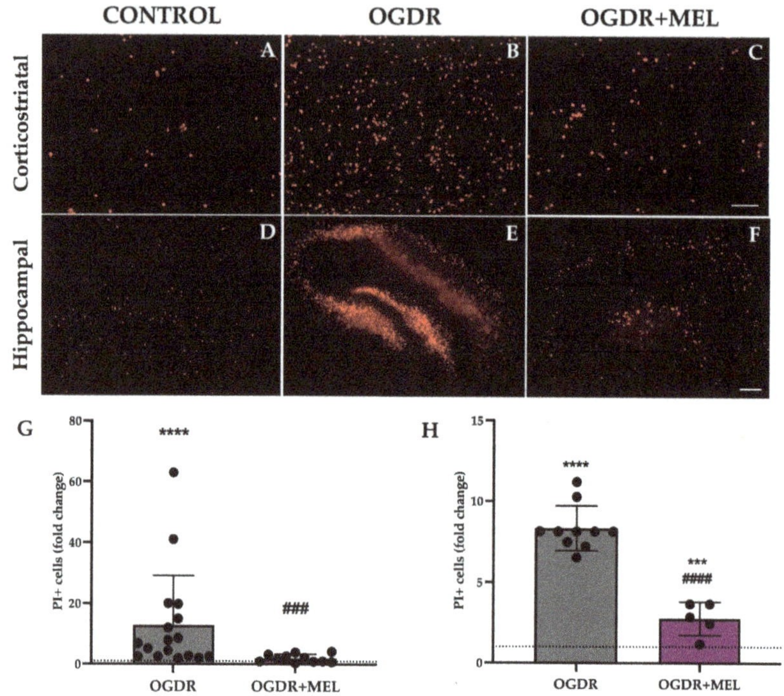

Figure 1. Cell death in corticostriatal and hippocampal slice cultures after OGDR and melatonin treatment. Cultures were subjected to 1 h (corticostriatal) or 45 min (hippocampal) of OGD and immediately treated with 50 µM melatonin followed by 24 h of reperfusion. Cell death was measured by PI uptake (**A–F**). Representative microphotographs of PI fluorescence in each of the experimental groups of the two regions ((**A–C**) Scale bar: 50 µm, (**D–F**) scale bar: 200 µm). (**G**,**H**). Graphs show PI+ cell counts in corticostriatal (**G**) and hippocampal (**H**) regions. n for control, OGDR and OGDR+MEL in the corticostriatal region was 16, 17 and 11 slices, respectively. n for control, OGDR and OGDR+MEL in the hippocampal region was 19, 10 and 5 slices, respectively. The dotted line at y = 1 represents the control values. *** $p < 0.001$ or **** $p < 0.0001$ vs. control group; ### $p < 0.001$ or #### $p < 0.0001$ vs. OGDR group.

In the hippocampus, cell death was also significantly increased after OGDR, reaching values 8 times higher than those observed in the control slices ($p < 0.0001$). Again, melatonin treatment after OGDR significantly decreased the number of PI+ cells from a fold change of 8 to 2.75 ($p < 0.0001$). However, unlike in the corticostriatal region, the treatment did not reduce dead cells to control-like values in the hippocampus, as we found nearly 3 times more dead cells in the OGDR+MEL group compared to the control group ($p < 0.001$). Results are shown in Figure 1.

https://doi.org/10.3390/antiox15010013

3.2. Oxidative Stress

To assess the redox state of the slices in the different experimental conditions, we measured the changes in the state of GSH and obtained the GSH/GSSG ratio, with decreased values as an indicator of cellular oxidative stress. These findings, obtained from pooled tissue samples analyzed in technical triplicates, should be interpreted as representative trends supporting the modulatory effect of melatonin on redox homeostasis.

In the corticostriatal region (Figure 2), GSH/GSSG ratio did not change after OGDR compared to control samples. However, after treating the slices with melatonin, the GSH/GSSG ratio clearly increased, suggesting that melatonin modulates the cellular antioxidant system beyond its role as a direct antioxidant.

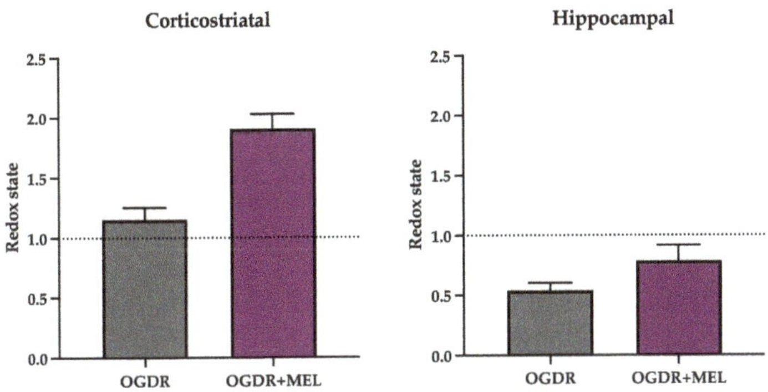

Figure 2. Redox state of the corticostriatal and hippocampal regions after OGDR and melatonin treatment. Cultures were subjected to 1 h (corticostriatal) or 45 min (hippocampal) of OGD and immediately treated with 50 μM melatonin followed by 24 h of reperfusion. GSH/GSSG ratio was calculated as an indicator of the redox state and is expressed relative to the control. n in the corticostriatal region was 6 slices in each experimental group. n in the hippocampal region was 9 slices in each experimental group. Slices were pooled and experiments were performed in triplicate. The dotted line at y = 1 represents the control values.

In the hippocampus (Figure 2), the redox state of the cells was markedly deteriorated following OGDR, as the GSH/GSSG ratio decreased to 0.53. After melatonin treatment, an improvement in redox status was found, as evidenced by a higher GSH/GSSG ratio in the OGDR+MEL group compared to OGDR. Nevertheless, this improvement did not reach control levels, thus suggesting that melatonin partially restores the cellular redox balance after OGDR.

3.3. Inflammatory Response

Following OGDR, several mediators of the inflammatory response were altered in the corticostriatal region (Figure 3). IP-10 (Interferon-γ-induced Protein-10), MIP-3α (Macrophage Inflammatory Protein-3α), CINC-1 (Cytokine-Induced Neutrophil Chemoattractant-1) and LIX (Lipopolysaccharide-Induced CXC chemokine) are chemokines involved in leukocyte recruitment, which help amplify the inflammatory response. After OGDR, all four appeared upregulated compared to controls, where very little or none of these markers was detected. Remarkably, after treating the slices with melatonin, the levels of these chemokines were strongly reduced, recovering control levels as in the case of IP-10 and MIP-3α. The expression of ICAM-1 (Intercellular Adhesion Molecule-1), a molecule associated with cell adhesion and transmigration, was also affected: while no ICAM-1 was detected in the control samples, OGDR clearly upregulated its levels. Again, melatonin treatment notably reduced the expression of this inflammatory molecule.

https://doi.org/10.3390/antiox15010013

Two markers related to tissue remodeling were also detected on corticostriatal samples. TIMP-1 (Tissue Inhibitor of Metalloproteinases-1), a tissue inhibitor of matrix metalloproteinases, showed elevated levels in all the experimental groups, with no differences between them. Compared to control samples, VEGF (Vascular Endothelial Growth Factor), a key regulator of angiogenesis and vascular remodeling, was strongly and mildly elevated after OGDR and OGDR+MEL, respectively.

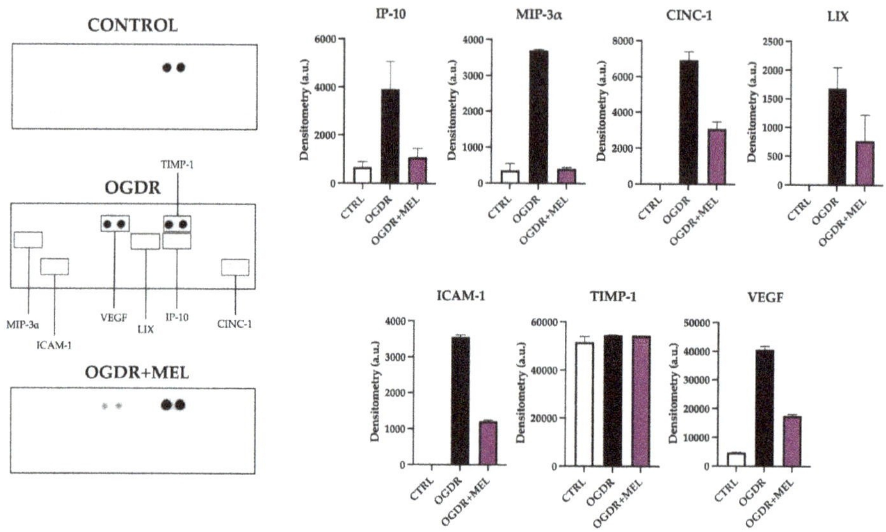

Figure 3. Proteome profile of the corticostriatal region after OGDR and melatonin treatment. Corticostriatal cultures were subjected to 1 h of OGD and immediately treated with 50 μM melatonin followed by 24 h of reperfusion. The proteome profile of supernatants was assessed using a membrane-based sandwich immunoassay (left side of the figure). The following were detected (graphs, right side of the figure): IP-10, MIP-3α, CINC-1, LIX, ICAM-1, TIMP-1 and VEGF. n was 6 slices in each experimental group. Samples were pooled and pixel densitometry was measured in duplicate.

In the hippocampus, the number of detected markers and their expression levels clearly varied compared to the corticostriatal region (Figure 4).

Regarding the recruitment phase of the inflammatory response, only IP-10 and CINC-1 were detected. The expression of IP-10 increased in OGDR slices compared to the control, where no IP-10 was detected. After treating cells with melatonin, the expression of this molecule did not differ from what was observed in OGDR slices. CINC-1, on the other hand, could not be detected in both the control and OGDR slices, but the supernatants of OGDR+MEL slices showed slightly increased levels of this marker. We did not detect MIP-3α, LIX or ICAM-1.

Tissue remodeling markers TIMP-1 and VEGF appeared again in hippocampi, but with a different profile compared to the corticostriatal region. Whereas TIMP-1 was not detected in control slices, its expression was strongly increased after OGDR and slightly reduced with melatonin. VEGF, unlike in the corticostriatal region, was only detected in the OGDR+MEL group, being absent in the non-treated OGDR group.

 https://doi.org/10.3390/antiox15010013

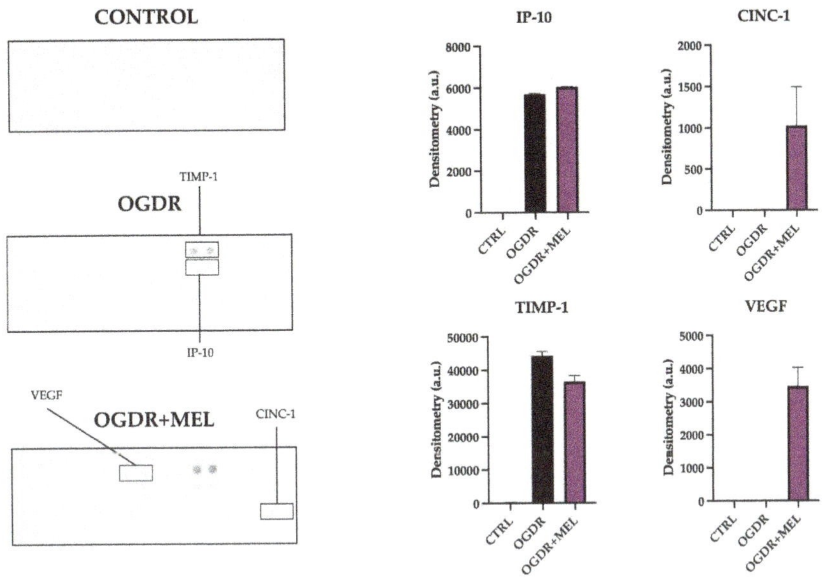

Figure 4. Proteome profile of the hippocampal region after OGDR and melatonin treatment. Hippocampal cultures were subjected to 45 min of OGD and immediately treated with 50 μM melatonin followed by 24 h of reperfusion. The proteome profile of supernatants was assessed using a membrane-based sandwich immunoassay (left side of the figure). The following were detected (graphs, right side of the figure): IP-10, CINC-1, TIMP-1 and VEGF. n was 10 slices in each experimental group. Samples were pooled and pixel densitometry was measured in duplicate.

4. Discussion

Using organotypic brain slices from neonatal rats, we compared here the effects of HI and melatonin treatment in the corticostriatal and hippocampal regions, two brain areas particularly sensitive to asphyxia in the neonatal period. Our findings confirmed such sensitivity and similar extent in cell death but revealed inter-regional differences in oxidative stress and inflammation. Melatonin treatment was able to reduce cell death and oxidative stress in both brain areas, but its anti-inflammatory effects differed depending on the region.

When evaluating cell death, the degree of injury showed minor regional differences, with the corticostriatal area exhibiting slightly higher cell counts compared to the hippocampus. The latter is usually a more sensitive region to asphyxia [28], so this discrepancy may be attributed to the different oxygen-glucose deprivation durations used in this work (60 min for corticostriatal [23] and 45 min for hippocampus [18]), which were established to avoid irreversible cell death in the hippocampus. There are different factors that could be behind the regional differences observed in cell death. First, each region has different intrinsic properties, which are likely to contribute to different responses after damage. As observed in previous works, neurons and astrocytes from the cortex, hippocampus and striatum differ in the activity of superoxide dismutase and the expression levels of gluthatione or Bcl-xL, cellular components related to the regulation of the antioxidant and apoptotic responses [29]. Different cell types are also likely to have specific responses depending on the region, which is also probably behind the different patterns of injury. For example, astrocytes are usually more resistant than neurons to excitotoxic and oxidative stimuli, but in the striatum, astrocytes seem to be particularly vulnerable to OGD. A similar thing

https://doi.org/10.3390/antiox15010013

happens with cortical neurons, which are more sensitive than striatal and hippocampal neurons [29,30]. At the molecular level, brain regions can also differ in cell death pathways after HI, as CA1 neurons usually undergo caspase-dependent apoptosis [31], whereas cortical and striatal regions may show a combination of caspase-dependent apoptosis, necrosis and other forms of cell death [32,33].

As cell death usually correlates with (or is a consequence of) oxidative stress and inflammation [34], we next studied these parameters.

The corticostriatal region did not show major variations in oxidative stress (measured by the GSH/GSSG ratio), whereas data from the hippocampus revealed a significant drop in this ratio after OGDR, pointing to more severe oxidative damage in this region. These regional differences in oxidative stress may be due to several factors. Together with its high consumption of oxygen as one of the most metabolically active regions of the developing brain [35], the hippocampus contains a very high density of glutamatergic neurons compared to other regions [36], which makes this area especially vulnerable to excitotoxicity and the resulting overproduction of reactive oxygen species (ROS) [37,38]. Consistent with data from the present work, GSH levels of the CA1 pyramidal neurons significantly decreased after OGDR in slice cultures [39], leaving the hippocampus with a less robust antioxidant defense, especially during early development. For its part, the low impact of oxidative stress on the corticostriatal slices may be due to a greater redox capacity of the region or to a previously resolved oxidative peak, as inflammation was relevant in this area at 24 h after OGDR.

In the corticostriatal region, we observed increased values of IP-10, MIP-3α, CINC-1, LIX, ICAM-1 and VEGF. These results suggest a classical inflammatory cascade in this region, where leukocyte recruitment chemokines (IP-10, MIP-3α, CINC-1 and LIX) are predominant at the studied time point [40]. In corticostriatal primary cultures subjected to OGDR, Zhang and collaborators reported IP-10 was markedly increased, which contributed to neuronal injury [41]. Both CINC-1 and LIX are potent neutrophil-attracting signals that aggravate the outcomes of neurological damage [42,43]. Neutrophils contribute to the damage caused by ischemia by impairing vascular remodeling, worsening blood–brain barrier integrity, increasing microvascular occlusions and ultimately promoting neuronal death [44–47], which correlates with the cell death observed in our work after OGDR. Similarly to our data, Terao et al. also reported an increase in MIP-3α in a rat model of ischemia, which was associated with exacerbation of the injury [48]. Molecules involved in endothelial adhesion (ICAM-1) and tissue remodeling (VEGF) were also detected in the corticostriatal region, which suggests that although the region appears to be primarily at the leukocyte recruitment stage, the tissue might be initiating the next phases of inflammation, which involve vascular adaptation and structural reorganization [49,50]. Further, the induction of ICAM-1 in the cortex and the activation of VEGF after ischemia are well documented in rodents and both markers appear to be upregulated in response to damage [51,52]. To sum up, corticostriatal-augmented inflammation may amplify tissue injury, microvascular dysfunction and blood–brain barrier disruption [53–55].

In contrast, the hippocampal inflammatory profile was much milder, with only two markers (IP-10 and TIMP-1) upregulated at 24 h post-OGDR, both markers correlating with brain injury and long-term sequelae, as reported before [41,56].

These region-dependent differences in inflammation may be due to several factors, like the spatial heterogeneity of microglia and the temporal heterogeneity in cytokine kinetics, among others. Microglia plays a leading role in inflammation, and its density, morphology and gene expression is not uniform across all brain areas. Further, its diversity peaks at the developmental stage [57,58], so this may influence a region-specific response to damage.

https://doi.org/10.3390/antiox15010013

Another factor to consider is the different temporal window of inflammatory molecules depending on the region. In neonatal HI, canonical proinflammatory cytokines often increase within 3–6 h post-injury and return to baseline after 24 h in the hippocampus [59], while in the cortex or striatum, signals of these cytokines can be detected across the 12–48 h window post-insult [60].

Our work also studied the therapeutic capacity of melatonin in the two regions. Melatonin significantly reduced cell death in both corticostriatal and hippocampal areas, a neuroprotective effect in line with prior data using primary neuronal cultures [61,62], organotypic hippocampal slices [18,63] and in vivo HI models [32,33].

Melatonin also improved GSH/GSSG ratio, with similar increases in both brain areas. These data align with previous works reporting the antioxidant capacity of the molecule, enhancing the synthesis of GSH, as well as scavenging free radicals [64–66]. Here, OGDR especially affected the redox state of the hippocampus, so the observed redox rebalancing after the treatment could contribute to the observed decrease in hippocampal cell death [13,18,63].

As discussed previously, the main process behind damage of the corticostriatal region in our work appears to be inflammation. The inflammatory mediators upregulated after OGDR in this region (IP-10, MIP-3α, CINC-1, LIX, ICAM, and VEGF) were all reduced after melatonin treatment, thus confirming a potent anti-inflammatory effect of the treatment, described in other models [67].

A different response in inflammation was observed in the hippocampus after the treatment. The upregulated markers after OGDR (IP-10 and TIMP-1) were not significantly reduced and two new markers (CINC-1 and VEGF) were detected. Some of these changes might contribute to the reduction in cell death; for instance, the appearance of VEGF after the treatment might indicate tissue regeneration [68].

This study has some limitations. First, all measurements were obtained at a single time point (24 h post-OGD). This time-point corresponds to a window in which cell death, inflammation and oxidative stress are clearly detectable in our model, but it does not allow us to characterize their temporal dynamics. Earlier sampling could help define the onset of the oxidative and inflammatory responses, whereas later sampling would probably help clarify whether these processes persist or resolve in the longer term, leading to a more robust understanding of the evolution of deleterious and/or repairing mechanisms in the studied brain areas. While organotypic slices preserve local architecture and neuron–glia interactions, being a highly representative model for the study of new therapies in HI, they do not preserve the full neurovascular unit present in vivo. As commented before, OGD duration differed between regions to avoid irreversible cell death in the hippocampus, and slices had to be pooled for some analyses, which limits some statistical comparisons. Future work incorporating multi-timepoint sampling, study of the upstream processes behind the immune response and complementary in vivo validation will help refine the interpretation of these findings.

In summary, our work suggests that organotypic corticostriatal and hippocampal slices respond differently to OGDR, with inflammation playing a predominant role in the corticostriatal region and oxidative stress in the hippocampus. Melatonin treatment reduced cell death in both regions, targeting both harmful mechanisms. From a translational perspective, our findings reinforce the therapeutic potential of melatonin as a neuroprotective strategy in neonatal hypoxic–ischemic injury and highlight the importance of considering region-specific pathophysiology when designing and evaluating new treatments. Future studies should determine whether similar patterns of vulnerability occur in vivo and evaluate longer-term structural and functional outcomes. Such

https://doi.org/10.3390/antiox15010013

work will help bring our results a step closer to clinical application in neonates with hypoxic–ischemic encephalopathy.

Author Contributions: Conceptualization, D.A.-A.; methodology, G.B., U.M., M.C. and J.A.A.; investigation, G.B., U.M., M.C. and J.A.A.; resources, D.A.-A.; data curation, G.B., U.M., M.C. and J.A.A.; writing—original draft preparation, G.B. and D.A.-A.; writing—review and editing, G.B., U.M., M.C., J.A.A., A.Á. and D.A.-A.; supervision, A.Á. and D.A.-A.; project administration, D.A.-A.; funding acquisition, D.A.-A. All authors have read and agreed to the published version of the manuscript.

Funding: This research was funded by University of the Basque Country (UPV/EHU; GIU21/054) and by the "Programa Investigo" by the European Union-Next Generation EU, grant PIFINVE22/15.

Institutional Review Board Statement: The animal study protocol was approved by the Institutional Review Board (or Ethics Committee) of the University of the Basque Country (protocol code M20/2024/042, approval date: 04/11/2025).

Informed Consent Statement: Not applicable.

Data Availability Statement: The original contributions presented in this study are included in the article. Further inquiries can be directed to the corresponding author.

Conflicts of Interest: The authors declare no conflicts of interest.

References

1. Gunn, A.J.; Thoresen, M. Neonatal Encephalopathy and Hypoxic–Ischemic Encephalopathy. In *Handbook of Clinical Neurology*; Elsevier: Amsterdam, The Netherlands, 2019; Volume 162, pp. 217–237. ISBN 978-0-444-64029-1.
2. Kurinczuk, J.J.; White-Koning, M.; Badawi, N. Epidemiology of Neonatal Encephalopathy and Hypoxic–Ischaemic Encephalopathy. *Early Hum. Dev.* **2010**, *86*, 329–338. [CrossRef]
3. Yıldız, E.P.; Ekici, B.; Tatlı, B. Neonatal Hypoxic Ischemic Encephalopathy: An Update on Disease Pathogenesis and Treatment. *Expert Rev. Neurother.* **2017**, *17*, 449–459. [CrossRef] [PubMed]
4. Edwards, A.D.; Brocklehurst, P.; Gunn, A.J.; Halliday, H.; Juszczak, E.; Levene, M.; Strohm, B.; Thoresen, M.; Whitelaw, A.; Azzopardi, D. Neurological Outcomes at 18 Months of Age after Moderate Hypothermia for Perinatal Hypoxic Ischaemic Encephalopathy: Synthesis and Meta-Analysis of Trial Data. *BMJ* **2010**, *340*, c363. [CrossRef] [PubMed]
5. Tagin, M.A.; Woolcott, C.G.; Vincer, M.J.; Whyte, R.K.; Stinson, D.A. Hypothermia for Neonatal Hypoxic Ischemic Encephalopathy: An Updated Systematic Review and Meta-Analysis. *Arch. Pediatr. Adolesc. Med.* **2012**, *166*, 558–566. [CrossRef]
6. Wassink, G.; Davidson, J.O.; Dhillon, S.K.; Zhou, K.; Bennet, L.; Thoresen, M.; Gunn, A.J. Therapeutic Hypothermia in Neonatal Hypoxic-Ischemic Encephalopathy. *Curr. Neurol. Neurosci. Rep.* **2019**, *19*, 2. [CrossRef]
7. Greco, P.; Nencini, G.; Piva, I.; Scioscia, M.; Volta, C.A.; Spadaro, S.; Neri, M.; Bonaccorsi, G.; Greco, F.; Cocco, I.; et al. Pathophysiology of Hypoxic–Ischemic Encephalopathy: A Review of the Past and a View on the Future. *Acta Neurol. Belg.* **2020**, *120*, 277–288. [CrossRef]
8. Martini, S.; Castellini, L.; Parladori, R.; Paoletti, V.; Aceti, A.; Corvaglia, L. Free Radicals and Neonatal Brain Injury: From Underlying Pathophysiology to Antioxidant Treatment Perspectives. *Antioxidants* **2021**, *10*, 2012. [CrossRef] [PubMed]
9. Odorcyk, F.K.; Kolling, J.; Sanches, E.F.; Wyse, A.T.S.; Netto, C.A. Experimental Neonatal Hypoxia Ischemia Causes Long Lasting Changes of Oxidative Stress Parameters in the Hippocampus and the Spleen. *J. Perinat. Med.* **2018**, *46*, 433–439. [CrossRef]
10. Jan, M.M.S. Melatonin for the Treatment of Handicapped Children with Severe Sleep Disorders. *Pediatr. Neurol.* **2000**, *23*, 229–232. [CrossRef]
11. Claustrat, B.; Leston, J. Melatonin: Physiological Effects in Humans. *Neurochirurgie* **2015**, *61*, 77–84. [CrossRef]
12. Alonso-Alconada, D.; Alvarez, A.; Lacalle, J.; Hilario, E. Histological Study of the Protective Effect of Melatonin on Neural Cells after Neonatal Hypoxia-Ischemia. *Histol. Histopathol.* **2012**, *27*, 771–783. [CrossRef] [PubMed]
13. Carloni, S.; Perrone, S.; Buonocore, G.; Longini, M.; Proietti, F.; Balduini, W. Melatonin Protects from the Long-term Consequences of a Neonatal Hypoxic-ischemic Brain Injury in Rats. *J. Pineal Res.* **2008**, *44*, 157–164. [CrossRef]
14. Robertson, N.J.; Faulkner, S.; Fleiss, B.; Bainbridge, A.; Andorka, C.; Price, D.; Powell, E.; Lecky-Thompson, L., Thei, L., Chandrasekaran, M.; et al. Melatonin Augments Hypothermic Neuroprotection in a Perinatal Asphyxia Model. *Brain* **2013**, *136*, 90–105. [CrossRef]

 https://doi.org/10.3390/antiox15010013

15. Welin, A.-K.; Svedin, P.; Lapatto, R.; Sultan, B.; Hagberg, H.; Gressens, P.; Kjellmer, I.; Mallard, C. Melatonin Reduces Inflammation and Cell Death in White Matter in the Mid-Gestation Fetal Sheep Following Umbilical Cord Occlusion. *Pediatr. Res.* **2007**, *61*, 153–158. [CrossRef]
16. Humpel, C. Organotypic Brain Slice Cultures: A Review. *Neuroscience* **2015**, *305*, 86–98. [CrossRef] [PubMed]
17. Landucci, E.; Pellegrini-Giampietro, D.E.; Facchinetti, F. Experimental Models for Testing the Efficacy of Pharmacological Treatments for Neonatal Hypoxic-Ischemic Encephalopathy. *Biomedicines* **2022**, *10*, 937. [CrossRef]
18. Carloni, S.; Facchinetti, F.; Pelizzi, N.; Buonocore, G.; Balduini, W. Melatonin Acts in Synergy with Hypothermia to Reduce Oxygen-Glucose Deprivation-Induced Cell Death in Rat Hippocampus Organotypic Slice Cultures. *Neonatology* **2018**, *114*, 364–371. [CrossRef]
19. Lana, D.; Gerace, E.; Magni, G.; Cialdai, F.; Monici, M.; Mannaioni, G.; Giovannini, M.G. Hypoxia/Ischemia-Induced Rod Microglia Phenotype in CA1 Hippocampal Slices. *Int. J. Mol. Sci.* **2022**, *23*, 1422. [CrossRef] [PubMed]
20. Landucci, E.; Mazzantini, C.; Lana, D.; Davolio, P.L.; Giovannini, M.G.; Pellegrini-Giampietro, D.E. Neuroprotective Effects of Cannabidiol but Not Δ9-Tetrahydrocannabinol in Rat Hippocampal Slices Exposed to Oxygen-Glucose Deprivation: Studies with Cannabis Extracts and Selected Cannabinoids. *Int. J. Mol. Sci.* **2021**, *22*, 9773. [CrossRef]
21. Landucci, E.; Filippi, L.; Gerace, E.; Catarzi, S.; Guerrini, R.; Pellegrini-Giampietro, D.E. Neuroprotective Effects of Topiramate and Memantine in Combination with Hypothermia in Hypoxic-Ischemic Brain Injury in Vitro and in Vivo. *Neurosci. Lett.* **2018**, *668*, 103–107. [CrossRef]
22. Gerace, E.; Landucci, E.; Scartabelli, T.; Moroni, F.; Pellegrini-Giampietro, D.E. Rat Hippocampal Slice Culture Models for the Evaluation of Neuroprotective Agents. In *Neurotrophic Factors*; Skaper, S.D., Ed.; Methods in Molecular Biology; Humana Press: Totowa, NJ, USA, 2012; Volume 846, pp. 343–354. ISBN 978-1-61779-535-0.
23. McKenna, M.; Filteau, J.R.; Butler, B.; Sluis, K.; Chungyoun, M.; Schimek, N.; Nance, E. Organotypic Whole Hemisphere Brain Slice Models to Study the Effects of Donor Age and Oxygen-Glucose-Deprivation on the Extracellular Properties of Cortical and Striatal Tissue. *J. Biol. Eng.* **2022**, *16*, 14. [CrossRef]
24. Parmentier, C.E.J.; De Vries, L.S.; Groenendaal, F. Magnetic Resonance Imaging in (Near-)Term Infants with Hypoxic-Ischemic Encephalopathy. *Diagnostics* **2022**, *12*, 645. [CrossRef]
25. Wisnowski, J.L.; Wintermark, P.; Bonifacio, S.L.; Smyser, C.D.; Barkovich, A.J.; Edwards, A.D.; De Vries, L.S.; Inder, T.E.; Chau, V. Neuroimaging in the Term Newborn with Neonatal Encephalopathy. *Semin. Fetal Neonatal Med.* **2021**, *26*, 101304. [CrossRef]
26. Soria, F.N.; Fernandez-Ballester, M. *Rat Organotypic Cultures for AAV-Mediated Vital Labeling of the Extracellular Matrix* (Version 2). protocols.io, 2024. Available online: https://dx.doi.org/10.17504/protocols.io.e6nvwdnzzlmk/v2 (accessed on 18 December 2025).
27. Allen, D. *Protein Array Tool* (Version 2.0.0.1). MATLAB Central File Exchange, 2020. Available online: https://www.mathworks.com/matlabcentral/fileexchange/35128-protein-array-tool (accessed on 18 December 2025).
28. Towfighi, J.; Mauger, D.; Vannucci, R.C.; Vannucci, S.J. Influence of Age on the Cerebral Lesions in an Immature Rat Model of Cerebral Hypoxia–Ischemia: A Light Microscopic Study. *Dev. Brain Res.* **1997**, *100*, 149–160. [CrossRef] [PubMed]
29. Xu, L.; Sapolsky, R.M.; Giffard, R.G. Differential Sensitivity of Murine Astrocytes and Neurons from Different Brain Regions to Injury. *Exp. Neurol.* **2001**, *169*, 416–424. [CrossRef]
30. Ianevski, A.; Cámara-Quílez, M.; Wang, W.; Suganthan, R.; Hildrestrand, G.; Grini, J.V.; Døskeland, D.S.; Ye, J.; Bjørås, M. Early Transcriptional Responses Reveal Cell Type-Specific Vulnerability and Neuroprotective Mechanisms in the Neonatal Ischemic Hippocampus. *Acta Neuropathol. Commun.* **2025**, *13*, 147. [CrossRef]
31. Nakajima, W.; Ishida, A.; Lange, M.S.; Gabrielson, K.L.; Wilson, M.A.; Martin, L.J.; Blue, M.E.; Johnston, M.V. Apoptosis Has a Prolonged Role in the Neurodegeneration after Hypoxic Ischemia in the Newborn Rat. *J. Neurosci.* **2000**, *20*, 7994–8004. [CrossRef] [PubMed]
32. Graham, E.M.; Sheldon, R.A.; Flock, D.L.; Ferriero, D.M.; Martin, L.J.; O'Riordan, D.P.; Northington, F.J. Neonatal Mice Lacking Functional Fas Death Receptors Are Resistant to Hypoxic–Ischemic Brain Injury. *Neurobiol. Dis.* **2004**, *17*, 89–98. [CrossRef] [PubMed]
33. Yigitkanli, K.; Zheng, Y.; Pekcec, A.; Lo, E.H.; Van Leyen, K. Increased 12/15-Lipoxygenase Leads to Widespread Brain Injury Following Global Cerebral Ischemia. *Transl. Stroke Res.* **2017**, *8*, 194–202. [CrossRef]
34. Alonso-Alconada, D.; Hilario, E.; Álvarez, F.J.; Álvarez, A. Apoptotic Cell Death Correlates With ROS Overproduction and Early Cytokine Expression After Hypoxia–Ischemia in Fetal Lambs. *Reprod. Sci.* **2012**, *19*, 754–763. [CrossRef] [PubMed]
35. Meibach, R.C.; Ross, D.A.; Cox, R.D.; Glick, S.D. The Ontogeny of Hippocampal Energy Metabolism. *Brain Res.* **1981**, *204*, 431–435. [CrossRef] [PubMed]
36. Khakpai, F.; Zarrindast, M.R.; Nasehi, M.; Haeri-Rohani, A.; Eidi, A. The Role of Glutamatergic Pathway between Septum and Hippocampus in the Memory Formation. *EXCLI J.* **2013**, *12*, 41–51.

37. Schmidt-Kastner, R.; Freund, T.F. Selective Vulnerability of the Hippocampus in Brain Ischemia. *Neuroscience* **1991**, *40*, 599–636. [CrossRef]
38. Wang, X.; Michaelis, E.K. Selective Neuronal Vulnerability to Oxidative Stress in the Brain. *Front. Aging Neurosci.* **2010**, *2*, 12. [CrossRef]
39. Yin, B.; Barrionuevo, G.; Weber, S.G. Optimized Real-Time Monitoring of Glutathione Redox Status in Single Pyramidal Neurons in Organotypic Hippocampal Slices during Oxygen–Glucose Deprivation and Reperfusion. *ACS Chem. Neurosci.* **2015**, *6*, 1838–1848. [CrossRef]
40. Ziemka-Nalecz, M.; Jaworska, J.; Zalewska, T. Insights Into the Neuroinflammatory Responses After Neonatal Hypoxia-Ischemia. *J. Neuropathol. Exp. Neurol.* **2017**, *76*, 644–654. [CrossRef] [PubMed]
41. Zhang, Y.; Gao, Z.; Wang, D.; Zhang, T.; Sun, B.; Mu, L.; Wang, J.; Liu, Y.; Kong, Q.; Liu, X.; et al. Accumulation of Natural Killer Cells in Ischemic Brain Tissues and the Chemotactic Effect of IP-10. *J. Neuroinflamm.* **2014**, *11*, 79. [CrossRef]
42. Alam, A.; Thelin, E.P.; Tajsic, T.; Khan, D.Z.; Khellaf, A.; Patani, R.; Helmy, A. Cellular Infiltration in Traumatic Brain Injury. *J. Neuroinflamm.* **2020**, *17*, 328. [CrossRef]
43. Wang, L.-Y.; Tu, Y.-F.; Lin, Y.-C.; Huang, C.-C. CXCL5 Signaling Is a Shared Pathway of Neuroinflammation and Blood–Brain Barrier Injury Contributing to White Matter Injury in the Immature Brain. *J. Neuroinflamm.* **2016**, *13*, 6. [CrossRef]
44. Kang, L.; Yu, H.; Yang, X.; Zhu, Y.; Bai, X.; Wang, R.; Cao, Y.; Xu, H.; Luo, H.; Lu, L.; et al. Neutrophil Extracellular Traps Released by Neutrophils Impair Revascularization and Vascular Remodeling after Stroke. *Nat. Commun.* **2020**, *11*, 2488. [CrossRef]
45. Laridan, E.; Denorme, F.; Desender, L.; François, O.; Andersson, T.; Deckmyn, H.; Vanhoorelbeke, K.; De Meyer, S.F. Neutrophil Extracellular Traps in Ischemic Stroke Thrombi. *Ann. Neurol.* **2017**, *82*, 223–232. [CrossRef] [PubMed]
46. Rosell, A.; Cuadrado, E.; Ortega-Aznar, A.; Hernández-Guillamon, M.; Lo, E.H.; Montaner, J. MMP-9–Positive Neutrophil Infiltration Is Associated to Blood–Brain Barrier Breakdown and Basal Lamina Type IV Collagen Degradation During Hemorrhagic Transformation After Human Ischemic Stroke. *Stroke* **2008**, *39*, 1121–1126. [CrossRef]
47. Kim, S.-W.; Lee, H.; Lee, H.-K.; Kim, I.-D.; Lee, J.-K. Neutrophil Extracellular Trap Induced by HMGB1 Exacerbates Damages in the Ischemic Brain. *Acta Neuropathol. Commun.* **2019**, *7*, 94. [CrossRef]
48. Terao, Y.; Ohta, H.; Oda, A.; Nakagaito, Y.; Kiyota, Y.; Shintani, Y. Macrophage Inflammatory Protein-3alpha Plays a Key Role in the Inflammatory Cascade in Rat Focal Cerebral Ischemia. *Neurosci. Res.* **2009**, *64*, 75–82. [CrossRef]
49. Yilmaz, G.; Granger, D.N. Leukocyte Recruitment and Ischemic Brain Injury. *Neuromol. Med.* **2010**, *12*, 193–204. [CrossRef]
50. Ramakrishnan, S.; Anand, V.; Roy, S. Vascular Endothelial Growth Factor Signaling in Hypoxia and Inflammation. *J. Neuroimmune Pharmacol.* **2014**, *9*, 142–160. [CrossRef]
51. Wang, X.; Siren, A.-L.; Liu, Y.; Yue, T.-L.; Barone, F.C.; Feuerstein, G.Z. Upregulation of Intercellular Adhesion Molecule 1 (ICAM-1) on Brain Microvascular Endothelial Cells in Rat Ischemic Cortex. *Mol. Brain Res.* **1994**, *26*, 61–68. [CrossRef]
52. Zhang, Z.G.; Zhang, L.; Tsang, W.; Soltanian-Zadeh, H.; Morris, D.; Zhang, R.; Goussev, A.; Powers, C.; Yeich, T.; Chopp, M. Correlation of VEGF and Angiopoietin Expression with Disruption of Blood–Brain Barrier and Angiogenesis after Focal Cerebral Ischemia. *J. Cereb. Blood Flow Metab.* **2002**, *22*, 379–392. [CrossRef] [PubMed]
53. Jickling, G.C.; Liu, D.; Ander, B.P.; Stamova, B.; Zhan, X.; Sharp, F.R. Targeting Neutrophils in Ischemic Stroke: Translational Insights from Experimental Studies. *J. Cereb. Blood Flow Metab.* **2015**, *35*, 888–901. [CrossRef] [PubMed]
54. Yang, C.; Hawkins, K.E.; Doré, S.; Candelario-Jalil, E. Neuroinflammatory Mechanisms of Blood-Brain Barrier Damage in Ischemic Stroke. *Am. J. Physiol.-Cell Physiol.* **2019**, *316*, C135–C153. [CrossRef] [PubMed]
55. Gidday, J.M.; Gasche, Y.G.; Copin, J.-C.; Shah, A.R.; Perez, R.S.; Shapiro, S.D.; Chan, P.H.; Park, T.S. Leukocyte-Derived Matrix Metalloproteinase-9 Mediates Blood-Brain Barrier Breakdown and Is Proinflammatory after Transient Focal Cerebral Ischemia. *Am. J. Physiol.-Heart Circ. Physiol.* **2005**, *289*, H558–H568. [CrossRef] [PubMed]
56. Bednarek, N.; Svedin, P.; Garnotel, R.; Favrais, G.; Loron, G.; Schwendiman, L.; Hagberg, H.; Morville, P.; Mallard, C.; Gressens, P. Increased MMP-9 and TIMP-1 in Mouse Neonatal Brain and Plasma and in Human Neonatal Plasma after Hypoxia–Ischemia: A Potential Marker of Neonatal Encephalopathy. *Pediatr. Res.* **2012**, *71*, 63–70. [CrossRef]
57. Abellanas, M.A.; Zamarbide, M.; Basurco, L.; Luquin, E.; Garcia-Granero, M.; Clavero, P.; San Martin-Uriz, P.; Vilas, A.; Mengual, E.; Hervas-Stubbs, S.; et al. Midbrain Microglia Mediate a Specific Immunosuppressive Response under Inflammatory Conditions. *J. Neuroinflamm.* **2019**, *16*, 233. [CrossRef]
58. Grabert, K.; Michoel, T.; Karavolos, M.H.; Clohisey, S.; Baillie, J.K.; Stevens, M.P.; Freeman, T.C.; Summers, K.M.; McColl, B.W. Microglial Brain Region−dependent Diversity and Selective Regional Sensitivities to Aging. *Nat. Neurosci.* **2016**, *19*, 504–516. [CrossRef]
59. Szaflarski, J.; Burtrum, D.; Silverstein, F.S. Cerebral Hypoxia-Ischemia Stimulates Cytokine Gene Expression in Perinatal Rats. *Stroke* **1995**, *26*, 1093–1100. [CrossRef]

 https://doi.org/10.3390/antiox15010013

60. Liu, T.; McDonnell, P.C.; Young, P.R.; White, R.F.; Siren, A.L.; Hallenbeck, J.M.; Barone, F.C.; Feurestein, G.Z. Interleukin-1 Beta mRNA Expression in Ischemic Rat Cortex. *Stroke* **1993**, *24*, 1746–1750. [CrossRef]
61. Wang, X.; Figueroa, B.E.; Stavrovskaya, I.G.; Zhang, Y.; Sirianni, A.C.; Zhu, S.; Day, A.L.; Kristal, B.S.; Friedlander, R.M. Methazolamide and Melatonin Inhibit Mitochondrial Cytochrome C Release and Are Neuroprotective in Experimental Models of Ischemic Injury. *Stroke* **2009**, *40*, 1877–1885. [CrossRef]
62. Lin, Y.; Chen, T.; Hung, C.; Tai, S.; Huang, S.; Chang, C.; Hung, H.; Lee, E. Melatonin Protects Brain against Ischemia/Reperfusion Injury by Attenuating Endoplasmic Reticulum Stress. *Int. J. Mol. Med.* **2018**, *42*, 182–192. [CrossRef]
63. Luchetti, F.; Nasoni, M.G.; Burattini, S.; Mohammadi, A.; Pagliarini, M.; Canonico, B.; Ambrogini, P.; Balduini, W.; Reiter, R.J.; Carloni, S. Melatonin Attenuates Ischemic-like Cell Injury by Promoting Autophagosome Maturation via the Sirt1/FoxO1/Rab7 Axis in Hippocampal HT22 Cells and in Organotypic Cultures. *Cells* **2022**, *11*, 3701. [CrossRef] [PubMed]
64. Ding, K.; Wang, H.; Xu, J.; Li, T.; Zhang, L.; Ding, Y.; Zhu, L.; He, J.; Zhou, M. Melatonin Stimulates Antioxidant Enzymes and Reduces Oxidative Stress in Experimental Traumatic Brain Injury: The Nrf2–ARE Signaling Pathway as a Potential Mechanism. *Free Radic. Biol. Med.* **2014**, *73*, 1–11. [CrossRef] [PubMed]
65. Yan, L.; Han, X.; Zhang, M.; Kou, H.; Liu, H.; Cheng, T. Melatonin Exerts Neuroprotective Effects in Mice with Spinal Cord Injury by Activating the Nrf2/Keap1 Signaling Pathway via the MT2 Receptor. *Exp. Ther. Med.* **2023**, *27*, 37. [CrossRef] [PubMed]
66. Ikram, M.; Park, H.Y.; Ali, T.; Kim, M.O. Melatonin as a Potential Regulator of Oxidative Stress, and Neuroinflammation: Mechanisms and Implications for the Management of Brain Injury-Induced Neurodegeneration. *J. Inflamm. Res.* **2021**, *14*, 6251–6264. [CrossRef]
67. Alonso-Alconada, D.; Álvarez, A.; Arteaga, O.; Martínez-Ibargüen, A.; Hilario, E. Neuroprotective Effect of Melatonin: A Novel Therapy against Perinatal Hypoxia-Ischemia. *Int. J. Mol. Sci.* **2013**, *14*, 9379–9395. [CrossRef]
68. Feng, Y.; Rhodes, P.G.; Bhatt, A.J. Neuroprotective Effects of Vascular Endothelial Growth Factor Following Hypoxic Ischemic Brain Injury in Neonatal Rats. *Pediatr. Res.* **2008**, *64*, 370–374. [CrossRef] [PubMed]

 https://doi.org/10.3390/antiox15010013

antioxidants

MDPI

Article

Caffeine Protects Against Hyperoxia-Induced Structural Lung Injury and Restores Alveolar Development in Neonatal Rats

Stefanie Endesfelder * and Christoph Bührer

Department of Neonatology, Charité—Universitätsmedizin Berlin, Augustenburger Platz 1, 13353 Berlin, Germany; christoph.buehrer@charite.de

* Correspondence: stefanie.endesfelder@charite.de; Tel.: +49-030-450-559-548

Abstract: In the developing lung, oxidative stress caused by relative hyperoxia constitutes a central pathogenic mechanism of neonatal lung injury resulting in bronchopulmonary dysplasia (BPD). The immature postnatal lung is highly susceptible to oxidative damage due to incomplete antioxidant defenses and ongoing alveolar and vascular maturation. In a postnatal high-oxygen-induced rat model of BPD-associated lung injury, three or five days of exposure to 80% oxygen was found to disrupt developmental signaling pathways, downregulating genes essential for alveolarization and angiogenesis while inducing profibrotic mediators and collagen expression (Sirius Red staining). These changes resulted in simplified alveolar architecture, as quantified by toluidine blue staining and mean linear intercept analysis of normalized volumes of parenchyma, non-parenchyma, airspaces, septa, and edema. Acting as a multifunctional antioxidant with antifibrotic activity, caffeine mitigated structural lung damage and normalized the transcription of angiogenic and fibrotic genes. It counteracted TGF-β/CTGF-driven fibrogenic signaling and promoted recovery of normal lung morphology following hyperoxic injury. Under normoxic conditions, however, caffeine transiently upregulated profibrotic mediators. Overall, caffeine mitigates hyperoxia-induced lung injury and may actively promote physiological lung maturation, warranting future studies to define optimal dosing windows, clarify context-dependent fibrotic signaling, and translate gene-level effects into long-term clinical outcomes.

Keywords: hyperoxia; oxygen; postnatal immature lung; caffeine; morphology; fibrosis

1. Introduction

The prevention of bronchopulmonary dysplasia (BPD) and the reduction in respiratory instability remain major challenges in neonatal intensive care. Improved survival rates are accompanied by a higher rate of chronic respiratory complications [1,2]. Extremely and very preterm infants (<28 weeks of gestational age) are especially susceptible to respiratory instability evolving into (BPD) [3]. Despite advances in neonatal intensive care medicine, such as the increased use of non-invasive ventilation and optimized surfactant therapy, the incidence of BPD remains high or is even increasing, while mortality is declining [4,5].

BPD is one of the most common respiratory complications in premature infants. Similarly, preterm infants without BPD, as well as moderate and late preterm infants, may also exhibit relevant and sometimes persistent impairments in lung function and an increased risk of respiratory morbidity [6–8]. Lifelong morbidities affecting various aspects of quality of life can result from respiratory diseases following premature birth. Cohort studies show that preterm infants with very low gestational age (<32 weeks) have a significantly increased risk of persistent and progressive lung function deficits, asthma, and the development of chronic obstructive pulmonary disease in adulthood [9–11]. These morbidities are reflected in higher hospitalization rates, greater medication requirements, and an increased prevalence of respiratory symptoms and infections already in childhood, often with functional limitations persisting into adulthood [12].

Current intervention options for the treatment of BPD primarily involve a combination of non-invasive ventilation, targeted pharmacotherapy, and supportive measures. The most important evidence-based therapies include the early application of non-invasive ventilation immediately after birth, volume-controlled ventilation, and the minimization of oxygen toxicity to prevent invasive ventilation and the associated lung injury [13–15], as well as surfactant therapy, preferably using less invasive surfactant administration [16]. Corticosteroids are employed to reduce the incidence of BPD; however, the optimal timing, choice of agent, and dosage remain debated [17,18]. Vitamin A supplementation may lower the risk of BPD but is not routinely used due to uncertain cost–benefit considerations. Diuretics and bronchodilators are applied symptomatically in established BPD, improving lung function in the short term but showing no preventive efficacy [14]. Experimental approaches include clinical and preclinical studies investigating mesenchymal stromal cells (MSCs), IGF-1/IGFBP-3, and IL-1 receptor antagonists [13,19,20]. Addionally, new therapeutic concepts, such as systems pharmacology therapeutics, aim to interrupt the complex, self-reinforcing cycle of BPD progression, comprising oxidative stress, inflammation, and downstream endoplasmic reticulum (ER) stress, to promote proper lung regeneration rather than defective repair [21]. While preclinical data indicate improved lung development and reduced inflammation, clinical benefits remain to be confirmed. The prophylactic and therapeutic use of caffeine is well established and significantly reduces the risk of BPD [14,15,22].

BPD is characterized by a complex interplay of impaired lung development and dysregulated processes of injury and repair [23]. Morphologically, this is manifested by a simplified alveolar architecture with reduced septation and enlarged alveoli, resulting in a diminished surface area for gas exchange [24–26]. In parallel, pulmonary microvascular development is impaired, leading to dysmorphic angiogenesis with reduced capillary density and an increased risk of pulmonary hypertension [24,25,27]. The small airways also exhibit pathological alterations, including wall thickening, smooth muscle hyperplasia, and enhanced reactivity, which contribute to obstructive ventilatory disorders and increased susceptibility to infection [24,27]. Interstitial remodeling processes involving fibrosis and chronic inflammation further compromise lung elasticity and compliance [25,26].

Ventilation-induced damage aggravates these changes by causing endothelial and epithelial dysfunction and disrupting the integrity of alveolar-capillary barrier. This results in pulmonary edema, increased protein permeability and diffuse alveolar damage, which progresses through exudative, proliferative, and fibrotic phases characterized by distinct patterns of tissue remodeling. In premature infants, the lungs are particularly vulnerable as the structural and functional maturation of both endothelium and epithelium is incomplete [28,29]. Even short periods of mechanical ventilation can lead to significant parenchymal and bronchiolar injury. Although repair mechanisms are activated, the risk of developing chronic lung disease such as BPD remains high [30].

Oxidative stress resulting from hyperoxia and the formation of reactive oxygen species (ROS) is central to BPD pathogenesis [31,32]. Preterm infants are particularly susceptible due to immature antioxidant defense systems [33,34]. Reduced glutathione concentrations limit free radical detoxification, promote inflammation and apoptosis, and disrupt alveolar and vascular development [35,36]. Elevated proinflammatory cytokines such as IL-6, IL-

8, and TNF-α contribute to chronic tissue damage and impaired alveolarization [37,38], with an increased neutrophil-to-lymphocyte ratio indicating systemic inflammation [38]. Additional mechanisms include impaired VEGF signaling and angiogenesis [39], epigenetic modifications [38], and programmed cell death pathways such as apoptosis, autophagy, and ferroptosis [32]. Genetic predispositions and dysregulated immune responses, including Th1/Th2 imbalance, further increase risk [23,38]. Overall, oxidative injury constitutes the major molecular risk landscape BPD progression.

Given the central role in premature lung injury, antioxidant strategies, including potential effects of caffeine, may hold therapeutic relevance. The interaction between oxidative stress and inflammation, although critical, remains incompletely understood in BPD an is an important area of ongoing [40].

Animal models, particularly mouse and rat models, are essential for investigating these mechanisms under controlled conditions [41]. They allow targeted analysis of oxygen toxicity, antioxidant system dysregulation, and BPD development [42]. The lungs of rodents in the late fetal and early neonatal periods are comparable to human preterm lungs in terms of development and antioxidant capacity [14,42]. Similarities include immature surfactant production, reduced antioxidant enzyme activity, and heightened vulnerability to oxidative stress after birth and oxygen exposure [42,43]. Changes observed in animal models after hyperoxia or intermittent hypoxia, such as alveolar hypoplasia, inflammation, and impaired vascular development, mirror those seen in human preterm infants [42–44]. Despite these similarities, differences in lung development, genetics, and immune response limit clinical translatability, so findings must be critically evaluated [42,43,45].

Against this background, animal models are particularly valuable for investigating antioxidant therapies such as caffeine. In mouse and rat models, caffeine exhibits pronounced antioxidant and anti-inflammatory effects. It reduces oxidative DNA damage, modulates antioxidant enzyme expression, inhibits NLRP3 inflammasome and NF-κB activation, and decreases alveolar epithelial cell apoptosis [46–52]. Additionally, caffeine promotes angiogenesis and alveolar development [53,54]. These effects are dose-dependent, as excessive caffeine concentrations may induce cellular damage [55]. The protective effects observed in animal studies correspond to clinical findings of reduced BPD incidence in preterm infants.

This study aims to investigate the effects of neonatal hyperoxia on postnatal lung development and to explore whether caffeine can modulate these outcomes. We focus on structural, molecular, and remodeling changes, including alveolar simplification, angiogenesis, and expression of key developmental and pro-fibrotic markers. Specifically, we are investigating the acute phase of oxygen exposure within the saccular stage and the transition from the saccular to the alveolar stage of lung development, as well as the subsequent effects in adolescent rats during the alveolar stage without further insults, in order to assess both immediate and longer-term impacts on lung maturation. Furthermore, we aim to determine whether caffeine exerts protective effects under oxidative stress and whether its impact is context-dependent, potentially influencing lung development differently under normoxic versus hyperoxic conditions. The findings of this study are expected to clarify caffeine's role as a potential modulator of postnatal lung development and its therapeutic relevance for preterm infants at risk of BPD.

2. Materials and Methods

2.1. Animal Welfare

After an acclimatization period, the pregnant *Wistar* rats were housed individually for two days before labor under controlled environmental conditions with a constant 12 h/12 h light/dark cycle, room temperature, and 60% relative humidity and had access to food and water at all times. In a 12 h birth window, pups were randomized by sex and number and assigned to lactating dams according to the experimental group. The animal experimental procedures were evaluated and approved by the local animal welfare authority (LAGeSo, authorization number G-0088/16), including a prior case number determination. All procedures performed on the animal, as well as the husbandry conditions, were in accordance with institutional and ARRIVE guidelines. For a comprehensive evaluation of the data collected, reference is made to the previously published studies with identical experimental cohorts [50,51,56,57], which have focused on pulmonary and neurodevelopmental analyses.

2.2. Experimental Design and Caffeine Administration

As described previously [50,51,56,57], all pups born within a 12 h time window were randomized by sex across litters and assigned to lactating dams according to litter size. Sample size was calculated using G∗Power V3.1.2 [58] for local authority approval. Dams with assigned rat pups were randomly divided into the different test groups. Exposure to two oxygen concentrations, either room air (normoxia, NO) or an 80% oxygen enriched atmosphere (hyperoxia, HY), was performed. Normoxic conditions were maintained in the animal housing, while hyperoxia-exposed pups remained with their dams in an 80% oxygen environment in an incubator (OxyCycler BioSpherix, Lacona, NY, USA). The duration of the modified oxygen environment was applied for two distinct periods: from postnatal day (P)0 to P3 (n = 6–8) or from P0 to P5 (n = 6–8) for both normoxia and hyperoxia conditions. Compared to neonatal pups, adult rats have limited tolerance to high oxygen concentrations and may stop feeding the pups. To prevent oxygen toxicity in the dams, they were alternated between hyperoxic and normoxic litters every 24 h. Pups were randomized into experimental groups on the day of birth (P0). In addition to the oxygen concentration, the treatment of the pups varied in the application of either caffeine or vehicle. Caffeine or vehicle was administered intraperitoneally (i.p.), correlated to body weight (100 µL/10 g). Crucially, the first application was administered on P0 prior to the onset of the respective oxygen exposure. The subsequent dosing schedule depended on the exposure duration: the P0-P3 subset received a total of two applications (on P0 and P2), while the P0–P5 subset received a total of three applications (on P0, P2, and P4). Validation of plasma caffeine levels confirmed that the dosing regimen was sufficient to achieve concentrations in the pups (8.8–11.2 µg/mL, [50]) that fell within the clinical therapeutic range (5.5–23.7 µg/mL, [59]). Importantly, experimental groups maintained until P15 received no further applications of caffeine or vehicle after the initial exposure period (P3 or P5), thus remaining untreated during the subsequent recovery phase under room air/normoxia. The dose of 10 mg/kg pure caffeine is equivalent to 20 mg/kg caffeine citrate, which is commonly used clinically [60]. This protocol resulted in four distinct groups at each exposure duration: (a) normoxia (NO, control group): application of 21% oxygen with vehicle (phosphate-buffered saline, PBS, Thermo Fisher Scientific Inc., Waltham, MA, USA), (b) normoxia with caffeine (NOC): 21% oxygen with caffeine (10 mg/kg, Sigma-Aldrich, Steinheim, Germany), (c) hyperoxia (HY): 80% oxygen with vehicle (PBS) and (d) hyperoxia with caffeine (HYC): 80% oxygen with caffeine (10 mg/kg). The study relevant examinations were performed immediately after the end of the oxygen exposure (P3, P5) or after recovery in room air at P15 (P3_P15, P5_P15). The experimental design with the timeline, the experimental groups and the outcome parameters is shown below in Figure 1.

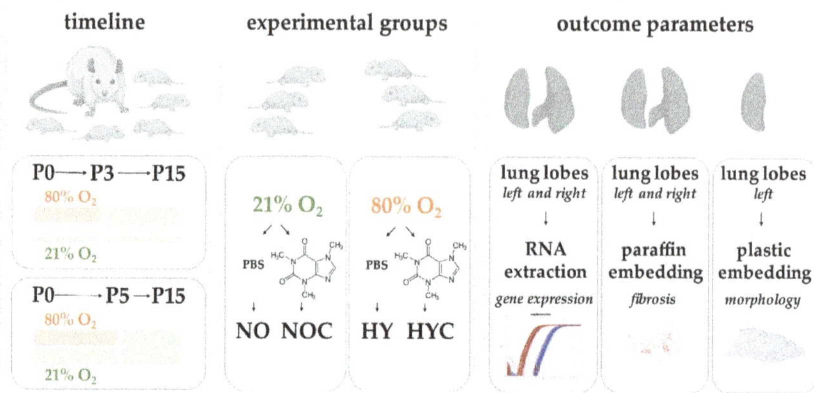

Figure 1. Schematic of the experimental design showing oxygen exposure (normoxia or hyperoxia, **left box**) with or without caffeine treatment from P0 to P3 or P5, followed by analyses either immediately (P3, P5) or after recovery in room air (P15). Four groups were studied at each time point: normoxia with vehicle (NO), normoxia with caffeine (NOC), hyperoxia with vehicle (HY), and hyperoxia with caffeine (HYC, **middle box**). Outcome parameters (**right box**) included RNA gene expression, Sirius Red staining (paraffin sections), and morphology (plastic embedding).

2.3. *Sampling and Processing Lung Tissue*

2.3.1. Tissue for RNA Extraction and Paraffin Embedding

At the time of targeted analyses (P3, P5, P3_P15 and P5_P15), as previously described [50,51], pups received i.p. injection of anesthetics (ketamine (100 mg/kg), xylazine (20 mg/kg), and acepromazine (3 mg/kg)). After opening the thorax, the pups were perfused transcardially as described previously. Cold PBS (pH 7.4) was used to perfuse the lung tissue for further transcript analysis. To prepare the whole lung for paraffin embedding, the PBS perfusion was followed by a further perfusion step with cold paraformaldehyde (PFA, 4% (*w*/*v*), pH 7.4, Merck KGaA, Darmstadt, Germany). After exsanguination of the deeply anesthetized rat pups, the lungs were removed and either snap-frozen in liquid nitrogen for gene expression studies and stored at −80 °C until further use, or dehydrated in ethanol for embedding after one day of post-fixation in PFA at 4 °C and then embedded in paraffin.

2.3.2. Tissue for Glycol Methacrylate Embedding

Sedation and deep anesthesia were achieved with an i.p. injection of ketamine, xylazine and acepromazine (see below 2.3.1). The abdomen and thorax were opened and the trachea was dissected free. A tube was inserted into the trachea via the oral cavity and fixed with a ligature in order to start transcranial perfusion. This was performed under constant pressure, first with cold isotonic Ringer's solution mixed with procaine and heparin, and then with cold glutaraldehyde-formaldehyde solution (4% (*w*/*v*) PFA and 0.1% (*v*/*v*) glutaraldehyde (GA, Carl Roth GmbH, Karlsruhe, Germany) in 0.2 M HEPES, Merck KGaA). Meanwhile, a recruitment maneuver (10 cm H_2O to 30 cm H_2O) was performed twice via a connected ventilation system (MidiVent Ventilator Model 849, Harvard Apparatus GmbH, March-Hugstetten, Germany) to optimally inflate the lungs. The tube was then removed and the heart-lung package carefully removed and post-fixed in the perfusion solution (PFA/GA/HEPES) for a further 24 h. Until further use and after removal of the heart, the lungs were stored at 4 °C in 0.15 M HEPES. The volume of the whole lungs was determined according to Archimedes' principle [61].

Plastic embedding of the left and the right lung lobes in glycol methacrylate (GMA, Technovit 7100; Heraeus Kulzer, Wehrheim, Germany) was performed over three consecutive days. For the embedding procedure, the lungs were washed twice for 10 min each in 0.15 M HEPES (pH 7.4) and then four times for 5 min each with 0.1 M sodium cacodylate buffer. The samples then remained in distilled water for about one hour. Dehydration was carried out using an ascending acetone dilution series with incubation times of 20 min twice each (70%, 80%, 90% and 100%), with the last step being repeated three times. A mixture of the base solution (Technovit 7100) with 100% acetone was added to the tissue for preinfiltration and incubated overnight at 4 °C. On the second day, the preinfiltration medium was replaced by freshly prepared infiltration medium (100 mL Technovit 7100 with 1 g hardener I). The samples were stored overnight at room temperature and shaken gently. Plastic embedding was carried out on the third day. Infiltration medium was mixed 15:1 with Hardener II, mixed well and placed in embedding cassettes for standardized positioning of the lungs. For optimal polymerization, the samples were left at room temperature for at least 24 h. For later sample preparation for cutting, they were glued to plexiglass blocks with superglue and stored cool. Using a rotary microtome, the glycol methacrylate-embedded lungs were cut into 1.5 μm thick sections and mounted on uncoated slides.

2.4. *RNA Extraction and qPCR*

The procurement of lung tissue for the transcript analyses with underlying RNA isolation and subsequent transcription into cDNA has already been described in the work carried out with the same animal cohorts [50,51]. Briefly described again, total RNA was isolated from frozen tissue by acid phenol/chloroform extraction (peqGOLD RNAPure™, PEQLAB Biotechnologie, Erlangen, Germany). The isolated and DNase-treated RNA (2 μg) was reverse-transcribed, then amplified and quantified in a real-time qPCR with the target gene sequences summarized in Supplementary Table S1, with detailed abbreviations for the amplified genes. Amplification of target genes was performed using the mastermix qPCR BIO Mix Hi-ROX (NIPPON Genetics Europe, Düren, Germany) in the QuantStudio™ 3 Real-Time PCR System (Applied Biosystems, Carlsbad, CA, USA, Thermo Fisher Scientific Inc.). Using standardized HPRT as an internal reference, expression analyses of selected target genes were performed using the $2^{-\Delta\Delta CT}$ method [62].

2.5. *Sirius Red and Toluidine Blue Staining*

The paraffin-embedded lungs were deparaffinized and then hydrated with descending alcohol concentrations (EtOH; 96%, 80%, 70%). After the sections were rinsed in distilled water, they were incubated with a saturated aqueous picric acid solution (Sigma-Aldrich) for at least 1 h. The slides were then washed twice with freshly prepared acidified water (0.5% (*v*/*v*) glacial acetic acid in ddH_2O) and hydrated again with an ascending concentration of alcohol (EtOH; 70%, 80%, 96%) and cleared with xylene substitute (ROTI®Histol, Carl Roth GmbH) before mounted on the slide with aqueous mounting medium.

The GMA-embedded lung sections from the young rats were incubated for 15 min at room temperature with an aqueous toluidine blue solution (1% (*w*/*v*) toluidine blue O (Carl Roth GmbH), 2.5% (*w*/*v*) sodium bicarbonate). The sections were then differentiated in deionized water and mounted in a resin-containing medium.

2.6. Aspects of Morphologic Analysis

The left toluidine blue-stained lung lobes of all experimental animals were used for quantitative analysis of morphological characteristics. The stereological analyses were performed according to the recommendations of the American Thoracic Society/European Respiratory Society [63] for the quantitative assessment of lung structure, whereby all lungs were embedded in the same orientation and serially cut into 1.5 μm thick sections. From the beginning of the tissue structure in the GMA block, the sections were numbered and selected for further analyses in comparable areas. For analysis, three left lung lobe sections from each animal were imaged at 10× and 20× magnification with the BZX800 microscope (Keyence, Osaka, Japan) using light microscopy with stitching and Z-stack. The images were taken under identical conditions and then analyzed blindly.

The mean linear intercept (MLI) of lung tissue sections was quantified using ImageJ (version 1.53, National Institutes of Health, Bethesda, MD, USA). Toluidine blue-stained images were acquired at 20× magnification, saved in TIFF format, and standardized to identical resolution. Image segmentation was performed by applying a uniform thresholding method across all samples [64]. To obtain chord lengths, horizontal and vertical test line grids were superimposed on the images. For each animal, ten fields per orientation were analyzed, corresponding to a defined test line length. Image processing was automated through a dedicated macro executed in batch mode, which included background correction, thresholding, chord isolation, and export of measurement data. The results were saved in Excel-compatible format. The MLI was calculated as the sum of the lengths [μm] of all counting lines divided by the total number of intercepts with alveolar septa. The mean intercept value per sample was obtained by averaging the values of all sections from the same animal. These means were then used to compare the intercepts of treated animals with those of control animals.

For each left lung examined in this study, a set of predefined morphological parameters was assessed, including the volume of parenchyma (V_V (par, lung), [mm^3]) and non-parenchyma (V_V (non-par, lung) [mm^3]), the volume of alveolar airspace (V_V (alv, lung), [mm^3]), the volume of alveolar septa (V_V (sept, lung), [mm^3]), the absolute surface area of the alveolar septum (S (sept, lung), [cm^2]), the mean thickness of the alveolar septum (τ (sept), [μm]), and the volume of alveolar edema (V_V (edema, lung), [mm^3]). As an initial step, the total lung volume was determined, since this value served as the essential reference volume against which all subsequent stereological volume estimations were normalized [65]. Histological sections were stained with toluidine blue and imaged at 10× magnification. Images were acquired in JPEG format and standardized to an identical resolution to ensure comparability across samples. The quantification of volume densities was performed using the point-counting method, a well-established stereological technique that allows unbiased estimation of three-dimensional structures from two-dimensional sections. For this purpose, the software STEPanizer (version 1) was employed, which facilitates stereological evaluation by superimposing grids of test points and lines on digital images [66]. The number of points hitting the respective tissue components was recorded, and volume densities were calculated relative to the reference lung volume. From these volume densities, absolute compartment volumes were derived. For each animal, the estimations were based on a systematic sampling of ten images, ensuring a representative coverage of the lung tissue. The measurements were blinded. All values determined for each sample were first averaged across all sections obtained from the same animal. These animal-level mean values were then used for group-level analyses by calculating the average of all animals within the respective experimental groups. Comparisons between treated and control groups were subsequently based on these group means. All stereological results were exported in an Excel-compatible format to enable structured data storage and facilitate subsequent statistical analysis.

Quantification of the red-stained collagen of paraffin-embedded rat lungs was performed by Sirius Red staining using ImageJ (version 1.53, [67]). For the fibrosis analyses, two complete left and two complete right lung lobes were used for each animal, recorded using a 10× magnification with stitching and Z-stack in the light field (BZX800 microscope). To determine the fibrosis per area, a calibration scale was first created for all subsequent analyses. For this purpose, an object of known size was measured and stored in ImageJ as a reference. The threshold value of the Sirius Red-stained areas was determined via 'colour convolution2' and the data were collected in a standardized manner. The percentage of the fibrotic area in the image section and the area of the corresponding lung were recorded, so that final fibrotic areas [%] per area [mm^2] were included in the evaluation. The measurements were blinded. The mean values per sample were calculated by averaging the values of all sections from the same animal and used to compare the cell counts of the treated animals with those of the control animals. The mean fibrosis percentages of the control animals were used as 100% values, as indicated in the figure legends.

2.7. Statistic

Box and whisker plots were created for the statistical evaluation, identical to the previous analyses of these cohorts. [50,51,56,57] The interquartile range (box) was shown, with the line representing the median, while the whiskers show the data variability outside the upper and lower quartiles. As already described, the groups were compared using a one-way analysis of variance (ANOVA). In the case of a partially non-Gaussian distribution, the analysis was carried out using the Kruskal–Wallis test or, assuming that the groups do not have the same variances, using the Brown–Forsythe test. Post hoc tests were matched to the respective omnibus test with Bonferroni for ANOVA, Dunn's for Kruskal–Wallis, and Dunnett's T3 for the Brown–Forsythe test. A p-value of <0.05 was considered significant. All graphs and statistical analyses were performed using GraphPad Prism 8.0 software (GraphPad Software, La Jolla, CA, USA).

3. Results

3.1. Survival, Growth and Lung Volume

All rat pups survived exposure to high oxygen concentrations. The nursing dams were rotated between hyperoxia- and normoxia-exposed litters, as described, in order to avoid a possible undersupply of the pups. The weight at the time of analysis (P3, P5, and after survival to P15) increased over time according to postnatal development and significant differences between the oxygen and/or caffeine treated pups could not be detected within the different time points (Figure 2a). However, when looking at the percentage difference in weight gain from the day of birth in a randomized distribution of pups, there was a delayed weight gain in caffeine-treated pups with caffeine application under room air conditions at P3 and P5 (Figure 2b). Caffeine-treated rats that remained under 80% oxygen for the first five days of life also showed delayed weight gain compared to both the control and hyperoxia groups.

The lung volumes of the respective rat pups were determined using the fluid displacement method and are equivalent to the lung volume. The lung volumes increase over time due to development. A significant increase in lung volume was observed for the rats exposed to 80% oxygen for 3 and 5 days after recovery in room air (P3_P15, P5_P15) (Figure 2c). The lung volume stabilized at a comparable level to the control animals when rats were treated with caffeine during hyperoxia. This effect was confirmed when lung volume was related to body weight (Figure 2d).

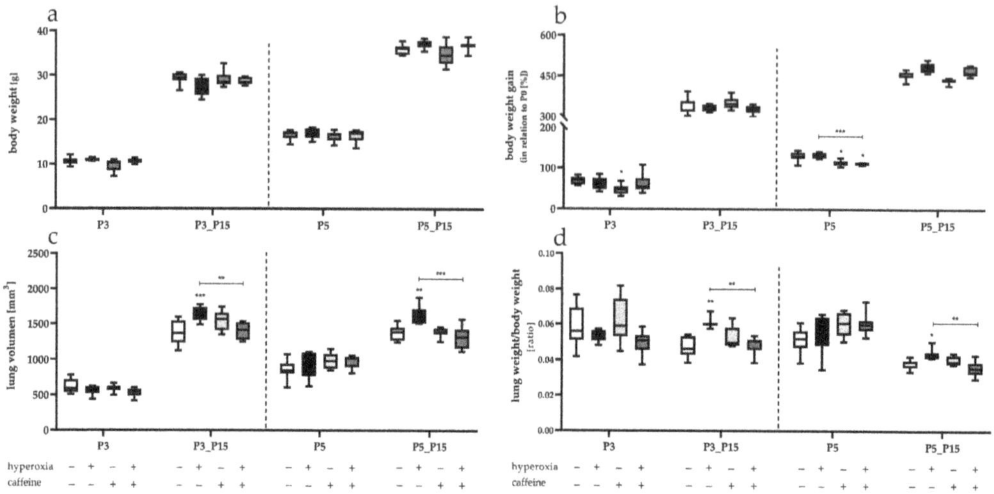

Figure 2. Quantification of (**a**) body weight on examination day, (**b**) body weight gain over time, (**c**) lung volume on examination day, and (**d**) the ratio of lung weight to body weight for 3 days postnatal oxygen exposure (P3) and recovery (P3_P15) as well as 5 days postnatal oxygen exposure (P5) and recovery (P5_P15). Data are presented as median of the experimental groups (normoxia, white; hyperoxia, black; normoxia with caffeine, light gray; hyperoxia with caffeine, dark gray) or normalized to the level of rat pups exposed to normoxia at each time point (body weight gain, control 100%, white bars). n = 7–8 per group; * $p < 0.05$, ** $p < 0.01$, *** $p < 0.001$ (ANOVA; Kruskal–Wallis; Brown–Forsythe).

3.2. Lung Morphology

3.2.1. Lung Development

Rat lung maturation occurs postnatally, with alveolarization taking place during the first two weeks of life [68,69]. Postnatal exposure of newborn rats (P0) to 80% oxygen (HY) for 3 (P3) or 5 days (P5) resulted in marked alterations in alveolar development (Figure 3). Compared to normoxic controls (NO), hyperoxia caused a simplified alveolar architecture with fewer alveoli and enlarged distal airspaces (HY at P3 and P5). These abnormalities persisted after recovery in room air until day 15 (HY at P3_P15, P5_P15), while caffeine treatment (HYC) mitigated these effects.

Quantitative assessment using the MLI (Figure 4a) revealed that hyperoxia increased MLI values at all time points (P3, P5, P3_P15, P5_P15), reflecting alveolar wall destruction and enlarged airspaces. Caffeine prevented this oxygen-induced increase at P3, P5, and after recovery from short-term exposure, although after 5 days of hyperoxia followed by recovery, MLI values remained slightly elevated. Under normoxia, caffeine had no structural effect.

Stereological analyses were performed to evaluate parenchymal and non-parenchymal lung compartments. Volume and surface densities are relative parameters influenced by changes in the reference space, all volume densities were related to total lung volume. The volume fraction of parenchyma (Vv (par, lung), Figure 4b), comprising alveoli, septa, and distal airways, was similar among groups at P3 and P5 but increased in hyperoxia-exposed pups after recovery (P3_P15, P5_P15). This increase was prevented by caffeine. Non-parenchymal volumes (Vv (non-par, lung)), including conducting airways and large vessels, remained unchanged (Figure 4c).

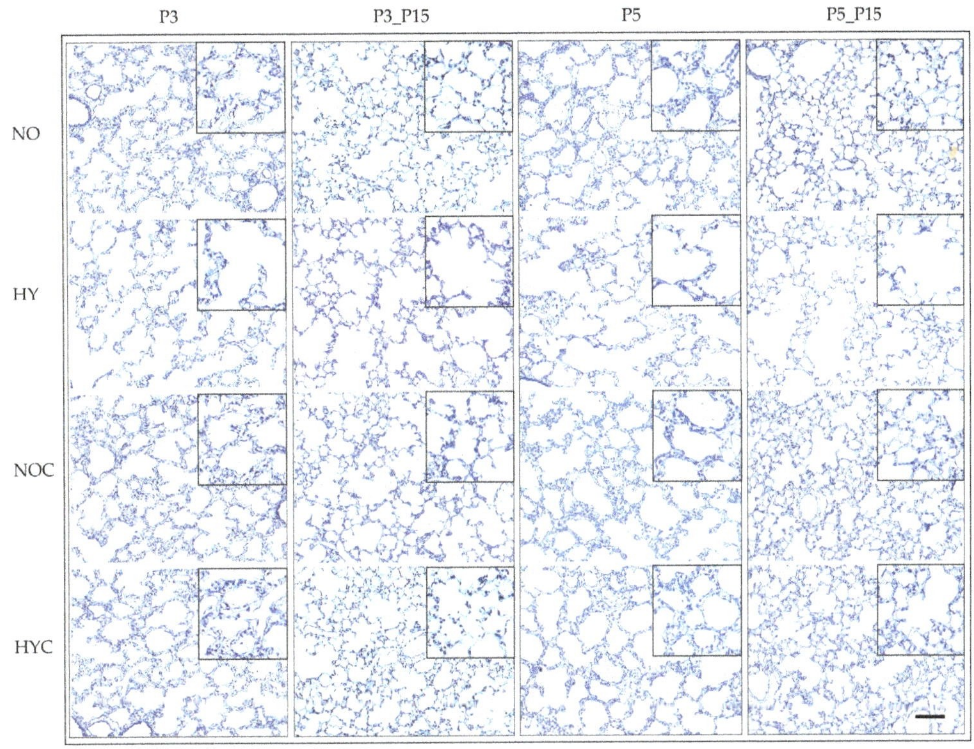

Figure 3. Representative toluidine stain imaging of left lung lobes for 3 days postnatal oxygen exposure (P3, **left**) and recovery (P3_P15, **middle left**) as well as 5 days postnatal oxygen exposure (P5, **middle right**) and recovery (P5_P15, **right**) of rats exposed to normoxia (NO) or hyperoxia (HY) compared to lung lobes of rats treated with caffeine (NOC, HYC). Scale bar 100 μm. The outlined regions (black bordered boxes) are magnified 1.8× compared to the original image.

The fraction of ventilated alveoli and alveolar ducts (Vv (alv, lung), Figure 4d) was unaffected by hyperoxia or caffeine at P3. After recovery, alveolar volume was significantly increased following both 3- and 5-day hyperoxia, an effect blocked by caffeine. Notably, caffeine under normoxia slightly increased alveolar volume at P5. Conversely, the volume fraction of alveolar septa (Vv (sept, lung)) decreased under hyperoxia, consistent with alveolar simplification. Caffeine restored septal volumes to near-control levels except at P3 (Figure 4e). The absolute surface area of the alveolar septum (S (sept, lung), Figure 4f) was markedly reduced by hyperoxia, with a near 50% decrease already evident at P3. At P5, hyperoxia induced only a trend toward reduction, yet both 3- and 5-day exposures resulted in a persistent decrease that remained detectable through P15. Caffeine did not modify these hyperoxia-induced reductions at any time point (Figure 4f). In contrast, the mean thickness of the alveolar septum (τ (sept), Figure 4g) was consistently diminished under hyperoxic conditions at all acute time points (P3, P5), and this reduction persisted after the recovery phase (P15). Caffeine mitigated septal thinning induced by hyperoxia and fully normalized septal thickness by P15. Under normoxic conditions, caffeine alone slightly decreased septal thickness at P5 (Figure 4g).

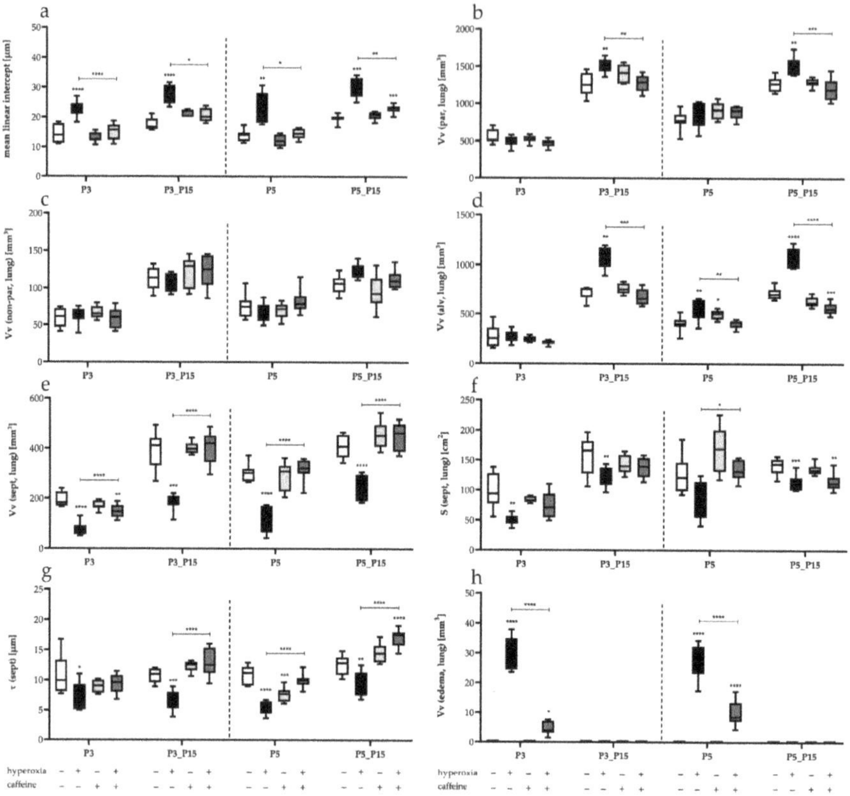

Figure 4. Quantification of (**a**) mean linear intercept (MLI), volume of (**b**) parenchyma (V_V (par, lung)), (**c**) non-parenchyma (V_V (non-par, lung)), (**d**) alveolar airspace (V_V (alv, lung)), (**e**) alveolar septa (V_V (sept, lung)), (**f**) absolute surface of the alveolar septum (S (sept, lung)), (**g**) mean thickness of the alveolar septum (τ (sept)), and (**h**) alveolar edema (V_V (edema, lung)) for 3 days postnatal oxygen exposure (P3) and recovery (P3_P15) as well as 5 days postnatal oxygen exposure (P5) and recovery (P5_P15). Data are presented as median of the experimental groups (normoxia, white; hyperoxia, black; normoxia with caffeine, light gray; hyperoxia with caffeine, dark gray). n = 6–8 per group; * $p < 0.05$, ** $p < 0.01$, *** $p < 0.001$, **** $p < 0.0001$ (ANOVA; Kruskal–Wallis; Brown–Forsythe).

Exposure to high oxygen concentrations may have increased the permeability of the alveolo-capillary barrier, promoting interstitial fluid accumulation. Hyperoxia increased alveolo-capillary permeability, leading to mild interstitial edema ((V_V (edema, lung) with 4.7% at P3; 3.2% at P5). Caffeine successfully reduced the edema to approximately 1% of total lung volume (Figure 4h), rendering this effect unmeasurable in the overall assessment of lung volume. However, separate assessment at the P15 time point, representing early hyperoxia followed by recovery in room air, showed an increased lung volume, both in absolute terms and relative to the pups' body weight (Figure 2c).

3.2.2. Gene Expression of Alveolarization- and Angiogenesis-Associated Mediators

The effect of hyperoxia on the developing lung in relation to alveolarization and angiogenesis was investigated by gene expression analysis of pulmonary transcripts.

CycD2 expression was reduced by 3-day and 5-day hyperoxia. This impairment in transcription persisted after recovery under normoxia (Figure 5a). Without persistent downregulation of gene expression, *FGF10*, *PDGFα*, *PDGFRα*, and *PDGFRβ* showed reduced transcription after acute hyperoxia at P3 and P5 (Figure 5c,e–g). *TGFβ* was down-regulated after neonatal hyperoxic insult only after recovery without high oxygen (P3_P15, P5_P15) (Figure 5h) after it had undergone a strong induction of transcription under acute hyperoxia (P3, P5). Excessive induction of gene expression after significant inhibition at P3 and P5 was seen at P15 for *PDGFRα* and *PDGFRβ* (Figure 5f). An increased transcription level could be detected for *FGF2* and Grem1 after a recovery period in room air (P15) without prior hyperoxia-induced downregulation (Figure 5b,d). Caffeine affected transcription of all target genes when administered concomitantly with 80% oxygen exposure and, with few exceptions, caused significant counter-regulation (*CycD2* at P3, P5 and P3_P15; *FGF2* at P3_P15 and P5_P15; *FGF10* at P3 and P5; *Grem1* at P3_P15 and P5_P15; *PDGFα* at P3; *PDGFRα* at P3 and P5; *PDGFRβ* at P3; *TGFβ* at P3, P5 and P5_P15; Figure 5a–h).

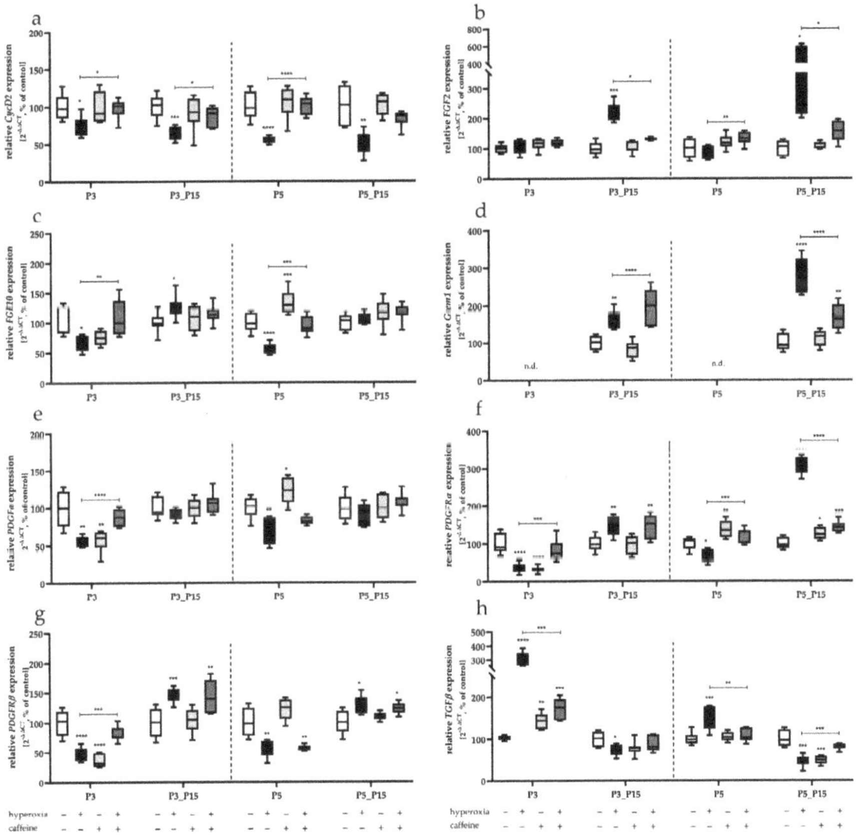

Figure 5. Quantification of alveolarization mediator's transcript expression of a whole lung lobe with (**a**) *CycD2*, (**b**) *FGF2*, (**c**) *FGF10*, (**d**) *Grem1*, (**e**) *PDGFα*, (**f**) *PDGFRα*, (**g**) *PDGFRβ*, and (**h**) *TGFβ* for 3 days postnatal oxygen exposure (P3) and recovery (P3_P15) as well as 5 days postnatal oxygen exposure (P5) and recovery (P5_P15). Data are normalized to the level of rat pups exposed to normoxia at each time point (control 100%, white bars) with verum groups hyperoxia (black), normoxia with caffeine (light gray), and hyperoxia with caffeine (dark gray). n = 7–8 per group; * $p < 0.05$, ** $p < 0.01$, *** $p < 0.001$, **** $p < 0.0001$ (ANOVA; Kruskal-Wallis; Brown–Forsythe).

When caffeine alone was applied under normoxia, various alveolarization-related factors were modulated. *FGF10*, *PDGFα* and *PDGFRα* were up-regulated at P5 compared to untreated controls (Figure 5c,e–f), with *PDGFRα* expression remaining elevated even after recovery in room air (Figure 5f). Caffeine had an inhibitory effect on the transcription of *PDGFα*, *PDGFRα* and *PDGFRβ* after treatment up to P3 (Figure 5e–g), while *TGFβ* was induced by caffeine at P3 (Figure 5h).

Exposure to 80% oxygen for three days decreased the expression of *Angpt1* (P3, Figure 6a) and was shown to be overexpressed after recovery below 21% (P3_P15). *Angpt2*, on the other hand, was strictly upregulated both after three (P3) and after 5 days of hyperoxia (P5, Figure 6b). P5_P15 then showed a drastic downregulation of *Angpt2* (Figure 6b). As an associated receptor to *Angpt1* and *Angpt2*, Tie2 was decreased under 80% oxygen and then increased at P5_P15 (Figure 6c). *VEGF* and *VEGFR2* were also inhibited in transcription under hyperoxia at P3 and P5, while VEGFR1 was induced at P3_P15 and P5 (Figure 6e). Caffeine affected transcription of all target genes when administered in parallel with 80% oxygen exposure and, with few exceptions, caused significant counter-regulation (*Angpt1* at P3 and P3_P15; *Angpt2* at P3, P5 and P5_P15; *Tie2* at P3, P5 and P5_P15; *VEGF* at P3, P5 and P5_P15; *VEGFR1* at P5; *VEGFR2* at P3 and P5_P15; Figure 6a–f).

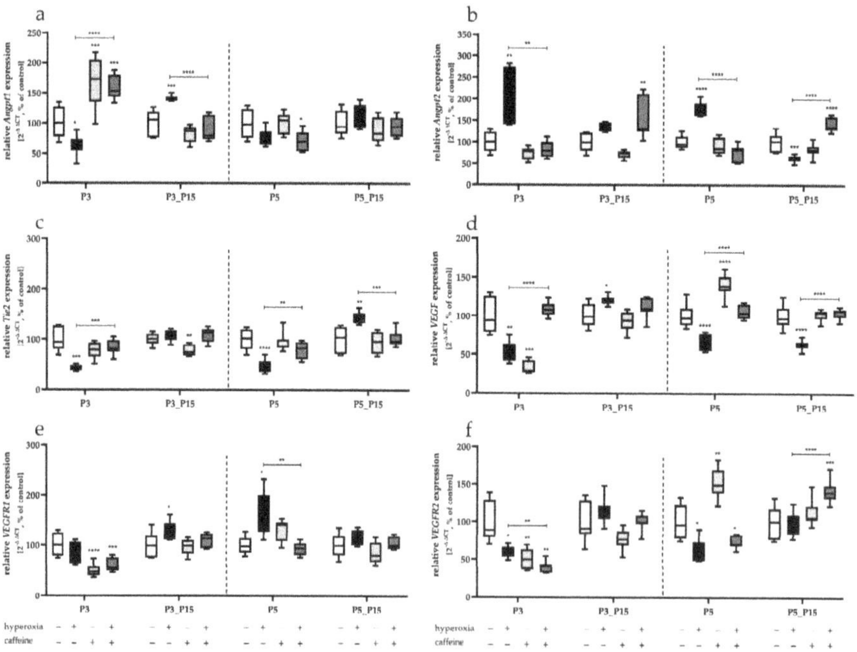

Figure 6. Quantification of angiogenesis mediator's transcript expression of a whole lung lobe with (**a**) *Angpt1*, (**b**) *Angpt2*, (**c**) *Tie2*, (**d**) *VEGF*, (**e**) *VEGFR1*, and (**f**) *VEGFR2* for 3 days postnatal oxygen exposure (P3) and recovery (P3_P15) as well as 5 days postnatal oxygen exposure (P5) and recovery (P5_P15). Data are normalized to the level of rat pups exposed to normoxia at each time point (control 100%, white bars) with verum groups hyperoxia (black), normoxia with caffeine (light gray), and hyperoxia with caffeine (dark gray). n = 7–8 per group; * $p < 0.05$, ** $p < 0.01$, *** $p < 0.001$, **** $p < 0.0001$ (ANOKruskal–Wallisllis; Brown–Forsythe).

When caffeine was administered under normoxia, the transcription of *Angpt1* at P3 and *VEGFR2* at P5 was induced (Figure 6a,f), and an inhibition of transcription by caffeine was detected for *Tie2* at P3_P15 (Figure 6c), *VEGF* at P3 (Figure 6d), for *VEGFR1* at P3 and for *VEGFR2* at P3 (Figure 6e,f).

3.3. *Fibrosis*

3.3.1. Collagen Staining

Histologic analysis of lung sections from normoxia (NO)- and hyperoxia (HY)-exposed rats showed no subjective difference between the groups in the percentage of area covered by collagen after 3-day (P3, top left) or 5-day (P5, top right) exposure (Figure 7), as well as in comparison to caffeine-treated pups under hyperoxia (HYC). However, the caffeine-treated rats (NOC) under room air showed intensified Sirius Red staining over the entire area. However, survival in room air up to P15 showed that 80% oxygen (HY) resulted in increased collagen production over the entire tissue area after both 3 (P3_P15, bottom left) and 5 days (P5_P15, bottom right). Caffeine treatment (HYC) attenuated this hyperoxia-induced effect, and was no longer elevated for the caffeine-treated group under 21% oxygen (NOC). After quantification of the collagen content over the area, these observations were confirmed for all time points and experimental groups (Figure 8).

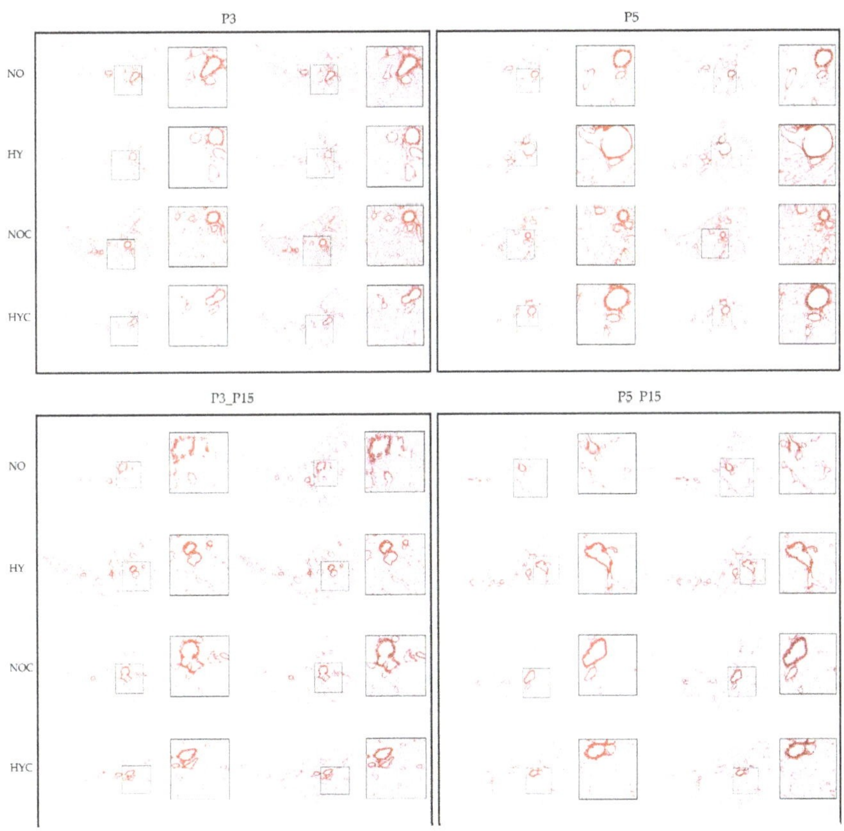

Figure 7. Sirius red stain imaging representative of general collagen for 3 days postnatal oxygen exposure (P3, **upper left**) and recovery (P3_P15, **bottom left**) as well as 5 days postnatal oxygen exposure (P5, **upper right**) and recovery (P5_P15, **bottom right**) in lung lobes of rats exposed to normoxia (NO) or hyperoxia (HY) compared to lung lobes of rats treated with caffeine (NOC, HYC). Images indicated stained Sirius Red areas (threshold, **left**) and merged with brightfield images (**right**).

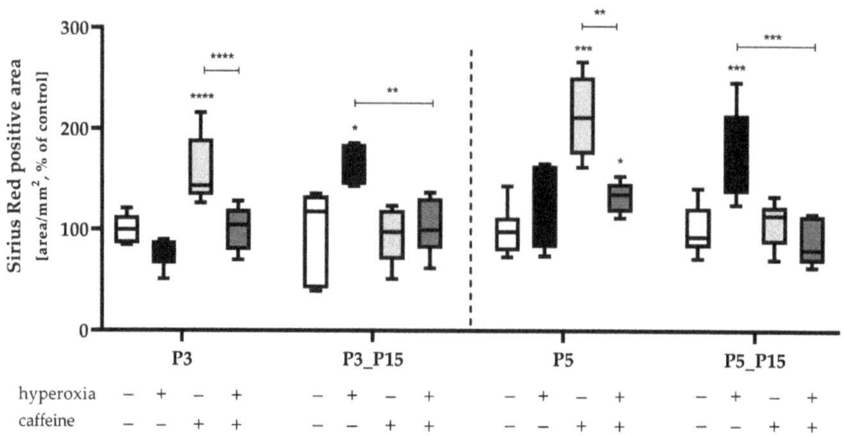

Figure 8. Quantification of Sirius Red positive areas for 3 days postnatal oxygen exposure (P3) and recovery (P3_P15) as well as 5 days postnatal oxygen exposure (P5) and recovery (P5_P15). Data are normalized to the level of rat pups exposed to normoxia at each time point (control 100%, white bars) and represented hyperoxia (black), normoxia with caffeine (light gray), and hyperoxia with caffeine (dark grey). n = 6–8 per group; * $p < 0.05$, ** $p < 0.01$, *** $p < 0.001$, **** $p < 0.0001$ (ANOVA; Kruskal–Wallis; Brown–Forsythe).

3.3.2. Gene Expression of Fibroblast-Associated Mediators

The effect of hyperoxia on the developing lung in relation to fibroblast-associated factors was investigated using gene expression analysis of pulmonary transcripts.

This indicated that some transcripts were reduced in expression to P3 and P5 by acute hyperoxia, such as *Acta2* (Figure 9a), *Col1a1* (Figure 9b), *Col3a1* (Figure 9c), *Csf2* (Figure 9d), and *Pu.1* (Figure 9i). Three days of hyperoxia reduced the transcription of *Fn1* (Figure 9h), whereas only 5 days of exposure to 80% oxygen impaired the transcription of *Timp1* and *Timp2* (Figure 9j,k). In some cases, down-regulated transcription persisted after recovery until postnatal day 15 (*Csf2* at P5_P15; *Pu.1* at P3_P15 and P5_P15; *Timp1* at P5_P15; Figure 9d,i,j). A drastically increased expression compared to a down-regulation after hyperoxia directly after termination at P3 and P5 was observed for some transcripts after recovery (*Acta2, Col1a1, Col3a1* at P3_P15 and P5_P15; *Fn1* at P3_P15; Figure 9a–c,h). A direct hyperoxia-induced increase in expression was seen at P3 and P5 for *Egr1* (Figure 9g) and *tPa* (Figure 9l), for *CTGF* at P5 (Figure 9e), as well as for *Timp1* and *Timp2* at P3 (Figure 9j,k), some of which remained elevated without further oxygen-induced noxae (*CTGF* at P5_P15, *Edn1* at P3_P15, *Egr1* at P3_P15, *tPa* at P3_P15 and P5_P15). Caffeine affected the transcription of all target genes when administered in parallel with 80% oxygen exposure and, with few exceptions, caused a significant counter-regulation (Figure 9a–l).

Caffeine under control conditions modulated, except *Col1a1* and *tPa*, all investigated factors, predominantly directly after the end of caffeine administration (*Acta2, Col1a1, CTGF, Edn1* at P3 and P5; *Fn1, Timp2* at P3; *Egr1, Pu.1* at P5).

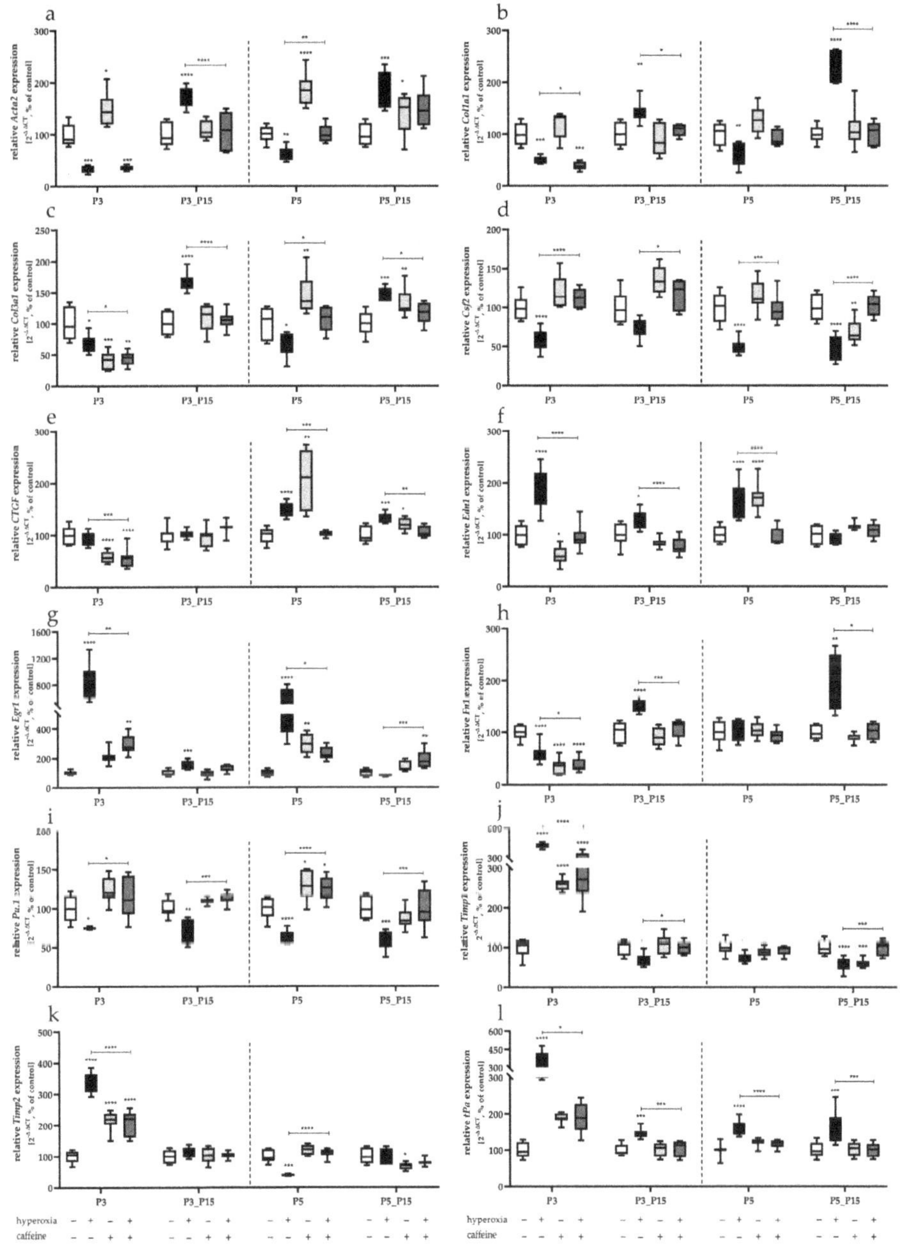

Figure 9. Quantification of fibroblast-associated mediator's transcript expression of a whole lung lobe with (**a**) *Acta2*, (**b**) *Col1a1*, (**c**) *Col3a1*, (**d**) *Csf2*, (**e**) *CTGF*, (**f**) *Edn1*, (**g**) *Egr1*, (**h**) *Fn1*, (**i**) *Pu.1*, (**j**) *Timp1*, (**k**) *Timp2*, and (**l**) *tPa* for 3 days postnatal oxygen exposure (P3) and recovery (P3_P15) as well as 5 days postnatal oxygen exposure (P5) and recovery (P5_P15). Data are normalized to the level of rat pups exposed to normoxia at each time point (control 100%, white bars) with verum groups hyperoxia (black), normoxia with caffeine (light gray), and hyperoxia with caffeine (dark grey). n = 7–8 per group; * $p < 0.05$, ** $p < 0.01$, *** $p < 0.001$, **** $p < 0.0001$ (ANOVA; Kruskal–Wallis; Brown–Forsythe).

4. Discussion

Exposure of neonatal rats to 80% oxygen during the early postnatal period resulted in marked alterations in lung development, architecture, and molecular pathways regulating alveolarization, angiogenesis, and fibrosis. Although survival and body weight gain were not significantly affected, hyperoxia induced classical BPD-like pathology characterized by simplified alveolar structure, enlarged distal airspaces, reduced septal volume, and increased alveolo-capillary permeability. Morphological and molecular impairments were examined both during the acute phase of oxygen exposure, corresponding to the saccular stage and the transition to the alveolar stage, and later in adolescent rats during the alveolar stage without further insults, allowing assessment of immediate and longer-term effects on lung maturation. These morphological impairments were accompanied by downregulation of key alveolo-angiogenic genes, alongside transient activation of fibroblast-associated mediators. Notably, the injury progressed in a two-phase manner, comprising an acute, reversible barrier disruption with edema followed by chronic fibrotic remodeling that persisted through P15, which corresponds to the infant developmental stage and falls within the early alveolar stage of lung development.

Caffeine administration during hyperoxia markedly mitigated these structural and transcriptional abnormalities, preserving lung volume and architecture and largely restoring physiological gene expression profiles. Importantly, caffeine also exerted intrinsic developmental effects under normoxia by enhancing mediators of alveolarization and angiogenesis and suppressing pro-fibrotic and tissue-remodeling signals. Thus, while caffeine counteracted hyperoxia-induced injury, it also displayed context-dependent regulatory effects on postnatal lung maturation; anti-fibrotic under oxidative stress, yet transiently pro-fibrotic under normoxic conditions.

Overall, this study provides comprehensive evidence that caffeine preserves lung structure and function under oxidative stress while actively shaping physiological lung maturation, extending its known antioxidant and anti-inflammatory properties and suggesting therapeutic benefit for preventing or ameliorating neonatal lung injury. Compared with previous preclinical studies investigating the effects of caffeine in neonatal hyperoxia [46,53,54,70–73], the present work offers several unique contributions, driven by an enhanced methodological rigor. Earlier studies often lacked systematic morphometric lung endpoints, stereological quantification, or detailed phase-specific analyses. Our study overcomes these limitations by integrating high-resolution stereological and transcriptional analyses across two distinct postnatal exposure phases, namely the saccular and the alveolar stages. This approach allowed us to systematically evaluate organ-level remodeling, capture the stage-specific impact of hyperoxia and caffeine, and define the context-dependent, pro- versus anti-fibrotic effects of caffeine over time, which were previously unresolved. This combination of temporal resolution and morphometric rigor establishes a novel framework with high translational relevance to preterm infants.

The postnatal lung is highly vulnerable to oxidative injury given immature antioxidant defenses and ongoing alveolar and vascular growth. Hyperoxia downregulated essential alveolo-angiogenic mediators such as cyclin D2, FGF10, PDGFα/β, VEGF, and Tie2, while also dysregulating fibrotic mediators including Col1a1, CTGF, and TIMP1/2 [42,74]. Downregulation of Acta2/α-SMA must be interpreted within neonatal pathobiology: unlike adult fibrosis, neonatal hyperoxia primarily disrupts fibroblast subtype dynamics and alveolar septation rather than inducing classical collagen-rich fibrosis [75]. Lipofibroblasts and matrix fibroblasts, critical for septation, are selectively reduced under hyperoxia [76,77], explaining the apparent reduction in Acta2 and pro-fibrotic markers as a loss of myofibroblast commitment and impaired septation rather than fibrosis resolution [75,78].

Hyperoxia produced the characteristic simplified alveolar architecture repeatedly described in neonatal models, including enlarged distal airspaces, reduced septal volume and surface area, and increased permeability [42,74,79,80]. Persistent impairment of gas-exchange surface area is supported by transcriptional evidence of disrupted AT1/AT2 cell dynamics [81], consistent with increased MLI at later ages (P60) [81,82] and the structural deficits observed here up to the alveolar phase in rat pups (P15). The absolute surface area and mean thickness of the alveolar septa are key determinants of gas-exchange capacity; destruction of septa reduces total surface area and diffusion capacity, ultimately limiting oxygen uptake [83,84]. Early hyperoxia induced acute barrier disruption with permeability increases and interstitial edema without changes in total lung volume [85]; edema resolved after return to normoxia [86], but structural remodeling, increased parenchymal volume, and collagen deposition indicated progression toward chronic, likely irreversible injury. This two-phase pathophysiology, an acute, reversible edema followed by chronic structural reorganization, fits the persistent simplification observed after neonatal hyperoxia.

Neonatal programming effects can have lifelong implications. Neonatal hyperoxia causes long-term depletion of an AT2 precursor subset, which potentially impairs adult lung regeneration [81]. This aligns with epidemiological evidence showing increased risk of asthma, obstructive disease, and respiratory infections among individuals born preterm, independent of atopy, with risk increasing as gestational age decreases [11,87–94]. Datta et al. [95] confirmed that early, but not late, postnatal hyperoxia disrupts alveolarization and pulmonary arterial remodeling, and that mitochondria-targeted antioxidants protect alveolar development, paralleling the protective effects of caffeine in our model.

Hyperoxia triggers excessive ROS production, causing inflammation, apoptosis, mitochondrial dysfunction, and altered ECM turnover [42,50,51,79,95–99]. These effects impair lung development and recapitulate central features of BPD. Consistent with extensive prior work from our group using the same experimental cohort, caffeine reduced ROS accumulation, apoptosis, inflammatory signaling, and immune cell recruitment, while modulating cell cycle and ECM dynamics [50,51,100]. Hyperoxia-induced inflammatory responses involve macrophages with mixed activation states, neutrophils, and epithelial and endothelial cell subsets [51,79], all of which contribute to the injury cascade.

Vascular development is integral to alveolar growth and BPD pathogenesis. Hyperoxia strongly downregulated *FGF10*, VEGF, PDGF, and Tie2, all critical for vascular stability and microvascular maturation [39,101–103]. Experimental and human BPD consistently show disrupted VEGF signaling. Recent studies highlight the complex, dynamic regulation of VEGF in BPD and its potential as a biomarker [104]. While preclinical models suggest therapeutic benefits of VEGF modulation, safety concerns limit direct application [105]. Hyperoxia acutely suppressed PDGFα and its receptors, with post-recovery PDGFRα/β overexpression, and disrupted Tie2 signaling, indicating impaired mesenchymal diversity and vascular stability in our study. FGF10 and VEGF were also reduced, compromising microvascularization and contributing to BPD-like pathology [25,33,106].

As shown in this study, caffeine preserves lung architecture, normalizes lung volume, and counteracts transcriptional disturbances induced by oxygen toxicity, supporting its therapeutic potential. Its protective profile combines anti-inflammatory, antioxidant, anti-fibrotic, and antiapoptotic effects, with additional regulation of angiogenesis [46,49–51,107]. These mechanisms, including antioxidant and antiapoptotic actions, are consistent with findings from our previous studies in the same experimental cohort [49–51]. Caffeine reduces proinflammatory cytokine release, inhibits NLRP3 inflammasome activation, suppresses NF-κB signaling, and antagonizes adenosine A2A receptors, thereby modulating apoptosis and vascular development [46,51,108–110]. It enhances antioxidant enzyme expression [31,50], regulates pro-fibrotic mediators such as CTGF [111], and may improve pulmonary vascular remodeling [53,54]. Effects are concentration-dependent and can be paradoxical under certain conditions [46,52,55]. Beyond oxidative stress, caffeine modulates lung maturation under normoxia by upregulating FGF10, PDGFα, angiopoietin 1, and VEGFR2, while suppressing remodeling- and fibrosis-associated genes such as CTGF and Timp2, demonstrating context-dependent, multimechanistic regulation of postnatal lung development. Dumpa et al. [54] demonstrated that caffeine stabilizes HIF-2α, modulates VEGFR1, and improves lung and microvascular architecture, supporting its multifaceted role in developmental signaling.

Given this impact on angiogenic signaling, it is important to consider how alterations in the angiopoietin/Tie2/VEGF axis may affect vascular integrity in hyperoxia-induced lung injury. Dysregulation of this axis has been consistently linked to severe impairment of pulmonary vasculature, even when PECAM-1/CD31 levels remain stable [112]. Other hyperoxia models report concurrent reductions in PECAM-1 alongside changes in these mediators [113], and in chronic lung injury, PECAM-1 downregulation is associated with endothelial barrier loss and vascular remodeling [114]. Although our study did not directly assess PECAM-1/CD31, changes in Angpt1, Angpt2, Tie2, VEGF, VEGFR1, and VEGFR2 provide meaningful insight into microvessel density and endothelial function. Interpreting our molecular findings in the context of established PECAM/CD31 variability [112,115–118] acknowledges this limitation while underscoring the relevance of our approach for evaluating pulmonary microvascular integrity.

Caffeine significantly mitigates hyperoxia-induced lung injury. Hyperoxia increases ROS, DNA, and lipid damage and activates endogenous antioxidant pathways, including Nrf2/Keap1, superoxide dismutase, and heme oxygenase-1 [47,48,50–52,55]. Consistent with previous findings in the identical experimental cohort of newborn rat pups, our data demonstrate that caffeine reduces oxidative stress in neonatal lungs, reflected by decreased DNA damage, lower hydrogen peroxide levels, and modulation of antioxidant enzyme expression [50]. Mechanistically, this involves antagonism of adenosine A2A receptors, inhibition of cAMP/PKA-MAPK signaling, and attenuation of ER stress, thereby reducing alveolar apoptosis [47,48]. Consequently, caffeine acts as a potent antioxidant at clinically relevant doses.

Beyond oxidative stress, hyperoxia promotes fibrosis in the developing lung, particularly in preterm infants, via ROS-driven inflammation, TGF-β signaling, fibroblast–myofibroblast differentiation, and ECM accumulation, contributing to BPD pathogenesis [31,33,42,119–127]. Key profibrotic mediators, including Acta2/α-SMA, Col1a1, Col3a1, Fn1, Timp1/2, CTGF, and Gremlin-1, were induced by hyperoxia, with Gremlin-1 at P15 differentially modulated by caffeine in our study [128–136]. Caffeine exhibits antifibrotic effects by inhibiting TGF-β/CTGF signaling, reducing myofibroblast differentiation, ECM deposition, and airway remodeling [47,111,137–142]. Its mechanisms include blockade of TGF-β signaling, Nrf2 activation, and Snail1 suppression, with limited impact on Timp1/2, and modulation of MMP activity [99,141–143].

Caffeine's regulation of fibrosis is context- and time-dependent. Under normoxia, we observed transient upregulation of TGF-β at P3 and CTGF at P5, indicating potential early pro-fibrotic signaling, whereas under hyperoxic stress, antioxidant and antifibrotic effects predominate. This reflects the complex, multi-stage process of pulmonary fibrosis, in which TGF-β drives myofibroblast differentiation and ECM accumulation [111,138,144]. Caffeine also modulates Smad signaling, reducing CTGF expression and downstream ECM production, thereby attenuating fibrotic remodeling. Furthermore, caffeine directly affects fibroblasts by influencing Acta2/α-SMA and other ECM-associated gene expression, limiting myofibroblast activation in injured lungs.

These effects are particularly relevant in the immature lung. In early postnatal development, excessive TGF-β and CTGF promote pathological matrix deposition, impaired alveolarization, and vascular malformation. In this context, caffeine can antagonize profibrotic signaling under oxidative stress, supporting structural lung maturation. Overall, our findings indicate that caffeine's impact on the TGF-β/CTGF axis is strongly dependent on oxygen status and timing: under normoxia, transient pro-fibrotic effects may occur, while under hyperoxic stress, antioxidant and antifibrotic effects predominate, highlighting its dual, context-dependent role in postnatal lung development.

Caffeine therapy in preterm infants is well established for apnea of prematurity and is consistently associated with reduced BPD risk and improved respiratory outcomes [15,145–148]. Early initiation within the first two days of life further decreases BPD, mortality, and need for prolonged respiratory support [149–151]. Benefits may partly reflect reduced duration of mechanical ventilation and lower cumulative oxygen exposure [107,149,152]. Higher-dose regimens may provide additional pulmonary advantages, although long-term safety in extremely preterm infants requires caution [15,146].

Although caffeine citrate is an established and effective drug for the treatment of AOP, its dosage and application in clinical practice are still discussed, and further research is needed to determine the optimal doses and timing for its multiple beneficial effects on various organs [60]. It is important to note that while the benefits of caffeine are well documented in animal models, its effects outside the context of oxidative stress may have undesirable consequences, so its use beyond the recommended indications must be carefully considered [52].

5. Conclusions

Neonatal hyperoxia induced a biphasic, BPD-like lung injury, with an early, reversible barrier disruption followed by sustained fibrotic remodeling into adolescence, accompanied by impaired alveolo-angiogenic signaling and transient fibroblast activation. Caffeine counteracted these acute and long-term effects by preserving alveolar structure, maintaining angiogenic gene expression, and limiting fibrosis via TGF-β/CTGF modulation. Its effects were context-dependent, being largely protective under hyperoxia but briefly pro-fibrotic under normoxia. These findings highlight caffeine's capacity to modulate postnatal lung maturation and its therapeutic promise for preterm infants at risk of BPD.

Supplementary Materials: The following supporting information can be downloaded at: https://www.mdpi.com/article/10.3390/antiox14121497/s1, Table S1: Sequences of oligonucleotides.

Author Contributions: S.E. conceived the idea and designed the experiments. S.E. wrote the first draft of the manuscript and approved the final draft; S.E. performed the animal experiments and gene expression analyses; S.E. analyzed tissue stains; S.E. collected and analyzed the collected data; S.E. and C.B. discussed the results in the experimental and clinical context, C.B. critically revised the paper for important intellectual content. All authors have read and agreed to the published version of the manuscript.

Funding: This research was supported by our employer, the Department of Neonatology, Charité-Universitätsmedizin Berlin, Germany, and by the "Förderverein für frühgeborene Kinder an der Charité e.V." and Sonnenfeld Stiftung, Berlin, Germany.

Institutional Review Board Statement: Animal care and the general protocols for animal use were approved by the local animal welfare authorities (LAGeSo, approval number G-0088/16 on 30 June 2016) and followed institutional guidelines.

Informed Consent Statement: Not applicable.

Data Availability Statement: The data used to support the findings of this study are available from the corresponding author upon request. The oligonucleotides used for qPCR and their target gene sequences are summarized in the Supplementary Table S1.

Acknowledgments: We would like to thank Evy Strauß, who passed away before this work was completed, for her assistance with the elaborate animal preparations and the preparation of paraffin sections. Many thanks to Nora Künzel for preparing the plastic-embedded tissue sections.

Conflicts of Interest: The authors declare no conflicts of interest.

References

1. Regin, Y.; Gie, A.; Eerdekens, A.; Toelen, J.; Debeer, A. Ventilation and respiratory outcome in extremely preterm infants: Trends in the new millennium. *Eur. J. Pediatr.* **2022**, *181*, 1899–1907. [CrossRef]
2. Davis, B.E.; Leppert, M.O.; German, K.; Lehmann, C.U.; Adams-Chapman, I. Primary Care Framework to Monitor Preterm Infants for Neurodevelopmental Outcomes in Early Childhood. *Pediatrics* **2023**, *152*, e2023062511. [CrossRef]
3. Lee, S.M.; Sie, L.; Liu, J.; Profit, J.; Lee, H.C. Evaluation of Trends in Bronchopulmonary Dysplasia and Respiratory Support Practice for Very Low Birth Weight Infants: A Population-Based Cohort Study. *J. Pediatr.* **2022**, *243*, 47–52.e2. [CrossRef]
4. Orvay, J.A.C.; Baucells, B.J.; Aloy, J.F.; Echegoyen, X.M.; Sebastiani, G. Trends in bronchopulmonary dysplasia and 15-year evolution in respiratory and hemodynamic management of very premature infants. *Sci. Rep.* **2025**, *15*, 24987. [CrossRef]
5. Jeon, G.W.; Oh, M.; Chang, Y.S. Increased bronchopulmonary dysplasia along with decreased mortality in extremely preterm infants. *Sci. Rep.* **2025**, *15*, 8720; Correction in *Sci. Rep.* **2025**, *15*, 15304. https://doi.org/10.1038/s41598-025-00489-2. [CrossRef]
6. Hart, K.; Cousins, M.; Watkins, W.J.; Kotecha, S.J.; Henderson, A.J.; Kotecha, S. Association of early-life factors with prematurity-associated lung disease: Prospective cohort study. *Eur. Respir. J.* **2022**, *59*, 2101766. [CrossRef]
7. Tsang, K.D.; Tramper-Stranders, G.A.; Been, J.V.; Hoffmann-Haringsma, A.K.; Reiss, I.K.; Pijnenburg, M.W.H.; De Kleer, I.M. Moderate-to-late prematurity: Understanding respiratory consequences and modifiable risk factors. *Eur. Respir. Rev.* **2025**, *34*, 240267. [CrossRef] [PubMed]
8. Fierro, J.L.; Passarella, M.; Lorch, S.A. Prematurity as an Independent Risk Factor for the Development of Pulmonary Disease. *J. Pediatr.* **2019**, *213*, 110–114. [CrossRef] [PubMed]
9. Simpson, S.J.; Du Berry, C.; Evans, D.J.; Gibbons, J.T.D.; Vollsæter, M.; Halvorsen, T.; Gruber, K.; Lombardi, E.; Stanojevic, S.; Hurst, J.R.; et al. Unravelling the respiratory health path across the lifespan for survivors of preterm birth. *Lancet Respir. Med.* **2024**, *12*, 167–180. [CrossRef]
10. Bui, D.S.; Perret, J.L.; Walters, E.H.; Lodge, C.J.; Bowatte, G.; Hamilton, G.S.; Thompson, B.R.; Frith, P.; Erbas, B.; Thomas, P.S.; et al. Association between very to moderate preterm births, lung function deficits, and COPD at age 53 years: Analysis of a prospective cohort study. *Lancet Respir. Med.* **2022**, *10*, 478–484. [CrossRef] [PubMed]
11. Smith, E.F.; Hemy, N.R.; Hall, G.L.; Wilson, A.C.; Murray, C.P.; Simpson, S.J. Risk factors for poorer respiratory outcomes in adolescents and young adults born preterm. *Thorax* **2023**, *78*, 1223–1232. [CrossRef]
12. Richter, L.L.; Shen, Y.; Lisonkova, S.; Bone, J.N.; Albert, A.; Ho, M.S.P.; Kieran, E.; Chan, E.S.; Mammen, C.; Lam, C.; et al. Preterm Birth and Risk of Health Service Use, Morbidity, and Medication Needs at 5 Years. *JAMA Pediatr.* **2025**, *179*, 1172. [CrossRef]
13. Durlak, W.; Thébaud, B. BPD: Latest Strategies of Prevention and Treatment. *Neonatology* **2024**, *121*, 596–607. [CrossRef]
14. Gilfillan, M.A.; Kiladejo, A.; Bhandari, V. Current and Emerging Therapies for Prevention and Treatment of Bronchopulmonary Dysplasia in Preterm Infants. *Paediatr. Drugs* **2025**, *27*, 539–562. [CrossRef] [PubMed]
15. Abiramalatha, T.; Ramaswamy, V.V.; Bandyopadhyay, T.; Somanath, S.H.; Shaik, N.B.; Pullattayil, A.K.; Weiner, G.M. Interventions to Prevent Bronchopulmonary Dysplasia in Preterm Neonates: An Umbrella Review of Systematic Reviews and Meta-analyses. *JAMA Pediatr.* **2022**, *176*, 502–516. [CrossRef] [PubMed]
16. Gillot, D.; Milési, C.; Combes, C.; Brodeau, V.; Gavotto, A.; Cambonie, G. Surfactant treatment rate in very premature infants following change to less-invasive surfactant administration. *Eur. J. Pediatr.* **2025**, *184*, 563. [CrossRef] [PubMed]
17. Doyle, L.W.; Mainzer, R.; Cheong, J.L.Y. Systemic Postnatal Corticosteroids, Bronchopulmonary Dysplasia, and Survival Free of Cerebral Palsy. *JAMA Pediatr.* **2025**, *179*, 65–72. [CrossRef]
18. Cummings, J.J.; Pramanik, A.K. Postnatal Corticosteroids to Prevent or Treat Chronic Lung Disease Following Preterm Birth. *Pediatrics* **2022**, *149*, e2022057530. [CrossRef]
19. Strueby, L.; Thebaud, B. Novel therapeutics for bronchopulmonary dysplasia. *Curr. Opin. Pediatr.* **2018**, *30*, 378–383. [CrossRef]
20. Sehgal, A.; Garrick, S.P.; Nold, M.F.; Nold-Petry, C.A. Novel concepts of treating vascular inflammation underlying neonatal lung diseases. *J. Appl. Physiol.* **2023**, *135*, 174–182. [CrossRef]

21. Teng, M.; Wu, T.J.; Pritchard, K.A., Jr.; Day, B.W.; Naylor, S.; Teng, R.J. The Destructive Cycle in Bronchopulmonary Dysplasia: The Rationale for Systems Pharmacology Therapeutics. *Antioxidants* **2025**, *14*, 844. [CrossRef] [PubMed]
22. Schmidt, B.; Roberts, R.S.; Davis, P.; Doyle, L.W.; Barrington, K.J.; Ohlsson, A.; Solimano, A.; Tin, W. Caffeine therapy for apnea of prematurity. *N. Engl. J. Med.* **2006**, *354*, 2112–2121. [CrossRef] [PubMed]
23. Dankhara, N.; Holla, I.; Ramarao, S.; Kalikkot Thekkeveedu, R. Bronchopulmonary Dysplasia: Pathogenesis and Pathophysiology. *J. Clin. Med.* **2023**, *12*, 4207. [CrossRef] [PubMed]
24. Moschino, L.; Bonadies, L.; Baraldi, E. Lung growth and pulmonary function after prematurity and bronchopulmonary dysplasia. *Pediatr. Pulmonol.* **2021**, *56*, 3499–3508. [CrossRef]
25. Ambalavanan, N.; Deutsch, G.; Pryhuber, G.; Travers, C.P.; Willis, K.A. The Evolving Pathophysiology of Bronchopulmonary Dysplasia. *Physiol. Rev.* **2025**, *106*, 197–237. [CrossRef]
26. Lignelli, E.; Palumbo, F.; Myti, D.; Morty, R.E. Recent advances in our understanding of the mechanisms of lung alveolarization and bronchopulmonary dysplasia. *Am. J. Physiol. Lung Cell. Mol. Physiol.* **2019**, *317*, L832–L887. [CrossRef]
27. Schmidt, B. Methylxanthine therapy for apnea of prematurity: Evaluation of treatment benefits and risks at age 5 years in the international Caffeine for Apnea of Prematurity (CAP) trial. *Biol. Neonatol.* **2005**, *88*, 208–213. [CrossRef]
28. Kalikkot Thekkeveedu, R.; El-Saie, A.; Prakash, V.; Katakam, L.; Shivanna, B. Ventilation-Induced Lung Injury (VILI) in Neonates: Evidence-Based Concepts and Lung-Protective Strategies. *J. Clin. Med.* **2022**, *11*, 557. [CrossRef]
29. Williams, E.E.; Greenough, A. Lung Protection During Mechanical Ventilation in the Premature Infant. *Clin. Perinatol.* **2021**, *48*, 869–880. [CrossRef]
30. van Kaam, A.H. Optimal Strategies of Mechanical Ventilation: Can We Avoid or Reduce Lung Injury? *Neonatology* **2024**, *121*, 570–575. [CrossRef]
31. Wang, J.; Dong, W. Oxidative stress and bronchopulmonary dysplasia. *Gene* **2018**, *678*, 177–183. [CrossRef]
32. Deng, X.; Bao, Z.; Yang, X.; Mei, Y.; Zhou, Q.; Chen, A.; Yu, R.; Zhang, Y. Molecular mechanisms of cell death in bronchopulmonary dysplasia. *Apoptosis* **2023**, *28*, 39–54. [CrossRef] [PubMed]
33. Cannavò, L.; Perrone, S.; Viola, V.; Marseglia, L.; Di Rosa, G.; Gitto, E. Oxidative Stress and Respiratory Diseases in Preterm Newborns. *Int. J. Mol. Sci.* **2021**, *22*, 12504. [CrossRef] [PubMed]
34. Lembo, C.; Buonocore, G.; Perrone, S. Oxidative Stress in Preterm Newborns. *Antioxidants* **2021**, *10*, 1672. [CrossRef]
35. Jia, X.; Yang, M.; Wu, B.; You, K.; Xu, W. Integrated Transcriptome and Metabolome Analysis Reveals a Disturbance in Antioxidant Metabolism Associated With the Progression of Bronchopulmonary Dysplasia. *FASEB J. Off. Publ. Fed. Am. Soc. Exp. Biol.* **2025**, *39*, e70874. [CrossRef]
36. Course, C.W.; Lewis, P.A.; Kotecha, S.J.; Cousins, M.; Hart, K.; Heesom, K.J.; Watkins, W.J.; Kotecha, S. Evidence of abnormality in glutathione metabolism in the airways of preterm born children with a history of bronchopulmonary dysplasia. *Sci. Rep.* **2023**, *13*, 19465. [CrossRef] [PubMed]
37. Zhang, Z.; Wu, W.; Hou, L.; Jiang, J.; Wan, W.; Li, Z. Cytokines and Exhaled Nitric Oxide Are Risk Factors in Preterm Infants for Bronchopulmonary Dysplasia. *Biomed Res. Int.* **2021**, *2021*, 6648208. [CrossRef]
38. Cho, H.Y.; Wang, X.; Campbell, M.R.; Panduri, V.; Coviello, S.; Caballero, M.T.; Bennett, B.D.; Kleeberger, S.R.; Polack, F.P.; Ofman, G.; et al. Prospective epigenome and transcriptome analyses of cord and peripheral blood from preterm infants at risk of bronchopulmonary dysplasia. *Sci. Rep.* **2023**, *13*, 12262. [CrossRef]
39. Perrone, S.; Manti, S.; Buttarelli, L.; Petrolini, C.; Boscarino, G.; Filonzi, L.; Gitto, E.; Esposito, S.M.R.; Nonnis Marzano, F. Vascular Endothelial Growth Factor as Molecular Target for Bronchopulmonary Dysplasia Prevention in Very Low Birth Weight Infants. *Int. J. Mol. Sci.* **2023**, *24*, 2729. [CrossRef]
40. Teng, M.; Wu, T.J.; Jing, X.; Day, B.W.; Pritchard, K.A., Jr.; Naylor, S.; Teng, R.J. Temporal Dynamics of Oxidative Stress and Inflammation in Bronchopulmonary Dysplasia. *Int. J. Mol. Sci.* **2024**, *25*, 10145. [CrossRef]
41. Alva, R.; Mirza, M.; Baiton, A.; Lazuran, L.; Samokysh, L.; Bobinski, A.; Cowan, C.; Jaimon, A.; Obioru, D.; Al Makhoul, T.; et al. Oxygen toxicity: Cellular mechanisms in normobaric hyperoxia. *Cell Biol. Toxicol.* **2022**, *39*, 111–143. [CrossRef] [PubMed]
42. Choi, Y.; Rekers, L.; Dong, Y.; Holzfurtner, L.; Goetz, M.J.; Shahzad, T.; Zimmer, K.P.; Behnke, J.; Behnke, J.; Bellusci, S.; et al. Oxygen Toxicity to the Immature Lung-Part I: Pathomechanistic Understanding and Preclinical Perspectives. *Int. J. Mol. Sci.* **2021**, *22*, 11006. [CrossRef]
43. Tamatam, C.M.; Reddy, N.M.; Potteti, H.R.; Ankireddy, A.; Noone, P.M.; Yamamoto, M.; Kensler, T.W.; Reddy, S.P. Preconditioning the immature lung with enhanced Nrf2 activity protects against oxidant-induced hypoalveolarization in mice. *Sci. Rep.* **2020**, *10*, 19034. [CrossRef] [PubMed]
44. Wang, J.; Zhang, A.; Li, Y.; Xu, J.; Huang, F.; Zhao, M.; Wu, B.; He, S. Effect of intermittent hypoxia or hyperoxia on lung development in preterm rat neonates during constant oxygen therapy. *J. Cell. Biochem.* **2019**, *120*, 17545–17554. [CrossRef]

45. Rößler, G.; Labode, J.; Regin, Y.; Salaets, T.; Gie, A.; Toelen, J.; Mühlfeld, C. Prematurity and Hyperoxia Have Different Effects on Alveolar and Microvascular Lung Development in the Rabbit. *J. Histochem. Cytochem.* **2023**, *71*, 259–271. [CrossRef]
46. Chen, S.; Wu, Q.; Zhong, D.; Li, C.; Du, L. Caffeine prevents hyperoxia-induced lung injury in neonatal mice through NLRP3 inflammasome and NF-κB pathway. *Respir. Res.* **2020**, *21*, 140. [CrossRef]
47. Teng, R.J.; Jing, X.; Michalkiewicz, T.; Afolayan, A.J.; Wu, T.J.; Konduri, G.G. Attenuation of endoplasmic reticulum stress by caffeine ameliorates hyperoxia-induced lung injury. *Am. J. Physiol. Lung Cell. Mol. Physiol.* **2017**, *312*, L586–L598. [CrossRef] [PubMed]
48. Wang, X.; Lv, S.; Sun, J.; Zhang, M.; Zhang, L.; Sun, Y.; Zhao, Z.; Wang, D.; Zhao, X.; Zhang, J. Caffeine reduces oxidative stress to protect against hyperoxia-induced lung injury via the adenosine A2A receptor/cAMP/PKA/Src/ERK1/2/p38MAPK pathway. *Redox Rep.* **2022**, *27*, 270–278. [CrossRef]
49. Weichelt, U.; Cay, R.; Schmitz, T.; Strauss, E.; Sifringer, M.; Buhrer, C.; Endesfelder, S. Prevention of hyperoxia-mediated pulmonary inflammation in neonatal rats by caffeine. *Eur. Respir. J.* **2013**, *41*, 966–973. [CrossRef]
50. Endesfelder, S.; Strauß, E.; Scheuer, T.; Schmitz, T.; Bührer, C. Antioxidative effects of caffeine in a hyperoxia-based rat model of bronchopulmonary dysplasia. *Respir. Res.* **2019**, *20*, 88. [CrossRef]
51. Endesfelder, S.; Strauß, E.; Bendix, I.; Schmitz, T.; Bührer, C. Prevention of Oxygen-Induced Inflammatory Lung Injury by Caffeine in Neonatal Rats. *Oxidative Med. Cell. Longev.* **2020**, *2020*, 3840124. [CrossRef] [PubMed]
52. Endesfelder, S. Caffeine: The Story beyond Oxygen-Induced Lung and Brain Injury in Neonatal Animal Models-A Narrative Review. *Antioxidants* **2024**, *13*, 1076. [CrossRef] [PubMed]
53. Jing, X.; Huang, Y.W.; Jarzembowski, J.; Shi, Y.; Konduri, G.G.; Teng, R.J. Caffeine ameliorates hyperoxia-induced lung injury by protecting GCH1 function in neonatal rat pups. *Pediatr. Res.* **2017**, *82*, 483–489. [CrossRef] [PubMed]
54. Dumpa, V.; Nielsen, L.; Wang, H.; Kumar, V.H.S. Caffeine is associated with improved alveolarization and angiogenesis in male mice following hyperoxia induced lung injury. *BMC Pulm. Med.* **2019**, *19*, 138. [CrossRef]
55. Tiwari, K.K.; Chu, C.; Couroucli, X.; Moorthy, B.; Lingappan, K. Differential concentration-specific effects of caffeine on cell viability, oxidative stress, and cell cycle in pulmonary oxygen toxicity in vitro. *Biochem. Biophys. Res. Commun.* **2014**, *450*, 1345–1350. [CrossRef]
56. Giszas, V.; Strauß, E.; Bührer, C.; Endesfelder, S. The Conflicting Role of Caffeine Supplementation on Hyperoxia-Induced Injury on the Cerebellar Granular Cell Neurogenesis of Newborn Rats. *Oxidative Med. Cell. Longev.* **2022**, *2022*, 5769784. [CrossRef]
57. Heise, J.; Schmitz, T.; Bührer, C.; Endesfelder, S. Protective Effects of Early Caffeine Administration in Hyperoxia-Induced Neurotoxicity in the Juvenile Rat. *Antioxidants* **2023**, *12*, 295. [CrossRef]
58. Faul, F.; Erdfelder, E.; Lang, A.G.; Buchner, A. G*Power 3: A flexible statistical power analysis program for the social, behavioral, and biomedical sciences. *Behav. Res. Methods* **2007**, *39*, 175–191. [CrossRef]
59. Abdel-Hady, H.; Nasef, N.; Shabaan, A.E.; Nour, I. Caffeine therapy in preterm infants. *World J. Clin. Pediatr.* **2015**, *4*, 81–93. [CrossRef]
60. O'Shea, M.; Butler, L.; Holohan, S.; Healy, K.; O'Farrell, R.; Shamit, A.; Cusack, R.; Elhadi, M.; Lynch, S.; Gilcrest, M.; et al. Caffeine and preterm infants: Multiorgan effects and therapeutic creep: Scope to optimise dose and timing. *Pediatr. Res.* **2025**, 1–8. [CrossRef] [PubMed]
61. Mühlfeld, C.; Knudsen, L.; Ochs, M. Stereology and morphometry of lung tissue. *Methods Mol. Biol.* **2013**, *931*, 367–390. [CrossRef]
62. Livak, K.J.; Schmittgen, T.D. Analysis of relative gene expression data using real-time quantitative PCR and the 2(-Delta Delta C(T)) Method. *Methods* **2001**, *25*, 402–408. [CrossRef]
63. Hsia, C.C.; Hyde, D.M.; Ochs, M.; Weibel, E.R. An official research policy statement of the American Thoracic Society/European Respiratory Society: Standards for quantitative assessment of lung structure. *Am. J. Respir. Crit. Care Med.* **2010**, *181*, 394–418. [CrossRef]
64. Crowley, G.; Kwon, S.; Caraher, E.J.; Haider, S.H.; Lam, R.; Batra, P.; Melles, D.; Liu, M.; Nolan, A. Quantitative lung morphology: Semi-automated measurement of mean linear intercept. *BMC Pulm. Med.* **2019**, *19*, 206. [CrossRef]
65. Mühlfeld, C.; Ochs, M. Quantitative microscopy of the lung: A problem-based approach. Part 2: Stereological parameters and study designs in various diseases of the respiratory tract. *Am. J. Physiol. Lung Cell. Mol. Physiol.* **2013**, *305*, L205–L221. [CrossRef]
66. Tschanz, S.A.; Burri, P.H.; Weibel, E.R. A simple tool for stereological assessment of digital images: The STEPanizer. *J. Microsc.* **2011**, *243*, 47–59. [CrossRef]
67. Schneider, C.A.; Rasband, W.S.; Eliceiri, K.W. NIH Image to ImageJ: 25 years of image analysis. *Nat. Methods* **2012**, *9*, 671–675. [CrossRef]
68. Hilgendorff, A.; Reiss, I.; Ehrhardt, H.; Eickelberg, O.; Alvira, C.M. Chronic lung disease in the preterm infant. Lessons learned from animal models. *Am. J. Respir. Cell Mol. Biol.* **2014**, *50*, 233–245. [CrossRef]

69. O'Reilly, M.; Thebaud, B. Animal models of bronchopulmonary dysplasia. The term rat models. *Am. J. Physiol. Lung Cell. Mol. Physiol.* **2014**, *307*, L948–L958. [CrossRef]
70. Rath, P.; Nardiello, C.; Surate Solaligue, D.E.; Agius, R.; Mizikova, I.; Huhn, S.; Mayer, K.; Vadasz, I.; Herold, S.; Runkel, F.; et al. Caffeine administration modulates TGF-beta signaling but does not attenuate blunted alveolarization in a hyperoxia-based mouse model of bronchopulmonary dysplasia. *Pediatr. Res.* **2017**, *81*, 795–805. [CrossRef]
71. Nagatomo, T.; Jimenez, J.; Richter, J.; De Baere, S.; Vanoirbeek, J.; Naulaers, G.; Allegaert, K.; Croubels, S.; Deprest, J.A.; Toelen, J. Caffeine Prevents Hyperoxia-Induced Functional and Structural Lung Damage in Preterm Rabbits. *Neonatology* **2016**, *109*, 274–281. [CrossRef] [PubMed]
72. Dayanim, S.; Lopez, B.; Maisonet, T.M.; Grewal, S.; Londhe, V.A. Caffeine induces alveolar apoptosis in the hyperoxia-exposed developing mouse lung. *Pediatr. Res.* **2014**, *75*, 395–402. [CrossRef] [PubMed]
73. Mielgo, V.; Catozzi, C.; Ricci, F.; Villetti, G.; Gomez-Solaetxe, M.A.; Rey-Santano, C. Caffeine Protects Newborn Rat Lungs Against the Effects of Hyperoxia and Intra-Amniotic Enterotoxin. *Pediatr. Pulmonol.* **2025**, *60*, e71395. [CrossRef]
74. Wagenaar, G.T.; ter Horst, S.A.; van Gastelen, M.A.; Leijser, L.M.; Mauad, T.; van der Velden, P.A.; de Heer, E.; Hiemstra, P.S.; Poorthuis, B.J.; Walther, F.J. Gene expression profile and histopathology of experimental bronchopulmonary dysplasia induced by prolonged oxidative stress. *Free Radic. Biol. Med.* **2004**, *36*, 782–801. [CrossRef]
75. Zanini, F.; Che, X.; Suresh, N.E.; Knutsen, C.; Klavina, P.; Xie, Y.; Domingo-Gonzalez, R.; Liu, M.; Kum, A.; Jones, R.C.; et al. Hyperoxia prevents the dynamic neonatal increases in lung mesenchymal cell diversity. *Sci. Rep.* **2024**, *14*, 2033. [CrossRef]
76. Al Alam, D.; El Agha, E.; Sakurai, R.; Kheirollahi, V.; Moiseenko, A.; Danopoulos, S.; Shrestha, A.; Schmoldt, C.; Quantius, J.; Herold, S.; et al. Evidence for the involvement of fibroblast growth factor 10 in lipofibroblast formation during embryonic lung development. *Development* **2015**, *142*, 4139–4150. [CrossRef]
77. Zhang, X.; Shi, X.; Xie, F.; Liu, Y.; Wei, X.; Cai, Y.; Chao, J. Dissecting pulmonary fibroblasts heterogeneity in lung development, health and diseases. *Heliyon* **2023**, *9*, e19428. [CrossRef]
78. Riccetti, M.R.; Ushakumary, M.G.; Waltamath, M.; Green, J.; Snowball, J.; Dautel, S.E.; Endale, M.; Lami, B.; Woods, J.; Ahlfeld, S.K.; et al. Maladaptive functional changes in alveolar fibroblasts due to perinatal hyperoxia impair epithelial differentiation. *J. Clin. Investig.* **2022**, *7*, e152404. [CrossRef]
79. Hurskainen, M.; Mižíková, I.; Cook, D.P.; Andersson, N.; Cyr-Depauw, C.; Lesage, F.; Helle, E.; Renesme, L.; Jankov, R.P., Heikinheimo, M.; et al. Single cell transcriptomic analysis of murine lung development on hyperoxia-induced damage. *Nat. Commun.* **2021**, *12*, 1565. [CrossRef]
80. Schmiedl, A.; Roolfs, T.; Tutdibi, E.; Gortner, L.; Monz, D. Influence of prenatal hypoxia and postnatal hyperoxia on morphologic lung maturation in mice. *PLoS ONE* **2017**, *12*, e0175804. [CrossRef] [PubMed]
81. Scaffa, A.; Yao, H.; Oulhen, N.; Wallace, J.; Peterson, A.L.; Rizal, S.; Ragavendran, A.; Wessel, G.; De Paepe, M.E.; Dennery, P.A. Single-cell transcriptomics reveals lasting changes in the lung cellular landscape into adulthood after neonatal hyperoxic exposure. *Redox Biol.* **2021**, *48*, 102091. [CrossRef]
82. Guo, M.; Du, Y.; Gokey, J.J.; Ray, S.; Bell, S.M.; Adam, M.; Sudha, P.; Perl, A.K.; Deshmukh, H.; Potter, S.S.; et al. Single cell RNA analysis identifies cellular heterogeneity and adaptive responses of the lung at birth. *Nat. Commun.* **2019**, *10*, 37. [CrossRef]
83. Ahlfeld, S.K.; Gao, Y.; Conway, S.J.; Tepper, R.S. Relationship of structural to functional impairment during alveolar-capillary membrane development. *Am. J. Pathol.* **2015**, *185*, 913–919. [CrossRef] [PubMed]
84. van der Vliet, A.; Janssen-Heininger, Y.M.W.; Anathy, V. Oxidative stress in chronic lung disease: From mitochondrial dysfunction to dysregulated redox signaling. *Mol. Asp. Med.* **2018**, *63*, 59–69. [CrossRef]
85. Wang, S.; Chen, H.; Li, Z.; Xu, G.; Bao, X. Hyperbaric oxygen-induced acute lung injury: A mouse model study on pathogenic characteristics and recovery dynamics. *Front. Physiol.* **2024**, *15*, 1474933. [CrossRef]
86. Garat, C.; Meignan, M.; Matthay, M.A.; Luo, D.F.; Jayr, C. Alveolar epithelial fluid clearance mechanisms are intact after moderate hyperoxic lung injury in rats. *Chest* **1997**, *111*, 1381–1388. [CrossRef]
87. Pulakka, A.; Risnes, K.; Metsälä, J.; Alenius, S.; Heikkilä, K.; Nilsen, S.M.; Näsänen-Gilmore, P.; Haaramo, P.; Gissler, M.; Opdahl, S.; et al. Preterm birth and asthma and COPD in adulthood: A nationwide register study from two Nordic countries. *Eur. Respir. J.* **2023**, *61*, 2201763. [CrossRef] [PubMed]
88. Crump, C.; Sundquist, J.; Sundquist, K. Preterm or early term birth and long-term risk of asthma into midadulthood: A national cohort and cosibling study. *Thorax* **2023**, *78*, 653–660. [CrossRef]
89. Voraphani, N.; Klevebro, S.; Larsson, F.; Wang, G.; Merid, S.K.; Björkander, S.; Um-Bergström, P.; Bergström, A.; Kull, I.; Merritt, A.S.; et al. The role of CC16 in the associations of preterm birth with lung function and asthma in adult life. *J. Allergy Clin. Immunol.* **2025**, *in press*. [CrossRef] [PubMed]
90. Anderson, J.; Do, L.A.H.; Wurzel, D.; Licciardi, P.V. Understanding the increased susceptibility to asthma development in preterm infants. *Allergy* **2023**, *78*, 928–939. [CrossRef]

91. Caffarelli, C.; Gracci, S.; Giannì, G.; Bernardini, R. Are Babies Born Preterm High-Risk Asthma Candidates? *J. Clin. Med.* **2023**, *12*, 5400. [CrossRef]
92. Davidesko, S.; Wainstock, T.; Sheiner, E.; Pariente, G. Long-Term Infectious Morbidity of Premature Infants: Is There a Critical Threshold? *J. Clin. Med.* **2020**, *9*, 3008. [CrossRef]
93. Coathup, V.; Carson, C.; Kurinczuk, J.J.; Macfarlane, A.J.; Boyle, E.; Johnson, S.; Petrou, S.; Quigley, M.A. Associations between gestational age at birth and infection-related hospital admission rates during childhood in England: Population-based record linkage study. *PLoS ONE* **2021**, *16*, e0257341. [CrossRef] [PubMed]
94. Steiner, L.; Diesner, S.C.; Voitl, P. Risk of infection in the first year of life in preterm children: An Austrian observational study. *PLoS ONE* **2019**, *14*, e0224766. [CrossRef] [PubMed]
95. Datta, A.; Kim, G.A.; Taylor, J.M.; Gugino, S.F.; Farrow, K.N.; Schumacker, P.T.; Berkelhamer, S.K. Mouse lung development and NOX1 induction during hyperoxia are developmentally regulated and mitochondrial ROS dependent. *Am. J. Physiol. Lung Cell. Mol. Physiol.* **2015**, *309*, L369–L377. [CrossRef] [PubMed]
96. Behnke, J.; Dippel, C.M.; Choi, Y.; Rekers, L.; Schmidt, A.; Lauer, T.; Dong, Y.; Behnke, J.; Zimmer, K.P.; Bellusci, S.; et al. Oxygen Toxicity to the Immature Lung-Part II: The Unmet Clinical Need for Causal Therapy. *Int. J. Mol. Sci.* **2021**, *22*, 10694. [CrossRef]
97. Berkelhamer, S.K.; Kim, G.A.; Radder, J.E.; Wedgwood, S.; Czech, L.; Steinhorn, R.H.; Schumacker, P.T. Developmental differences in hyperoxia-induced oxidative stress and cellular responses in the murine lung. *Free Radic. Biol. Med.* **2013**, *61*, 51–60. [CrossRef]
98. Zou, D.; Li, J.; Fan, Q.; Zheng, X.; Deng, J.; Wang, S. Reactive oxygen and nitrogen species induce cell apoptosis via a mitochondria-dependent pathway in hyperoxia lung injury. *J. Cell. Biochem.* **2019**, *120*, 4837–4850. [CrossRef]
99. Wang, H.; Guan, W.; Yang, W.; Wang, Q.; Zhao, H.; Yang, F.; Lv, X.; Li, J. Caffeine inhibits the activation of hepatic stellate cells induced by acetaldehyde via adenosine A2A receptor mediated by the cAMP/PKA/SRC/ERK1/2/P38 MAPK signal pathway. *PLoS ONE* **2014**, *9*, e92482. [CrossRef]
100. Endesfelder, S.; Weichelt, U.; Strauss, E.; Schlor, A.; Sifringer, M.; Scheuer, T.; Buhrer, C.; Schmitz, T. Neuroprotection by Caffeine in Hyperoxia-Induced Neonatal Brain Injury. *Int. J. Mol. Sci.* **2017**, *18*, 187. [CrossRef]
101. Keenaghan, M.; Cai, C.L.; Kumar, D.; Valencia, G.B.; Rao, M.; Aranda, J.V.; Beharry, K.D. Response of vascular endothelial growth factor and angiogenesis-related genes to stepwise increases in inspired oxygen in neonatal rat lungs. *Pediatr. Res.* **2013**, *73*, 630–638. [CrossRef]
102. Chao, C.M.; Moiseenko, A.; Kosanovic, D.; Rivetti, S.; El Agha, E.; Wilhelm, J.; Kampschulte, M.; Yahya, F.; Ehrhardt, H.; Zimmer, K.P.; et al. Impact of Fgf10 deficiency on pulmonary vasculature formation in a mouse model of bronchopulmonary dysplasia. *Hum. Mol. Genet.* **2019**, *28*, 1429–1444. [CrossRef]
103. D'Angio, C.T.; Maniscalco, W.M. The role of vascular growth factors in hyperoxia-induced injury to the developing lung. *Front. Biosci.* **2002**, *7*, d1609–d1623. [CrossRef]
104. Lun, Y.; Gu, D.; Yang, Z. Association of early VEGF trajectories with bronchopulmonary dysplasia severity in preterm infants. *Sci. Rep.* **2025**, *15*, 37406. [CrossRef] [PubMed]
105. Bolte, C.; Ustiyan, V.; Ren, X.; Dunn, A.W.; Pradhan, A.; Wang, G.; Kolesnichenko, O.A.; Deng, Z.; Zhang, Y.; Shi, D.; et al. Nanoparticle Delivery of Proangiogenic Transcription Factors into the Neonatal Circulation Inhibits Alveolar Simplification Caused by Hyperoxia. *Am. J. Respir. Crit. Care Med.* **2020**, *202*, 100–111. [CrossRef]
106. van Horik, C.; Meyboom, J.; Boerema-de Munck, A.; Buscop-van Kempen, M.; Eenjes, E.; Edel, G.G.; Kortekaas, D.; Wijnen, R.M.H.; van, I.W.F.J.; Reiss, I.; et al. The impact of hyperoxia and antibiotics on lung mesenchymal cells in experimental bronchopulmonary dysplasia. *Am. J. Physiol. Lung Cell. Mol. Physiol.* **2025**, *329*, L440–L454. [CrossRef]
107. Yuan, Y.; Yang, Y.; Lei, X.; Dong, W. Caffeine and bronchopulmonary dysplasia: Clinical benefits and the mechanisms involved. *Pediatr. Pulmonol.* **2022**, *57*, 1392–1400. [CrossRef]
108. Zhao, W.; Ma, L.; Cai, C.; Gong, X. Caffeine Inhibits NLRP3 Inflammasome Activation by Suppressing MAPK/NF-κB and A2aR Signaling in LPS-Induced THP-1 Macrophages. *Int. J. Biol. Sci.* **2019**, *15*, 1571–1581. [CrossRef] [PubMed]
109. Rocha, G.; Proença, E.; Guedes, A.; Carvalho, C.; Areias, A.; Ramos, J.P.; Rodrigues, T.; Guimarães, H. Cord blood levels of IL-6, IL-8 and IL-10 may be early predictors of bronchopulmonary dysplasia in preterm newborns small for gestational age. *Dis. Markers* **2012**, *33*, 51–60. [CrossRef] [PubMed]
110. Kou, C.; Han, D.; Li, Z.; Wu, W.; Liu, Z.; Zhang, Y.; Gao, Z. Influence of prevention of caffeine citrate on cytokine profile and bronchopulmonary dysplasia in preterm infants with apnea. *Minerva Pediatr.* **2020**, *72*, 95–100. [CrossRef]
111. Fehrholz, M.; Speer, C.P.; Kunzmann, S. Caffeine and rolipram affect Smad signalling and TGF-β1 stimulated CTGF and transgelin expression in lung epithelial cells. *PLoS ONE* **2014**, *9*, e97357. [CrossRef]
112. Obst, S.; Serdar, M.; Herz, J.; Kempe, K.; Assili, M.; Rizazad, M.; Hirani, D.; Alejandre Alcazar, M.A.; Endesfelder, S.; Möbius, M.A.; et al. A Novel Model for Simultaneous Evaluation of Hyperoxia-Mediated Brain and Lung Injury in Neonatal Rats. *Cells* **2025**, *14*, 443. [CrossRef]

113. Lee, Y.; Lee, J.; Nam, S.K.; Hoon Jun, Y. S-endoglin expression is induced in hyperoxia and contributes to altered pulmonary angiogenesis in bronchopulmonary dysplasia development. *Sci. Rep.* **2020**, *10*, 3043. [CrossRef]
114. Lu, X.; Gong, J.; Dennery, P.A.; Yao, H. Endothelial-to-mesenchymal transition: Pathogenesis and therapeutic targets for chronic pulmonary and vascular diseases. *Biochem. Pharmacol.* **2019**, *168*, 100–107. [CrossRef]
115. DeLisser, H.M.; Helmke, B.P.; Cao, G.; Egan, P.M.; Taichman, D.; Fehrenbach, M.; Zaman, A.; Cui, Z.; Mohan, G.S.; Baldwin, H.S.; et al. Loss of PECAM-1 function impairs alveolarization. *J. Biol. Chem.* **2006**, *281*, 8724–8731. [CrossRef]
116. Gu, A.; Shively, J.E. Angiopoietins-1 and -2 play opposing roles in endothelial sprouting of embryoid bodies in 3D culture and their receptor Tie-2 associates with the cell-cell adhesion molecule PECAM1. *Exp. Cell Res.* **2011**, *317*, 2171–2182. [CrossRef]
117. Perveen, S.; Patel, H.; Arif, A.; Younis, S.; Codipilly, C.N.; Ahmed, M. Role of EC-SOD overexpression in preserving pulmonary angiogenesis inhibited by oxidative stress. *PLoS ONE* **2012**, *7*, e51945. [CrossRef] [PubMed]
118. Hosford, G.E.; Olson, D.M. Effects of hyperoxia on VEGF, its receptors, and HIF-2alpha in the newborn rat lung. *Am. J. Physiol. Lung Cell. Mol. Physiol.* **2003**, *285*, L161–L168. [CrossRef] [PubMed]
119. Makena, P.; Kikalova, T.; Prasad, G.L.; Baxter, S.A. Oxidative Stress and Lung Fibrosis: Towards an Adverse Outcome Pathway. *Int. J. Mol. Sci.* **2023**, *24*, 12490. [CrossRef] [PubMed]
120. Bik-Multanowski, M.; Revhaug, C.; Grabowska, A.; Dobosz, A.; Madetko-Talowska, A.; Zasada, M.; Saugstad, O.D. Hyperoxia induces epigenetic changes in newborn mice lungs. *Free Radic. Biol. Med.* **2018**, *121*, 51–56. [CrossRef]
121. Louzada, R.A.; Corre, R.; Ameziane El Hassani, R.; Meziani, L.; Jaillet, M.; Cazes, A.; Crestani, B.; Deutsch, E.; Dupuy, C. NADPH oxidase DUOX1 sustains TGF-β1 signalling and promotes lung fibrosis. *Eur. Respir. J.* **2021**, *57*, 1901949. [CrossRef] [PubMed]
122. Cui, Y.; Yang, Z.; Lv, Z.; Lei, J. Disruption of extracellular redox balance drives persistent lung fibrosis and impairs fibrosis resolution. *Biochim. Biophys. Acta Mol. Basis Dis.* **2023**, *1869*, 166842. [CrossRef]
123. Kondrikov, D.; Caldwell, R.B.; Dong, Z.; Su, Y. Reactive oxygen species-dependent RhoA activation mediates collagen synthesis in hyperoxic lung fibrosis. *Free Radic. Biol. Med.* **2011**, *50*, 1689–1698. [CrossRef] [PubMed]
124. Ni, J.; Dong, Z.; Han, W.; Kondrikov, D.; Su, Y. The role of RhoA and cytoskeleton in myofibroblast transformation in hyperoxic lung fibrosis. *Free Radic. Biol. Med.* **2013**, *61*, 26–39. [CrossRef]
125. You, K.; Parikh, P.; Khandalavala, K.; Wicher, S.A.; Manlove, L.; Yang, B.; Roesler, A.; Roos, B.B.; Teske, J.J.; Britt, R.D., Jr.; et al. Moderate hyperoxia induces senescence in developing human lung fibroblasts. *Am. J. Physiol. Lung Cell. Mol. Physiol.* **2019**, *317*, L525–L536. [CrossRef]
126. Parikh, P.; Britt, R.D., Jr.; Manlove, L.J.; Wicher, S.A.; Roesler, A.; Ravix, J.; Teske, J.; Thompson, M.A.; Sieck, G.C.; Kirkland, J.L.; et al. Hyperoxia-induced Cellular Senescence in Fetal Airway Smooth Muscle Cells. *Am. J. Respir. Cell Mol. Biol.* **2019**, *61*, 51–60. [CrossRef]
127. Koloko Ngassie, M.L.; Thompson, M.A.; Roos, B.B.; Ayyalasomayajula, S.; Lagnado, A.B.; Passos, J.F.; Pabelick, C.M.; Prakash, Y.S. Hyperoxia-induced senescence in fetal airway smooth muscle cells: Role of mitochondrial reactive oxygen species and endoplasmic reticulum stress. *Am. J. Physiol. Lung Cell. Mol. Physiol.* **2025**, *329*, L1–L18. [CrossRef]
128. Rockey, D.C.; Bell, P.D.; Hill, J.A. Fibrosis—A Common Pathway to Organ Injury and Failure. *N. Engl. J. Med.* **2015**, *372*, 1138–1149. [CrossRef]
129. Druso, J.E.; MacPherson, M.B.; Chia, S.B.; Elko, E.; Aboushousha, R.; Seward, D.J.; Abdelhamid, H.; Erickson, C.; Corteselli, E.; Tarte, M.; et al. Endoplasmic Reticulum Oxidative Stress Promotes Glutathione-Dependent Oxidation of Collagen-1A1 and Promotes Lung Fibroblast Activation. *Am. J. Respir. Cell Mol. Biol.* **2024**, *71*, 589–602. [CrossRef] [PubMed]
130. Tam, A.Y.Y.; Horwell, A.L.; Trinder, S.L.; Khan, K.; Xu, S.; Ong, V.; Denton, C.P.; Norman, J.T.; Holmes, A.M.; Bou-Gharios, G.; et al. Selective deletion of connective tissue growth factor attenuates experimentally-induced pulmonary fibrosis and pulmonary arterial hypertension. *Int. J. Biochem. Cell Biol.* **2021**, *134*, 105961. [CrossRef] [PubMed]
131. Blaauboer, M.E.; Boeijen, F.R.; Emson, C.L.; Turner, S.M.; Zandieh-Doulabi, B.; Hanemaaijer, R.; Smit, T.H.; Stoop, R.; Everts, V. Extracellular matrix proteins: A positive feedback loop in lung fibrosis? *Matrix Biol.* **2014**, *34*, 170–178. [CrossRef]
132. Kelly, M.; Kolb, M.; Bonniaud, P.; Gauldie, J. Re-evaluation of fibrogenic cytokines in lung fibrosis. *Curr. Pharm. Des.* **2003**, *9*, 39–49. [CrossRef] [PubMed]
133. Weng, C.M.; Yu, C.C.; Kuo, M.L.; Chen, B.C.; Lin, C.H. Endothelin-1 induces connective tissue growth factor expression in human lung fibroblasts by ETAR-dependent JNK/AP-1 pathway. *Biochem. Pharmacol.* **2014**, *88*, 402–411. [CrossRef]
134. Samarakoon, R.; Overstreet, J.M.; Higgins, P.J. TGF-β signaling in tissue fibrosis: Redox controls, target genes and therapeutic opportunities. *Cell. Signal.* **2013**, *25*, 264–268. [CrossRef]
135. O'Reilly, S. Gremlin: A complex molecule regulating wound healing and fibrosis. *Cell. Mol. Life Sci. CMLS* **2021**, *78*, 7917–7923. [CrossRef]
136. Mthunzi, L.; Rowan, S.C.; Kostyunina, D.S.; Baugh, J.A.; Knaus, U.G.; McLoughlin, P. Gremlin 1 is required for macrophage M2 polarization. *Am. J. Physiol. Lung Cell. Mol. Physiol.* **2023**, *325*, L270–L276. [CrossRef]

137. Calthorpe, R.J.; Poulter, C.; Smyth, A.R.; Sharkey, D.; Bhatt, J.; Jenkins, G.; Tatler, A.L. Complex roles of TGF-β signaling pathways in lung development and bronchopulmonary dysplasia. *Am. J. Physiol. Lung Cell. Mol. Physiol.* **2023**, *324*, L285–L296. [CrossRef] [PubMed]
138. Tatler, A.L.; Barnes, J.; Habgood, A.; Goodwin, A.; McAnulty, R.J.; Jenkins, G. Caffeine inhibits TGFβ activation in epithelial cells, interrupts fibroblast responses to TGFβ, and reduces established fibrosis in ex vivo precision-cut lung slices. *Thorax* **2016**, *71*, 565–567. [CrossRef] [PubMed]
139. Nilnumkhum, A.; Kanlaya, R.; Yoodee, S.; Thongboonkerd, V. Caffeine inhibits hypoxia-induced renal fibroblast activation by antioxidant mechanism. *Cell Adhes. Migr.* **2019**, *13*, 260–272. [CrossRef]
140. Arauz, J.; Zarco, N.; Segovia, J.; Shibayama, M.; Tsutsumi, V.; Muriel, P. Caffeine prevents experimental liver fibrosis by blocking the expression of TGF-β. *Eur. J. Gastroenterol. Hepatol.* **2014**, *26*, 164–173. [CrossRef]
141. Gordillo-Bastidas, D.; Oceguera-Contreras, E.; Salazar-Montes, A.; González-Cuevas, J.; Hernández-Ortega, L.D.; Armendáriz-Borunda, J. Nrf2 and Snail-1 in the prevention of experimental liver fibrosis by caffeine. *World J. Gastroenterol.* **2013**, *19*, 9020–9033. [CrossRef]
142. Kanlaya, R.; Subkod, C.; Nanthawuttiphan, S.; Thongboonkerd, V. Caffeine prevents oxalate-induced epithelial-mesenchymal transition of renal tubular cells by its anti-oxidative property through activation of Nrf2 signaling and suppression of Snail1 transcription factor. *Biomed. Pharmacother.* **2021**, *141*, 111870. [CrossRef]
143. Gressner, O.A.; Lahme, B.; Rehbein, K.; Siluschek, M.; Weiskirchen, R.; Gressner, A.M. Pharmacological application of caffeine inhibits TGF-beta-stimulated connective tissue growth factor expression in hepatocytes via PPARgamma and SMAD2/3-dependent pathways. *J. Hepatol.* **2008**, *49*, 758–767. [CrossRef]
144. Bhardwaj, S.; Gautam, R.K.; Kushwaha, S. From senescence to scarring: Exploring TGF-β signaling in cellular aging, fibrotic remodeling, and pulmonary fibrosis. *Cytokine Growth Factor Rev.* **2025**, *86*, 29–39. [CrossRef]
145. Marques, K.A.; Bruschettini, M.; Roehr, C.C.; Davis, P.G.; Fiander, M.; Soll, R. Methylxanthine for the prevention and treatment of apnea in preterm infants. *Cochrane Database Syst. Rev.* **2023**, *2025*, CD013830. [CrossRef]
146. Bruschettini, M.; Brattström, P.; Russo, C.; Onland, W.; Davis, P.G.; Soll, R. Caffeine dosing regimens in preterm infants with or at risk for apnea of prematurity. *Cochrane Database Syst. Rev.* **2023**, *2023*, CD013873. [CrossRef]
147. Oliphant, E.A.; Hanning, S.M.; McKinlay, C.J.D.; Alsweiler, J.M. Caffeine for apnea and prevention of neurodevelopmental impairment in preterm infants: Systematic review and meta-analysis. *J. Perinatol. Off. J. Calif. Perinat. Assoc.* **2024**, *44*, 785–801. [CrossRef]
148. Brattström, P.; Russo, C.; Ley, D.; Bruschettini, M. High-versus low-dose caffeine in preterm infants: A systematic review and meta-analysis. *Acta Paediatr.* **2019**, *108*, 401–410. [CrossRef]
149. Elmowafi, M.; Mohsen, N.; Nour, I.; Nasef, N. Prophylactic versus therapeutic caffeine for apnea of prematurity: A randomized controlled trial. *J. Matern. Fetal Neonatal Med.* **2022**, *35*, 6053–6061. [CrossRef] [PubMed]
150. Szatkowski, L.; Fateh, S.; Abramson, J.; Kwok, T.C.; Sharkey, D.; Budge, H.; Ojha, S. Observational cohort study of use of caffeine in preterm infants and association between early caffeine use and neonatal outcomes. *Arch. Dis. Child. Fetal Neonatal Ed.* **2023**, *108*, 505–510. [CrossRef]
151. Karlinski Vizentin, V.; Madeira de Sá Pacheco, I.; Fahel Vilas Bôas Azevêdo, T.; Florêncio de Mesquita, C.; Alvim Pereira, R. Early versus Late Caffeine Therapy Administration in Preterm Neonates: An Updated Systematic Review and Meta-Analysis. *Neonatology* **2024**, *121*, 7–16, Erratum in *Neonatology* **2024**, *121*, 536. https://doi.org/10.1159/000538725. [CrossRef] [PubMed]
152. Côté, A.; Savanh, P.; Shalish, W.; Bizgu, V.; Canakis, A.M. Caffeine for Persistent Intermittent Hypoxia in Newborns with and Without Chronic Lung Disease. *Pediatr. Pulmonol.* **2025**, *60*, e71312. [CrossRef] [PubMed]

antioxidants

MDPI

Article

The Vasomotor Impact of Cu/ZnSODs Is Higher in Arterial Smooth Muscle of Early Postnatal Rats Compared to Adult Animals

Anastasia A. Shvetsova [1], Valentina S. Shateeva [1], Dmitry S. Semenovich [2], Rudolf Schubert [3,*] and Dina K. Gaynullina [1]

1 Faculty of Biology, M.V. Lomonosov Moscow State University, Moscow 119234, Russia; GaynullinaDK@my.msu.ru (D.K.G.)
2 A.N. Belozersky Institute of Physico-Chemical Biology, Moscow State University, Moscow 119992, Russia
3 Physiology, Institute of Theoretical Medicine, Faculty of Medicine, University of Augsburg, 86135 Augsburg, Germany
* Correspondence: rudolf.schubert@med.uni-augsburg.de

Abstract: The mechanisms of vascular tone regulation during early postnatal ontogenesis differ considerably from those in adulthood. We tested the hypothesis that the vasomotor influence of superoxide dismutases is higher in arteries of rats shortly after birth compared to adult animals. Saphenous arteries of 10 to 15 day ("young") and 3 to 4 month ("adult") old rats were studied using quantitative PCR, spectrophotometry, Western blotting, and isometric myography. *Sod3* mRNA was most abundant in both adult and young saphenous arteries. Total SOD activity, Cu/ZnSODs activity, and SOD3 protein content were higher in young compared to adult arteries. H_2O_2 caused vessel contraction, while elimination of H_2O_2 weakened contractile responses of endothelium-denuded young arteries. The Cu/ZnSOD inhibitor DETC had no influence on contraction of adult arteries, but weakened contraction of endothelium-denuded, but not endothelium-intact arteries of young rats. The latter effect was observed in the presence of the NO-synthase inhibitor L-NNA, but not with the soluble guanylate cyclase inhibitor ODQ. DETC had no effect in the presence of sodium nitroprusside. Thus, Cu/ZnSODs promote contraction of saphenous arteries in the early postnatal period, but not in adult age. This influence of Cu/ZnSODs is counteracted by endothelial NO in early postnatal arteries.

Keywords: superoxide dismutase; reactive oxygen species; diethyldithiocarbamate; artery; smooth muscle; vascular tone; endothelium; early postnatal period

1. Introduction

Superoxide dismutases (SODs) are enzymes that catalyze the conversion of the highly reactive superoxide anion radical $O_2^{\bullet-}$ into a more stable and less aggressive molecule, hydrogen peroxide H_2O_2. H_2O_2 has a number of signaling functions in the cell, including vasomotor effects [1]. Notably, H_2O_2 may cause both relaxation or contraction of arterial smooth muscle, which depends on many factors, including the animal species, the artery type, the presence of endothelium, etc. [2–6].

There are three isoforms of SODs: cytoplasmic Cu/ZnSOD (SOD1), mitochondrial MnSOD (SOD2), and extracellular Cu/ZnSOD (SOD3) [1]. All three isoforms are expressed in blood vessels. However, among the isoforms of SOD, the functional roles of SOD1 and SOD3 in the regulation of arterial smooth muscle tone have been the most studied due to the existence of their inhibitor diethyldithiocarbamate (DETC) [7], which chelates Cu^{2+} ions in the active centers of the enzymes. Thus, in rat aorta, DETC did not affect phenylephrine-evoked contractile responses [8,9]. However, DETC caused an increase of norepinephrine-induced contractile responses of mouse aorta [10] and rat mesenteric arteries [11], as well as serotonin-induced contraction of mouse aorta [12], which may indicate an anticontractile role of Cu/ZnSOD-produced H_2O_2 in several arteries. It is worth noting, however, that such effects were reported for arteries from mature animals.

Nevertheless, the mechanisms of smooth muscle tone regulation, including those involving reactive oxygen species (ROS) [13], are considerably different between adulthood and the period of early postnatal ontogenesis. However, the functional contribution of SODs to the regulation of arterial tone in the early postnatal period has been poorly studied and the few existing studies focus mainly on the pulmonary circulation. For example, it was shown that DETC increased U46619-induced contractile responses of pulmonary arteries from newborn and two-week-old piglets [14]. At the same time, SOD content/activity in lung tissue of newborn or early postnatal organisms (human, rabbit, and mouse) were lower compared to adults [15–17] which might be associated with a smaller functional contribution of SODs in the pulmonary circulation at the early postnatal period compared to adulthood. However, the vasomotor impact of SODs on the regulation of smooth muscle contraction in systemic arteries, which play a key role in the regulation of systemic blood pressure and blood flow distribution during the early postnatal period, has not been studied yet, to the best of our knowledge. The need and significance to address this issue also follows from the widely used antioxidant therapy of newborns in the clinic [18], which does not take into account the probable consequences of the vasomotor role of SODs.

Therefore, in this study we hypothesized that the functional impact of SODs (mainly SOD1 and SOD3) is different in systemic arteries of early postnatal rats and adult animals.

2. Materials and Methods

2.1. Animals and the Main Object of the Study

Animal studies are reported in compliance with the ARRIVE guidelines 2.0 [19,20]. The use of laboratory animals and of all procedures used in this study were approved by the Biomedical Ethics Committee of M.V. Lomonosov Moscow State University (Protocol number 149-g). Wistar rats (obtained from the Institute of General Pathology and Pathophysiology, Russia) were used. Rats were maintained at a controlled temperature and a 12/12 h light/dark cycle in the laboratory animal unit of the Biological Faculty, M.V. Lomonosov Moscow State University with access to food and water ad libitum. To obtain the offspring, sexually mature male and female rats were placed together for 4 days. Pregnant females were housed individually 3–4 days before the expected delivery. The next day after birth, the litters were reduced to 8 pups. Experiments were carried out on male rats aged 10 to 15 days ("young" in the following text) and 3 to 4 months ("adult" in the following text). In total, 30 adult and 75 young rats (offspring of 20 females) were used in the study. Rats were killed by decapitation, adult rats were first anesthetized by CO_2.

The main object of the study was the saphenous artery. There are several reasons for this choice. The rat saphenous artery is a muscular-type artery, is densely innervated, and develops strong contractile responses to the stimulation of sympathetic nerves [21]. Therefore, along with its conduit function the saphenous artery participates in the control of cutaneous blood flow, which comprises up to 20% of the cardiac output in neonatal rats [22]. Importantly, according to our data, developmental alterations of several regulatory mechanisms observed in the saphenous artery are associated with corresponding changes at the level of blood pressure control [23]. Therefore, we consider the saphenous artery to be a representative artery from the peripheral circulation, i.e., the findings from the saphenous artery can be generalized to other vascular beds. In addition, the saphenous artery is large enough in 10–15-day-old rats to assess its responses by wire myography and to obtain a sufficient amount of tissue for qPCR, Western blot, and spectrophotometry analyses (described further).

2.2. *Measurement of mRNA Expression Levels in Arterial Tissue by qPCR*

Measurement of mRNA expression levels in arterial tissue by qPCR was performed as previously described [13,24]. Briefly, saphenous arteries were isolated, carefully cleaned from surrounding tissue and blood, and kept in RNAlater solution (Qiagen, Venlo, The Netherlands) at −20 °C pending further procedures. Each sample in the adult group included one artery and in the young group two arteries.

Arteries were homogenized in RLT lysis buffer (Qiagen, Venlo, The Netherlands) containing 1% β-mercaptoethanol and proteinase K (20 mg/mL, MP Biomedicals, Irvine, CA, USA). RNA was extracted using the ExtractRNA kit (cat. number BC032, Evrogen, Moscow, Russia) according to the manufacturer's instructions and then processed with DNase I (cat. number EN0525, Fermentas, Waltham, MA, USA, 1000 U/mL). The RNA concentration was measured by a NanoDrop 1000 (Thermo Scientific, Waltham, MA, USA), and then all samples were diluted to an equal concentration of 35 ng/μL. Reverse transcription was performed using the MMLV RT kit (Evrogen, Moscow, Russia) according to the manufacturer's manual. qPCR was run in the Bio-Rad CFX 96 Real-Time PCR System (Bio-Rad, Hercules, CA, USA), using qPCRmix-HS SYBR (Evrogen, Moscow, Russia). Primers used in this study were synthesized by Evrogen; their sequences are listed in Table 1. The mRNA expression level was calculated as E^(−Cq), where E is the primer efficiency, and Cq is the cycle number corresponding to the maximum of the second derivative of the fluorescence curve. Primer efficiency was identified using the LinRegPCR (version 2015.1) software [25]. All E values were close to 2.0. mRNA expression levels of investigated genes were normalized to the mRNA expression level of the housekeeping gene beta-actin (*Actb*), detected in the same sample. Importantly, mRNA expression values of *Actb* did not differ between the two age groups and were $2.2 \times 10^{-6} \pm 0.5 \times 10^{-6}$ in adult arteries (n = 9) and $1.6 \times 10^{-6} \pm 0.3 \times 10^{-6}$ in young arteries (n = 9, $p > 0.05$, unpaired Student's *t*-test).

Table 1. Gene specific primers used in the study.

Gene Name	Gene Accession Numbers	Forward	Reverse
Sod1	NM_017050.1	GTACCACTGCAGGACCTCAT	CCACCTTTGCCCAAGTCATC
Sod2	NM_017051.2	CCGTGGTGGGTGTTTTGTAG	CGTCCAAGCAATTCAAGCCT
Sod3	NM_012880.2	CGCCTCCAGTCATCCTAGAG	AAAGTGTCCTGGTCTCCGAG
Actb	NM_031144.3	CAGGGTGTGATGGTGGGTATGG	AGTTGGTGACAATGCCGTGTTC

2.3. *Measurement of Superoxide Dismutase and Catalase Activity in Arterial Tissue*

Two saphenous arteries from one adult rat or four saphenous arteries from two young rats were used for one sample. Arteries were homogenized with ice cold PBS and centrifuged at 1500× *g* over 10 min at +4 °C. After that, probes were incubated with ice for 15 min. Supernatants were collected and frozen at −80 °C till further analysis.

SOD activity was determined by a spectrophotometric method based on the inhibition of the superoxide anion radical production rate by SOD, formed in the reaction of quercetin autooxidation in a slightly alkaline medium [26]. To determine the total SOD activity, 5 μL of the homogenate was mixed with 150 μL of a buffer containing 20 mM potassium phosphate buffer pH 7.8, 8 mM TEMED, and 0.08 mM EDTA in a 96-well plate. The reaction was started by adding 5 μL of 0.15 mg/mL quercetin in DMSO. To determine MnSOD activity, 3 mM KCN was added to the buffer. The change in optical density at 405 nm was recorded on a plate spectrophotometer INNO LTEK Co., Ltd. Instruments, Seongnam-si, Republic of Korea. One unit (U) of SOD activity was defined as the rate of 50% inhibition of the quercetin autooxidation reaction. Total SOD activity was expressed as U/mg total protein. Cu/ZnSOD activity was calculated as the difference between total SOD and MnSOD activities. Total protein content was determined by the modified Lowry method [27].

Catalase activity was determined spectrophotometrically using a method based on the reaction of excess (unreacted) hydrogen peroxide with FOX reagent [28]. The substrate–buffer mixture for catalase contained 50 mM potassium phosphate buffer (pH 7.4) and 0.2 mM hydrogen peroxide. In a 96-well plate, the supernatant of vascular homogenates was mixed with the substrate–buffer mixture at a ratio of 1:20 (v/v) and incubated for 10 min at room temperature. The reaction was stopped by the addition of FOX reagent and further incubated for 30 min at room temperature until a stable violet coloration developed. Optical density was recorded at 560 nm using a microplate spectrophotometer (INNO LTEK Co., Ltd. Instruments, Seongnam-si, Republic of Korea). The concentration of decomposed hydrogen peroxide was calculated using the millimolar extinction coefficient of 235 $mM^{-1} \cdot cm^{-1}$. One unit (U) of catalase activity was defined as the amount of enzyme required to decompose 1 μmol of hydrogen peroxide per minute. Catalase activity in vascular homogenates was expressed as U/mg of total protein. Total protein concentration was determined using the modified Lowry method [27].

2.4. Measurement of Protein Abundance in Arterial Tissue by Western Blotting

Measurement of protein abundance in arterial tissue was performed as described previously [13,29]. Four saphenous arteries from two young rats or two saphenous arteries from one adult rat were used for one sample. The endothelium was rapidly removed after isolation of the artery with the use of a custom-made wire myograph analogue [30]. Thereafter, saphenous arteries were quickly frozen in liquid nitrogen and kept at −80 °C until further analysis. Samples were homogenized in ice-cold SDS buffer supplemented with protease inhibitor cocktail (Roche, Bazel, Switzerland) and phosphatase inhibitors (2 mg/mL NaF and 180 mg/mL Na_3VO_4), centrifuged at 16,900× g for 5 min; the supernatant was kept at −20 °C till further analysis. Proteins were separated by SDS-PAGE and transferred to nitrocellulose membranes (Santa Cruz, Dallas, TX, USA) using the Trans-Blot Turbo transfer system (Bio-Rad). The transfer was visualized with Ponceau S staining. Thereafter, membranes (each containing samples from adult and young groups) were cut into two pieces at the level of appr. 28 kDa based on the corresponding levels of protein markers (Abcam, Cambridge, UK, cat. number 116027). The membranes were blocked with 5% nonfat milk (AppliChem, Darmstadt, Germany) in TBS with 0.1% Tween 20 (TBSt). Then, two pieces of the first membrane were incubated overnight with antibodies against GAPDH (mouse, Abcam, cat. number 125247, 1:2000 in TBSt with 5% milk, molecular weight 37 kDa) or SOD1 (rabbit, Sigma, St. Louis, MO, USA, cat. number SAB5200083 1:4000 in TBSt with 5% milk, molecular weight 23 kDa), and one piece of the second membrane was incubated overnight with antibodies against SOD3 (rabbit, Enzo, New York, NY, USA, cat. number ADI-SOD-106-F 1:1000 in TBSt with 5% milk, molecular weight 35–40 kDa). The next day, the membranes were washed from the primary antibodies and incubated with appropriate secondary antibodies: anti-mouse (Cell Signaling, Boston, MA,

USA, cat. number 7076, 1:5000 in 5% milk) or anti-rabbit (Cell Signaling, Boston, MA, USA, cat. number 7074, 1:10,000 in 5% milk) for 1 h and visualized with SuperSignal West Dura Substrate (Thermo Fisher Scientific, Waltham, MA, USA) using a ChemiDoc (Bio-Rad, Hercules, CA, USA). Of note, since SOD3 and GAPDH have similar molecular weights (close to 40 kDa), membrane stained with antibodies against SOD3 was stripped using stripping buffer (for composition see Section 2.6), blocked again as described above, and incubated overnight with antibodies against GAPDH. Western blotting experiments were analyzed using the Image Lab software 6.0 (Bio-Rad, Hercules, CA, USA). SOD1 and SOD3 protein abundance was normalized to GAPDH in each sample—the rationale for using GAPDH as a reference protein for comparing arteries from adult and early postnatal animals was provided previously [31])—and then the average ratio of the «Adult» group was taken as 100%.

2.5. Wire Myograph Experiments

Saphenous arteries were carefully cleaned from surrounding tissue in PSS for vessel dissection (PSS I), cut into 2-mm-long segments, and mounted in a wire myograph (410 A or 620 M, DMT A/S, Aarhus, Denmark) to measure the isometric force. Experiments were carried out both on arteries with an intact endothelium and on arteries with the endothelium removed. The endothelium was removed mechanically using a rat whisker by performing rotational movements inside the artery.

During the subsequent experiments, arterial segments were kept in PSS II. The myograph chambers were heated to 37 °C and continuously bubbled with 5% CO_2 in O_2 to maintain pH at 7.4. Data were recorded at 10 Hz using an analogue-to-digital converter (E14-140M, L-CARD, Moscow, Russia) and the PowerGraph 3.3 software (DISoft, Moscow, Russia). Each arterial segment was stretched to 0.9 × d100 (90% of the inner diameter it would have at a transmural pressure of 100 mmHg), corresponding to maximum active force development [13,32].

At the beginning of each experiment, all vessels were exposed to a standard start-up procedure. This included the following: (a) application of noradrenaline (10 μM); (b) application of acetylcholine (10 μM) on the top of methoxamine-induced contraction (1 μM) to confirm the integrity or removal of the endothelium (by the presence or the absence of a relaxation response, respectively); (c) application of methoxamine (10 μM).

Two experimental protocols were used in the study. The first experimental protocol included two sequential concentration–response relationships to methoxamine (concentration range from 0.01 to 100 μM, for three minutes each). The first relationship was started 20 min after the end of the activation procedure. After washout, one arterial segment was incubated for 20 min with one of the following substances or their combinations: inhibitor of Cu/ZnSOD DETC (1 mM), NO-synthase inhibitor L-NNA (0.1 mM), soluble guanylate cyclase inhibitor ODQ (1 μM), NO-donor sodium nitroprusside SNP (0.1 μM), superoxide dismutase SOD (300 U/mL), catalase (1500 U/mL). The other segment was incubated with an equivalent volume of the solvent (DMSO or H_2O or their combination). Then, the second concentration–response relationship was obtained. The second experimental protocol also included sequential concentration–response relationships. The first one was to methoxamine (concentration range from 0.01 to 100 μM), which allowed us to determine the concentration of methoxamine that causes a contraction of approximately 60% of the maximum. After washout, contractions at the level of approximately 60% of the maximum in two arterial segments were induced, and then H_2O_2 (concentration range from 1 μM to 0.3 mM, for 2 min each) was added sequentially to one of the segments and an equivalent volume of H_2O was added to another segment (Time-control).

To calculate active force values at each methoxamine concentration, the force value at the fully relaxed state was subtracted from all recorded values. All active force values obtained during the second concentration–response relationship to methoxamine were expressed as the percentage of the maximum active force developed during the respective first concentration–response relationship or as a percentage of the precontraction level (for H_2O_2). Areas under the curves (AUCs) were calculated in several experimental series as an additional way to evaluate the effects of inhibitors (Table S1). The internal diameter (d100), maximal force and values of acetylcholine-induced relaxation of the arteries in different series of wire-myography experiments are presented in Table S2.

2.6. Solutions

1. PSS for vessel isolation (PSS I), in mM: 145 NaCl, 4.5 KCl, 1.2 NaH_2PO_4, 1 $MgSO_4$, 0.1 $CaCl_2$, 0.025 EDTA, 5 HEPES (pH 7.4).
2. PSS for myograph experiments (PSS II), in mM: 120 NaCl, 26 $NaHCO_3$, 4.5 KCl, 1.2 NaH_2PO_4, 1.0 $MgSO_4$, 1.6 $CaCl_2$, 5.5 D-glucose, 0.025 EDTA, 5 HEPES; equilibrated with 5% CO_2 in 95% O_2 (pH 7.4).
3. SDS-buffer: 0.0625 M Tris-HCl (pH 6.8), 2.5% SDS, 10% water-free glycerin, 2.47% dithiothreitol, 0.002% bromophenol blue.
4. TBS: 50 mM Tris-HCl, 150 mM NaCl, pH 7.6.
5. TBSt: TBS with 0.1% Tween.
6. Stripping buffer: 0.2 M glycine, 0.1% SDS, 1% Tween (pH 2.2).

2.7. Drugs

Noradrenaline, acetylcholine, methoxamine, DETC, catalase (all dissolved in H_2O) and ODQ (dissolved in DMSO) were obtained from Sigma (St. Louis, MO, USA). L-NNA (dissolved in H_2O) was obtained from Alexis Biochemicals (San Diego, CA, USA). SOD (dissolved in H_2O) was obtained from BiosynthCarbosynth (Staad, Switzerland).

2.8. Statistics

Statistical analysis was performed using GraphPad Prism 8.0. The normality of data distribution was checked using the Shapiro–Wilk test; data are presented as the mean and standard error of the mean or as the median and interquartile range, depending on the type of data distribution. Statistical analysis was performed using two-way repeated measures ANOVA, one-way ANOVA, the parametric Student's *t*-test, and the Mann–Whitney test. Differences were accepted as statistically significant if the p value was less than 0.05; n represents the number of animals, i.e., biological replicates.

3. Results

3.1. mRNA Expression of SOD Isoforms in Saphenous Arteries in Adult and Young Rats

First, we identified the expression profile of SOD isoforms in the saphenous artery of young and adult rats. For this purpose, we compared mRNA contents of *Sod1*, *Sod2*, and *Sod3* genes. mRNAs of all three isoforms were detected in arterial tissue from adult and young rats (Figure 1). Among these three isoforms, mRNA of *Sod3* was most abundant in both adult and young saphenous arteries (Figure 1).

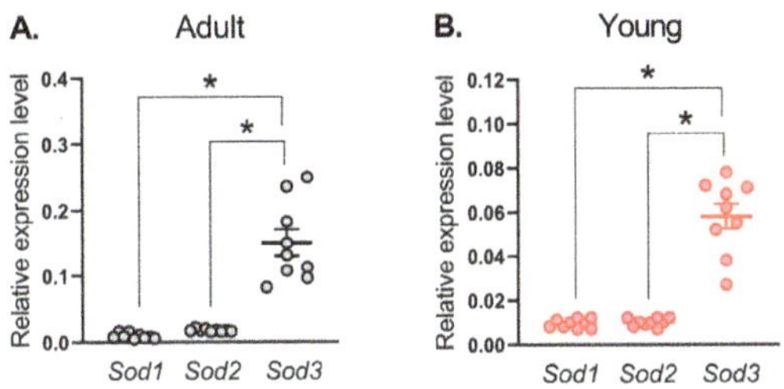

Figure 1. mRNA expression of superoxide dismutase isoforms in saphenous arteries of adult (**A**) and young (**B**) rats. Data are normalized to the beta-actin (*Actb*) expression level. The number of samples in each group is 9. * $p < 0.05$ (one-way ANOVA with Tukey's multiple comparisons test).

3.2. Activity of SODs and Catalase in Saphenous Arteries in Adult and Young Rats

Next, we evaluated the activity of SODs in saphenous arteries from adult and young rats. Total SOD activity was higher in arterial homogenates from young in comparison to adult rats (Figure 2A), which, apparently, is due to the higher activity of Cu/ZnSOD isoforms (SOD1 and SOD3) (Figure 2B), but not MnSOD (SOD2) (Figure 2C). Catalase activity was higher in arterial homogenates from young in comparison to adult rats (Figure 2D).

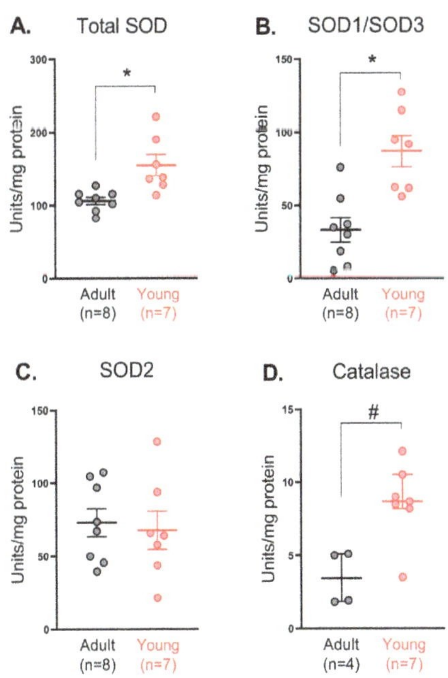

Figure 2. Activity of superoxide dismutases (total (**A**), Cu/ZnSOD (**B**) or MnSOD (**C**)) and catalase (**D**) in saphenous arteries of adult and young rats. Data are normalized to the total protein amount in the same sample; n is the number of animals. * $p < 0.05$ (unpaired Student's *t*-test), # $p < 0.05$ (Mann–Whitney test).

3.3. SOD1 and SOD3 Protein Abundances in Saphenous Arteries in Adult and Young Rats

Since Cu/ZnSODs demonstrated higher activity in arteries from young rats, we further compared the protein abundance of SOD1 and SOD3 between arteries of young and adult rats. We did not detect any difference in the protein amount of SOD1 in arterial tissue of adult and young rats (Figure 3A), while the protein amount of SOD3 was augmented in young rats (Figure 3B).

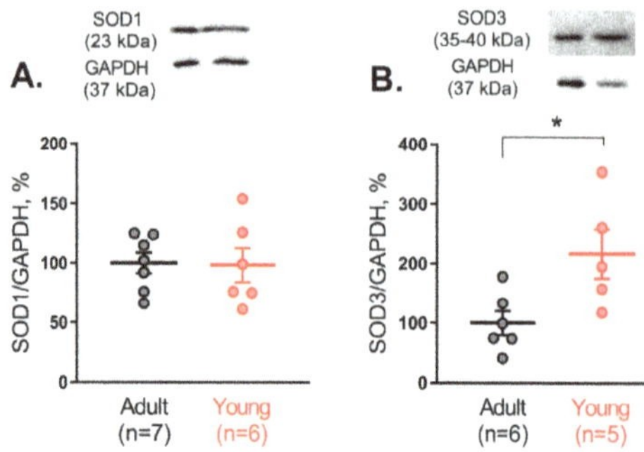

Figure 3. Protein abundance of SOD1 (**A**) and SOD3 (**B**) in saphenous arteries of adult and young rats. Data are normalized to GAPDH. The mean value of SOD1 or SOD3 protein content in the adult group is taken as 100%; n is the number of animals. * $p < 0.05$ (unpaired Student's *t*-test). The whole-length Western blot membranes are shown in Supplemetary Materials (Figure S1).

3.4. *Effect of H_2O_2 on Arterial Tone in Saphenous Arteries of Young Rats*

One of the main functions of SODs in the cells is the catalysis of $O_2^{\bullet-}$ to H_2O_2, which in its turn has a number of signaling functions, including vasomotor effects [1]. However, the vasomotor effect of H_2O_2 has been much less studied in arterial smooth muscle at early postnatal age. Therefore, we evaluated the effects of H_2O_2 on the tone of endothelium-denuded saphenous arteries of young rats.

First, we added H_2O_2 to precontracted (approximately 60% of maximal force) endothelium-denuded segments of saphenous arteries of young rats. In the concentration range used H_2O_2 caused additional contraction of the arteries in comparison to vehicle application (Figure 4A). Importantly, the initial values of precontraction did not differ between solvent- and H_2O_2-treated arteries and were 57 ± 7% and 64 ± 4% of maximal force, respectively, ($p > 0.05$, unpaired Student's *t*-test).

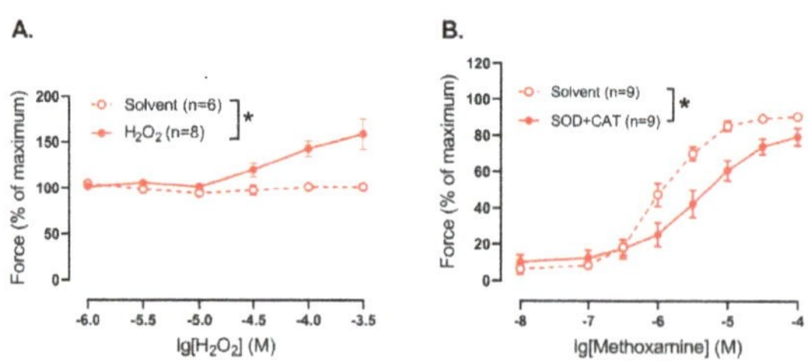

Figure 4. Effect of H_2O_2 addition and elimination on the tone of endothelium-denuded saphenous arteries from young rats. (**A**) Concentration–response relationship of precontracted by methoxamine arteries to exogenous H_2O_2. (**B**) Contractile responses to methoxamine in the presence of solvent or superoxide dismutase (SOD, 300 U/mL) together with catalase (CAT, 1500 U/mL); n is the number of animals. * $p < 0.05$ (two-way repeated measures ANOVA).

Second, when H_2O_2 was eliminated from the extracellular space by addition of superoxide dismutase (300 U/mL) together with catalase (1500 U/mL, promotes H_2O_2 breakdown), methoxamine-induced contractile responses of young saphenous arteries were reduced compared to untreated vessels (Figure 4B).

3.5. *Effect of a Cu/ZnSOD Inhibitor on Arterial Contractile Responses in Saphenous Arteries of Adult and Young Rats*

In the following experiments we focused on SOD1 and SOD3 since these superoxide dismutase isoforms (i) are abundant in the saphenous arteries in both age groups; (ii) demonstrate higher activity and amount in arteries of young rats, and (iii) can be blocked by the inhibitor DETC (unlike SOD2) [7], which allowed us to evaluate their role at the functional level, i.e., in the regulation of contractile responses of isolated arterial segments.

We did not detect any effect of pre-treatment with DETC (1 mM) on methoxamine-induced contractile responses in endothelium-denuded arteries of adult rats (Figure 5A). However, DETC considerably reduced contractile responses to methoxamine of endothelium-denuded arteries of young rats (Figure 5B).

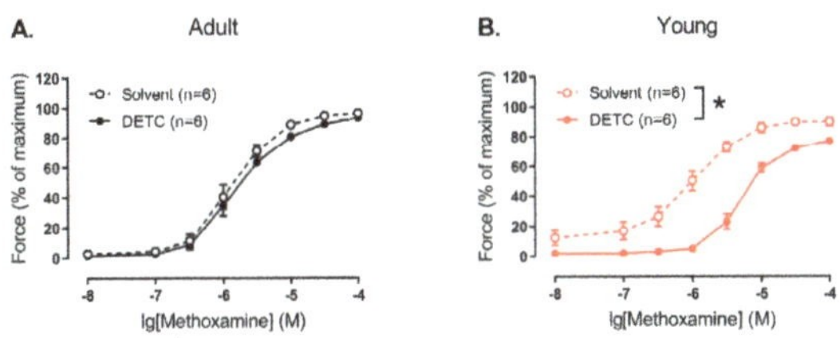

Figure 5. Effect of the Cu/ZnSOD inhibitor DETC on the contractile responses to methoxamine of endothelium-denuded arteries from young and adult rats. Concentration–response relationships to methoxamine in the presence of solvent (H_2O) or DETC (1 mM) of endothelium-denuded arteries from adult (**A**) and young (**B**) rats; n is the number of animals. * $p < 0.05$ (two-way repeated measures ANOVA).

In contrast, we did not detect any effect of DETC in endothelium-intact arteries of both adult and young rats (Figure 6).

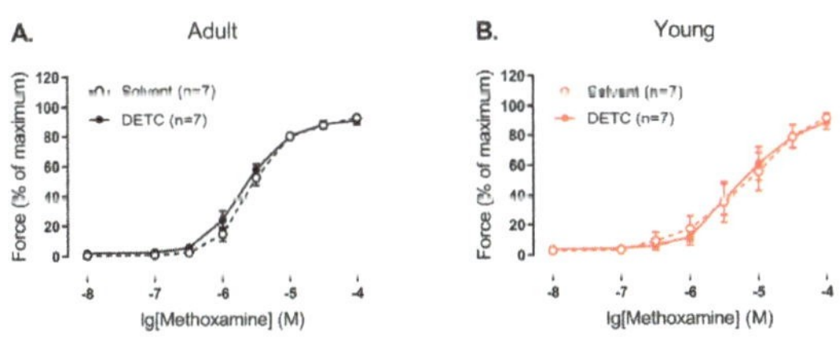

Figure 6. Effect of the Cu/ZnSOD inhibitor DETC on the contractile responses to methoxamine of endothelium-intact arteries from young and adult rats. Concentration–response relationships to methoxamine in the presence of solvent (H_2O) or DETC (1 mM) of endothelium-intact arteries from adult (**A**) and young (**B**) rats; n is the number of animals.

3.6. *Effect of a Cu/ZnSOD Inhibitor on Arterial Contractile Responses in Saphenous Arteries of Young Rats During Manipulation of NO-Dependent Signaling*

Next, we interrogated which endothelium-dependent pathway counteracts the functional impact of SOD1 and SOD3 in smooth muscle of saphenous arteries of young rats. Since the anticontractile effect of endothelial NO is particularly pronounced during early postnatal ontogenesis [33], we focused primarily on the NO pathway.

Thus, we evaluated the effect of DETC in the presence of the NO-synthase inhibitor L-NNA. L-NNA increased contractile responses of endothelium-intact arteries of young rats (Figure 7A, Table S1). Importantly, DETC decreased contractile responses of the arteries in the presence of L-NNA (Figure 7A, Table S1).

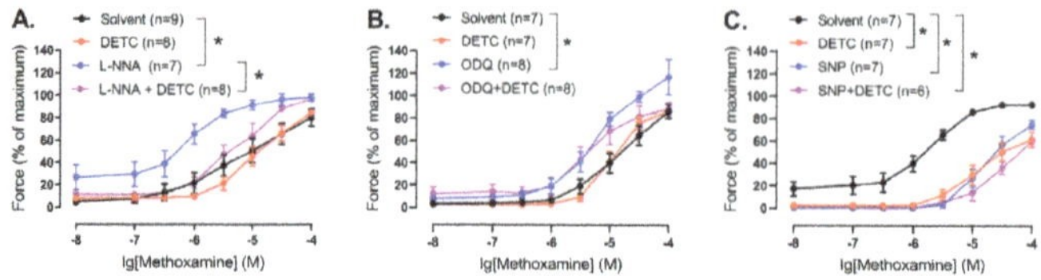

Figure 7. Effect of the Cu/ZnSOD inhibitor DETC on the contractile responses to methoxamine of arteries from young rats during manipulation of NO-dependent signaling. (**A**) Concentration–response relationships to methoxamine in the presence of solvent (H_2O) or DETC (1 mM), or the NO-synthase inhibitor L-NNA (100 µM), or L-NNA together with DETC of endothelium-intact saphenous arteries. (**B**) Concentration–response relationships to methoxamine in the presence of solvent (H_2O) or DETC (1 mM), or the soluble guanylate cyclase inhibitor ODQ (1 µM), or ODQ together with DETC of endothelium-intact saphenous arteries. (**C**) Concentration–response relationships to methoxamine in the presence of solvent (H_2O), or DETC (1 mM), or the NO donor sodium nitroprusside SNP (0.1 µM), or SNP together with DETC of endothelium-denuded saphenous arteries; n is the number of animals. * $p < 0.05$ (two-way repeated measures ANOVA, followed with Tukey's multiple comparisons test).

Next, we evaluated the effect of DETC in the presence of ODQ, an inhibitor of soluble guanylate cyclase, the intracellular target of NO. ODQ caused an increase in the contractile responses of endothelium-intact arteries of young rats (Figure 7B, Table S1). At the same time, we did not detect any effect of DETC in the presence of ODQ (Figure 7B, Table S1).

In order to test whether NO (or its derivatives) may neutralize SODs directly, we assessed the effects of DETC in endothelium-denuded arteries in the presence of the NO donor sodium nitroprusside (SNP). SNP attenuated arterial contractile responses (Figure 7C, Table S1). Importantly, we did not detect any effect of DETC in the presence of SNP (Figure 7C, Table S1).

4. Discussion

In the present study, for the first time, age-related changes of SOD1 and SOD3 vasomotor functional contribution in systemic arteries were evaluated. We showed that activities of Cu/ZnSODs (SOD1 and SOD3) as well as SOD3 protein content were considerably increased in arterial tissue from early postnatal rats in comparison to adult animals. In accordance with these results, functional experiments indicate that SOD1 and SOD3 functioning promotes contraction of arterial smooth muscle of young, but not adult rats. In addition, endothelial NO (or its derivatives) considerably counteracts the SOD1 and SOD3 influence in arterial smooth muscle of early postnatal rats.

Age-related changes in SOD expression and/or activity were addressed previously, but only in the pulmonary circulation, not in the systemic arteries. For example, the activity and protein amount of Cu/ZnSOD were significantly decreased in the lung tissue of newborn rabbits compared to adults [15]. Similarly, the expression of Cu/ZnSOD mRNA in human lung tissue increased from newborns to adults [16]. In addition, the protein content of SOD1 and SOD3 in the lung tissue of adult mice was higher than in 7-day-old mice [17]. Of note, lung tissue, used in the cited studies, includes many types of cells, not just vascular ones. Nevertheless, taken together, these data suggest that the activity and expression of SOD1 and SOD3 in the pulmonary circulation at the early stages of postnatal development is decreased, making newborn organisms more susceptible to persistent pulmonary hypertension development [18]. Our study, for the first time, addresses the expression pattern and age-related changes in SOD activity/protein content in systemic arteries. We found that the mRNA expression level of *Sod3* was the highest in both adult and young rats. At the same time, total SOD activity as well as Cu/ZnSOD activity was considerably increased in arterial tissue from early postnatal rats in comparison to adult. Importantly, although SOD1 protein content did not differ between arterial smooth muscle of adult and young rats, the protein amount of SOD3 in the arterial smooth muscle of 10–15-day old rats was increased in comparison to adults. Thus, in contrast to the pulmonary circulation, arteries of the systemic circulation demonstrate a higher activity and protein amount of Cu/ZnSODs during the early postnatal period.

As mentioned, SODs catalyze the conversion of $O_2^{\bullet-}$ into H_2O_2, which participates in cell signaling. H_2O_2 may cause both relaxation and contraction of arteries, as it was shown for different arteries from adult animals [2–6]. However, much less is known about vasomotor effects of H_2O_2 in arteries of newborn/early postnatal organisms. A few studies demonstrated transient H_2O_2-induced contractions of piglets' pial arterioles [34] as well as small pulmonary arteries of 1–2-week-old piglets [35]. In accordance with this, our data demonstrate that H_2O_2 addition promotes contraction, while H_2O_2 elimination, on the contrary, results in the weakening of contractile responses of endothelium-denuded saphenous arteries from 10–15 day-old rats, indicating a pro-contractile influence of H_2O_2 in arterial smooth muscle in the early postnatal period. Of note, previously we demonstrated that H_2O_2 elimination had no effect on methoxamine-induced contractile responses of saphenous arteries of adult rats [33], indicating that H_2O_2 does not have a pronounced vasomotor effect in this artery at adulthood.

Our next step was to evaluate the functional impact of H_2O_2, produced by Cu/ZnSODs, on the regulation of smooth muscle contraction in saphenous arteries from two age groups of rats. To do so, we used DETC, which chelates Cu^{2+}, leading to a decrease of enzyme activity [7] and, consequently, H_2O_2 content in the arterial wall [36]. In accordance with our molecular experiments, which demonstrated low activity and protein amount of Cu/ZnSODs in the arteries of adult rats, DETC had no influence on methoxamine-induced contractions of endothelium-denuded arteries in adult animals. Similarly, in previous studies, DETC had no effect on phenylephrine-induced contractile responses of rat aorta [8,9] and contractile responses to PGF2alpha of mouse aorta [12]. Along with that, DETC augmented vasocontraction of mouse aorta in response to serotonin [12] and norepinephrine [10] as well as induced further vasoconstriction of precontracted by norepinephrine rat mesenteric arteries [11], pointing to an anti-contractile influence of H_2O_2, produced by Cu/ZnSODs. Thus, the vasomotor contribution of SODs may depend on the artery type and/or the agonist used. Importantly, our data demonstrate that the vasomotor influence of SODs may also depend on the stage of the organism's development: DETC led to pronounced weakening of methoxamine-induced contraction of endothelium-denuded saphenous arteries of young rats. In accordance with the described above pro-contractile influence of H_2O_2 in endothelium-denuded saphenous arteries of young rats, the weakening of arterial contraction after Cu/ZnSOD inhibition indicates a pro-contractile role of H_2O_2, produced by these enzymes, in the smooth muscle of early postnatal saphenous arteries. Of note, the mechanism of such pro-contractile influence remains unclear. Previously we demonstrated, that ROS, produced by NADPH-oxidase, promote vasocontraction of saphenous artery in early postnatal rats by activation of LTCC [24]. Taking into account (1) the high production of $O_2^{\bullet-}$ (demonstrated previously [13]) and (2) high activity of Cu/ZnSODs in saphenous arteries of early postnatal rats (demonstrated in the present study), it is possible that H_2O_2 formed from $O_2^{\bullet-}$ activates LTCC. However, such a hypothesis should be tested in future studies.

It should be also noted that the activity of one of the H_2O_2 scavengers, catalase, was higher in arteries of young compared to adult rats. As already mentioned, we demonstrated previously increased $O_2^{\bullet-}$ production [13] in saphenous arteries of early postnatal rats.

Therefore, apparently, saphenous arteries of early postnatal rats exhibit both higher levels of ROS production and ROS degradation. Such an augmentation of catalase activity in young arteries may serve as an adjustment for increased ROS levels, where initially increased production leads to a rise in ROS levels and a secondary increase in ROS degradation then establishes a new, higher ROS steady state with balanced production and degradation. However, we do not know the kinetics of these processes and cannot say whether such a compensation occurs to the same extent as in adults. Determination of the kinetics of the production and degradation processes requires a large number of experiments that were beyond the scope of the present study. In addition, a set of data demonstrating (1) higher ROS production ($O_2^{\bullet-}$ [13]), (2) higher Cu/ZnSODs activity, (3) higher SOD3 protein content in saphenous arteries of early postnatal rats allows the suggestion that the differences in functional impact of Cu/ZnSODs between young and adult arteries are mainly associated with ROS generation than with their downstream signaling.

Importantly, vasomotor effects of many molecules, including H_2O_2, on smooth muscle tone may depend on the endothelium. For example, endothelium denudation eliminated contraction of pressurized gracilis muscle arterioles of adult rats in response to 0.1 mM of H_2O_2 [3]. In addition, incubation with SOD weakened phenylephrine-induced contraction of endothelium-intact pulmonary arteries of newborn rabbits, which was not observed after endothelium denudation [37]. Therefore, we evaluated the effects of the Cu/ZnSOD inhibitor DETC in endothelium-intact saphenous arteries. The presence of endothelium did not affect the effect of DETC in adult arteries: contractile responses to methoxamine were not affected by DETC. However, DETC did not weaken anymore the contraction of young arteries when the endothelium was left intact. Thus, endothelium is able to counteract the vasomotor impact of Cu/ZnSODs in the arteries of young rats. Since the anti-contractile effect of endothelial NO is particularly pronounced during early postnatal ontogenesis [33], we hypothesized that the lack of DETC effect in endothelium-intact arteries of young rats is primarily associated with the NO pathway. Indeed, DETC decreased contractile responses of young arteries in the presence of the NO-synthase inhibitor L-NNA, as we previously observed in vessels without endothelium (Figure 5B). This indicates that endothelial NO counteracts the vasomotor impact of Cu/ZnSODs in the arterial smooth muscle of young rats.

Further experiments were aimed to answer the question of how NO counteracts the effect of DETC in the arteries of young rats. According to the literature, the activity of Cu/ZnSODs can be altered by post-translational modifications via phosphorylation [38]. Therefore, we suggested that the signaling cascade activated by NO in the smooth muscle cell (activation of soluble guanylate cyclase followed by activation of PKG) may be involved. To answer this question, we evaluated the effect of DETC in the presence of the soluble guanylate cyclase inhibitor ODQ. Importantly, the effectiveness of using 1 μM of ODQ in the saphenous artery of rats was previously demonstrated [33]. Our data show that DETC had no effect in the presence of ODQ, indicating that the neutralizing effect of NO on the effects of DETC in endothelium-intact arteries of young rats is not linked with soluble guanylate cyclase activation, and, probably, its main target in smooth muscle cell—PKG. Of note, Cu/ZnSODs may be nitrosylated, which may also change their activity [38]. Therefore, we suggested that NO (or its derivatives, such as peroxynitrite ($ONOO^-$)) may neutralize SODs directly. To check this, we assessed the effects of DETC in endothelium-denuded arteries in the presence of the NO donor sodium nitroprusside (SNP), which mimics best the effect of naturally produced endothelial NO [39]. Importantly, incubation with DETC did not further attenuate the contractile response in the presence of SNP, similar to that which we observed in endothelium-intact arteries. Thus, NO (or its derivatives, such as peroxynitrite ($ONOO^-$)) counteracts the SOD1 and SOD3 functional impact in the arterial smooth muscle of young rats regardless of the classical signaling cascade of NO, in particular soluble guanylate cyclase activation.

Limitations

The present study has a number of limitations. First, the effect of exogenous H_2O_2 was not studied in the arteries of adult rats. Second, sex-dependent differences were not assessed, since only males were used. Third, H_2O_2 downstream signaling mechanisms were not explored.

5. Conclusions

In conclusion, our novel findings demonstrate that Cu/ZnSODs promote contraction of saphenous arteries, an artery of the systemic circulation, at the early postnatal period, but not at adult age. Moreover, endothelial NO, the anti-contractile influence of which is particularly high at the early stages of postnatal development, counteracts the pro-contractile influence of Cu/ZnSODs in early postnatal, but not in adult arteries. The data obtained should be taken into account at least for several reasons. First, our data complement and develop knowledge about the altered vasomotor contributions of ROS in early postnatal ontogenesis. In addition, the more pronounced functional contribution of Cu/ZnSODs, along with the previously demonstrated high vasomotor influence of NADPH oxidases [13,24], in arteries of young animals may be associated with the stimulation of smooth muscle differentiation [40,41] and angiogenesis [42] by ROS, which actively occur in the early postnatal period. Second, the direction of age-related changes in the functional contribution of Cu/ZnSODs in the systemic circulation appears to be opposite to that in the pulmonary circulation, as described in the literature. Therefore, we have received further evidence for the differences in the functioning between the systemic and pulmonary circulations. Finally, the data obtained need to be taken into account when developing therapeutic approaches in newborns, including those intended to treat "oxygen radical disease in the newborn" [18,43,44].

Supplementary Materials: The following supporting information can be downloaded at: https://www.mdpi.com/article/10.3390/antiox14101231/s1. Table S1: Area under the curve (AUC) in experimental series on evaluation of the Cu/ZnSOD inhibitor effects on the contractile responses to methoxamine of arteries from young rats during manipulation of NO-dependent signaling. Table S2: The internal diameter (d100), maximal force, and values of acetylcholine-induced relaxation of the arteries in different series of wire-myography experiments. Figure S1: Original unprocessed images of Western blot membranes.

Author Contributions: Conceptualization, A.A.S., R.S., and D.K.G.; investigation, A.A.S., V.S.S., D.S.S., and D.K.G.; data curation, A.A.S., and D.K.G.; writing—original draft preparation, A.A.S.; writing—review and editing, A.A.S., R.S., and D.K.G. All authors have read and agreed to the published version of the manuscript.

Funding: This research received no external funding.

Institutional Review Board Statement: The animal study protocol was approved by the Biomedical Ethics Committee of M.V. Lomonosov Moscow State University (Protocol number 149-g, 20 April 2023).

Informed Consent Statement: Not applicable.

Data Availability Statement: All data generated during this study are available from the corresponding author on reasonable request.

Conflicts of Interest: The authors declare no conflicts of interest.

Abbreviations

The following abbreviations are used in this manuscript:

ROS	Reactive oxygen species
SOD	Superoxide dismutase
DETC	Diethyldithiocarbamate
NO	Nitric oxide
SNP	Sodium nitroprusside

References

1. Fukai, T.; Ushio-Fukai, M. Superoxide dismutases: Role in redox signaling, vascular function, and diseases. *Antioxid. Redox Signal.* **2011**, *15*, 1583–1606. [CrossRef] [PubMed]
2. Santiago, E.; Contreras, C.; García-Sacristán, A.; Sánchez, A.; Rivera, L.; Climent, B.; Prieto, D. Signaling pathways involved in the H2O2-induced vasoconstriction of rat coronary arteries. *Free Radic. Biol. Med.* **2013**, *60*, 136–146. [CrossRef]
3. Csató, V.; Peto, A.; Koller, Á.; Édes, I.; Tóth, A.; Papp, Z. Hydrogen peroxide elicits constriction of skeletal muscle arterioles by activating the arachidonic acid pathway. *PLoS ONE* **2014**, *9*, e103858. [CrossRef]
4. Thengchaisri, N.; Kuo, L. Hydrogen peroxide induces endothelium-dependent and -independent coronary arteriolar dilation: Role of cyclooxygenase and potassium channels. *Am. J. Physiol. -Hear. Circ. Physiol.* **2003**, *285*, H2255–H2263. [CrossRef]
5. Ardanaz, N.; Beierwaltes, W.H.; Pagano, P.J. Comparison of H_2O_2-induced vasoconstriction in the abdominal aorta and mesenteric artery of the mouse. *Vascul. Pharmacol.* **2007**, *47*, 288–294. [CrossRef] [PubMed]
6. Puri, N.; Zhang, F.; Monu, S.R.; Sodhi, K.; Bellner, L.; Lamon, B.D.; Zhang, Y.; Abraham, N.G.; Nasjletti, A. Antioxidants condition pleiotropic vascular responses to exogenous H_2O_2: Role of modulation of vascular TP receptors and the heme oxygenase system. *Antioxid. Redox Signal.* **2013**, *18*, 471–480. [CrossRef]
7. Wambi-kiéssé, C.O.; Katusic, Z.S.; Patel, K.; Chen, Y.; Dennehy, K.; Blau, J.; Connors, S.; Tarpey, M.; Krishna, M.; Mitchell, J.B.; et al. Inhibition of copper/zinc superoxide dismutase impairs NO·-mediated endothelium-dependent relaxations. *Nitric Oxide* **1999**, *276*, H1043–H1048. [CrossRef]
8. Nunes, K.Z.; Nunes, D.O.; Silveira, E.A.; Pereira, C.A.C.; Filho, G.B.B.; Vassallo, D.V.; Fioresi, M. Chronic lead exposure decreases the vascular reactivity of rat aortas: The role of hydrogen peroxide. *PLoS ONE* **2015**, *10*, e0120965. [CrossRef] [PubMed]
9. Nunes, K.Z.; Fioresi, M.; Marques, V.B.; Vassallo, D.V. Acute copper overload induces vascular dysfunction in aortic rings due to endothelial oxidative stress and increased nitric oxide production. *J. Toxicol. Environ. Heal.-Part. A Curr. Issues* **2018**, *81*, 218–228. [CrossRef]
10. Kanie, N.; Kamata, K. Contractile responses in spontaneously diabetic mice: I. Involvement of superoxide anion in enhanced contractile response of aorta to norepinephrine in C57BL/KsJ(db/db) mice. *Gen. Pharmacol. Vasc. Syst.* **2000**, *35*, 311–318. [CrossRef]
11. Wang, Y.C.; Hu, C.W.; Liu, M.Y.; Jiang, H.C.; Huo, R.; Dong, D.L. Copper induces vasorelaxation and antagonizes noradrenaline -Induced vasoconstriction in rat mesenteric artery. *Cell. Physiol. Biochem.* **2013**, *32*, 1247–1254. [CrossRef]
12. Andresen, J.J.; Faraci, F.M.; Heistad, D.D. Vasomotor responses in MnSOD-deficient mice. *Am. J. Physiol. Circ. Physiol.* **2004**, *287*, H1141–H1148. [CrossRef] [PubMed]
13. Shvetsova, A.A.; Khlystova, M.A.; Makukha, Y.A.; Shateeva, V.S.; Borzykh, A.A.; Gaynullina, D.K.; Tarasova, O.S. Reactive oxygen species augment contractile responses of saphenous artery in 10-15-day-old but not adult rats: Substantial role of NADPH oxidases. *Free Radic. Biol. Med.* **2024**, *216*, 24–32. [CrossRef]
14. Villamor, E.; Kessels, C.G.A.; Fischer, M.A.J.; Bast, A.; De Mey, J.G.R.; Blanco, C.E. Role of superoxide anion on basal and stimulated nitric oxide activity in neonatal piglet pulmonary vessels. *Pediatr. Res.* **2003**, *54*, 372–381. [CrossRef]
15. Nozik-Grayck, E.V.A.; Dieterle, C.S.; Piantadosi, C.A.; Enghild, J.J.; Oury, T.D. Secretion of extracellular superoxide dismutase in neonatal lungs. *Am. J. Physiol.-Lung Cell. Mol. Physiol.* **2000**, *279*, 977–984. [CrossRef] [PubMed]
16. Asikainen, T.M.; Raivio, K.O.; Saksela, M.; Kinnula, V.L. Expression and developmental profile of antioxidant enzymes in human lung and liver. *Am. J. Respir. Cell Mol. Biol.* **1998**, *19*, 942–949. [CrossRef]
17. Berkelhamer, S.K.; Kim, G.A.; Radder, J.E.; Wedgwood, S.; Czech, L.; Steinhorn, R.H.; Schumacker, P.T. Developmental differences in hyperoxia-induced oxidative stress and cellular responses in the murine lung. *Free Radic. Biol. Med.* **2013**, *61*, 51–60. [CrossRef]
18. Perez, M.; Robbins, M.E.; Revhaug, C.; Saugstad, O.D. Oxygen radical disease in the newborn, revisited: Oxidative stress and disease in the newborn period. *Free Radic. Biol. Med.* **2019**, *142*, 61–72. [CrossRef] [PubMed]

19. Lilley, E.; Stanford, S.C.; Kendall, D.E.; Alexander, S.P.H.; Cirino, G.; Docherty, J.R.; George, C.H.; Insel, P.A.; Izzo, A.A.; Ji, Y.; et al. ARRIVE 2.0 and the British Journal of Pharmacology: Updated guidance for 2020. *Br. J. Pharmacol.* **2020**, *177*, 3611–3616. [CrossRef]
20. Percie du Sert, N.; Hurst, V.; Ahluwalia, A.; Alam, S.; Avey, M.T.; Baker, M.; Browne, W.J.; Clark, A.; Cuthill, I.C.; Dirnagl, U.; et al. The ARRIVE guidelines 2.0: Updated guidelines for reporting animal research. *Br. J. Pharmacol.* **2020**, *177*, 3617–3624. [CrossRef]
21. Tarasova, O.; Sjöblom-Widfeldt, N.; Nilsson, H. Transmitter characteristics of cutaneous, renal and skeletal muscle small arteries in the rat. *Acta Physiol. Scand.* **2003**, *177*, 157–166. [CrossRef]
22. Štulcová, B. Postnatal Development of Cardiac Output Distribution Measured by Radioactive Microspheres in Rats. *Neonatology* **1977**, *32*, 119–124. [CrossRef]
23. Shvetsova, A.A.; Gaynullina, D.K.; Schmidt, N.; Bugert, P.; Lukoshkova, E.V.; Tarasova, O.S.; Schubert, R. TASK-1 channel blockade by AVE1231 increases vasocontractile responses and BP in 1- to 2-week-old but not adult rats. *Br. J. Pharmacol.* **2020**, *177*, 5148–5162. [CrossRef]
24. Shvetsova, A.A.; Shateeva, V.S.; Khlystova, M.A.; Makukha, Y.A.; Tarasova, O.S.; Gaynullina, D.K. NADPH oxidase derived ROS promote arterial contraction in early postnatal rats by activation of L-type voltage-gated Ca^{2+} channels. *Free Radic. Res.* **2025**, *59*, 49–60. [CrossRef] [PubMed]
25. Ruijter, J.M.; Ramakers, C.; Hoogaars, W.M.H.; Karlen, Y.; Bakker, O.; van den Hoff, M.J.B.; Moorman, A.F.M. Amplification efficiency: Linking baseline and bias in the analysis of quantitative PCR data. *Nucleic Acids Res.* **2009**, *37*, e45. [CrossRef] [PubMed]
26. Kostyuk, V.A.; Potapovich, A.I. Superoxide--driven oxidation of quercetin and a simple sensitive assay for determination of superoxide dismutase. *Biochem. Int.* **1989**, *19*, 1117–1124. [PubMed]
27. Fryer, H.J.L.; Davis, G.E.; Manthorpe, M.; Varon, S. Lowry protein assay using an automatic microtiter plate spectrophotometer. *Anal. Biochem.* **1986**, *153*, 262–266. [CrossRef]
28. Shangari, N.; O'Brien, P.J. Catalase activity assays. *Curr. Protoc. Toxicol.* **2006**, *27*, 7.7.1–7.7.16. [CrossRef]
29. Shateeva, V.S.; Simonenko, S.D.; Khlystova, M.A.; Selivanova, E.K.; Borzykh, A.A.; Gaynullina, D.K.; Shvetsova, A.A. Perinatal hypoxia augments contractile impact of NADPH oxidase-derived ROS in early postnatal rat arteries. *Pediatr. Res.* **2024**, *97*, 1220–1226. [CrossRef]
30. Schubert, R.; Gaynullina, D.; Shvetsova, A.; Tarasova, O.S. Myography of isolated blood vessels: Considerations for experimental design and combination with supplementary techniques. *Front. Physiol.* **2023**, *14*, 1–24. [CrossRef]
31. Puzdrova, V.A.; Kudryashova, T.V.; Gaynullina, D.K.; Mochalov, S.V.; Aalkjaer, C.; Nilsson, H.; Vorotnikov, A.V.; Schubert, R.; Tarasova, O.S. Trophic action of sympathetic nerves reduces arterial smooth muscle Ca^{2+} sensitivity during early post-natal development in rats. *Acta Physiol.* **2014**, *212*, 128–141. [CrossRef]
32. Mulvany, M.J.; Halpern, W. Contractile properties of small arterial resistance vessels in spontaneously hypertensive and normotensive rats. *Circ. Res.* **1977**, *41*, 19–26. [CrossRef]
33. Gaynullina, D.; Lubomirov, L.T.; Sofronova, S.I.; Kalenchuk, V.U.; Gloe, T.; Pfitzer, G.; Tarasova, O.S.; Schubert, R. Functional remodelling of arterial endothelium during early postnatal development in rats. *Cardiovasc. Res.* **2013**, *99*, 612–621. [CrossRef] [PubMed]
34. Leffler, C.W.; Busija, D.W.; Armstead, W.M.; Mirro, R. H_2O_2 effects on cerebral prostanoids and pial arteriolar diameter in piglets. *Am. J. Physiol.-Hear. Circ. Physiol.* **1990**, *258*, H1382–H1387. [CrossRef] [PubMed]
35. Fike, C.D.; Aschner, J.L.; Slaughter, J.C.; Kaplowitz, M.R.; Zhang, Y.; Pfister, S.L. Pulmonary arterial responses to reactive oxygen species are altered in newborn piglets with chronic hypoxia-induced pulmonary hypertension. *Pediatr. Res.* **2011**, *70*, 136–141. [CrossRef] [PubMed]
36. Ahmad, M.; Zhao, X.; Kelly, M.R.; Kandhi, S.; Perez, O.; Abraham, N.G.; Wolin, M.S. Heme oxygenase-1 induction modulates hypoxic pulmonary vasoconstriction through upregulation of ecSOD. *Am. J. Physiol.-Hear. Circ. Physiol.* **2009**, *297*, 1453–1461. [CrossRef]
37. Morecroft, I.; MacLean, M.R. Developmental changes in endothelium-dependent vasodilation and the influence of superoxide anions in perinatal rabbit pulmonary arteries. *Br. J. Pharmacol.* **1998**, *125*, 1585–1593. [CrossRef]
38. Banks, C.J.; Andersen, J.L. Mechanisms of SOD1 regulation by post-translational modifications. *Redox Biol.* **2019**, *26*, 101270. [CrossRef]
39. Shvetsova, A.A.; Gaynullina, D.K.; Winkler, P.; Wohlfart, P.; Schubert, R. Potassium channel-mediated NO-induced vasodilation during maturation: Dominance of Kv7 channels. *FASEB BioAdvances* **2025**, *7*, e1490. [CrossRef]
40. Zhou, Y.; Yan, H.; Guo, M.; Zhu, J.; Xiao, Q.; Zhang, L. Reactive oxygen species in vascular formation and development. *Oxid. Med. Cell. Longev.* **2013**, *2013*, 374963. [CrossRef]

41. Clempus, R.E.; Sorescu, D.; Dikalova, A.E.; Pounkova, L.; Jo, P.; Sorescu, G.P.; Schmidt, H.H.H.; Lassègue, B.; Griendling, K.K. Nox4 is required for maintenance of the differentiated vascular smooth muscle cell phenotype. *Arterioscler. Thromb. Vasc. Biol.* **2007**, *27*, 42–48. [CrossRef]
42. Yang, J. The role of reactive oxygen species in angiogenesis and preventing tissue injury after brain ischemia. *Microvasc. Res.* **2019**, *123*, 62–67. [CrossRef] [PubMed]
43. Jankov, R.P.; Negus, A.; Keith Tanswell, A. Antioxidants as therapy in the newborn: Some words of caution. *Pediatr. Res.* **2001**, *50*, 681–687. [CrossRef] [PubMed]
44. Lee, J.W.; Davis, J.M. Future applications of antioxidants in premature infants. *Curr. Opin. Pediatr.* **2011**, *23*, 161–166. [CrossRef]

antioxidants

MDPI

Article

Prenatal Melatonin Therapy Enhances Postnatal Lung Development in a Mouse Model of Inflammation-Induced Preterm Birth

So Hee Park [1], Hee Young Cho [2], Jin Hyun Jun [3], Haengseok Song [1,4,*] and Ji Yeon Lee [5,*]

1 Departments of Life Science, Graduate School, CHA University, Seongnam 13488, Republic of Korea; psh@chauniv.ac.kr
2 Department of Obstetrics and Gynecology, Seoul National University Hospital, Seoul National University College of Medicine, Seoul 03080, Republic of Korea; hycho.md@snu.ac.kr
3 Department of Biomedical Laboratory Science, College of Health Science, Eulji University, Seongnam 13135, Republic of Korea; junjh55@eulji.ac.kr
4 Division of Life Science, CHA University, Pocheon 11160, Republic of Korea
5 Department of Obstetrics and Gynecology, CHA Bundang Medical Center, CHA University School of Medicine, Seongnam 13496, Republic of Korea
* Correspondence: hssong@cha.ac.kr (H.S.); lenna@hanmail.net (J.Y.L.); Tel.: +82-31-881-7150 (H.S.); +82-31-780-5290 (J.Y.L.); Fax: +82-31-881-7249 (H.S.)

Abstract: Inflammation-induced preterm birth (PTB) significantly impacts neonatal development, particularly due to fetal lung immaturity. The lungs undergo critical development both in utero and postnatally, and PTB disrupts this process, leading to impaired pulmonary function. Current treatments for promoting lung maturation in preterm infants have limited efficacy and safety. Melatonin, known for its potent antioxidant and anti-inflammatory properties, has shown promise in preventing PTB, but its effects on fetal and postnatal lung maturation remain unclear. This study evaluated the therapeutic efficacy of melatonin in a mouse model of intrauterine inflammation-induced PTB (IPTB). Pregnant mice (Pregnancy Day 17, [PD17]) were assigned to control, lipopolysaccharide (LPS), and LPS + melatonin groups. LPS (25 μg) was injected into the right uterine horn, with melatonin (10 mg/kg) administered intraperitoneally 30 min prior. Uterine tissues were collected at 6 and 24 h post-LPS administration for molecular and histological analyses. PTB occurred in seven out of eleven (63.6%) IPTB mice within 24 h of LPS injection, whereas melatonin significantly reduced this rate to 25% (2/8). In melatonin-treated mice, the downregulation of pro-inflammatory genes in uterine tissues, restoration of placental blood flow, increased lamellar body counts, and prevention of LPS-induced vacuolation in PD18 fetal lungs were observed. Furthermore, melatonin administration enhanced surfactant protein B expression and improved lung structure. In the offspring of IPTB mice that survived, melatonin further suppressed pro-inflammatory markers and promoted lung septal thickening at postnatal day 3. In conclusion, melatonin prevents PTB, mitigates inflammation, and supports fetal lung maturation in IPTB mice, highlighting its therapeutic potential for improving neonatal pulmonary outcomes.

Keywords: preterm birth; melatonin; fetal lung development; inflammation; postnatal lung

1. Introduction

Preterm birth (PTB), defined as delivery before 37 weeks of gestation, has various causes and poses significant health risks [1,2], particularly due to immature organ development, especially the lungs. The earlier PTB occurs, the higher the risk of severe respiratory complications, as underdeveloped lungs can lead to oxygen insufficiency, triggering systemic dysfunction, including central nervous system impairments [3,4].

Among the causes of PTB, intrauterine inflammation is a major contributor [2]. Elevated expression of pro-inflammatory mediators such as cyclooxygenase (COX)-2, interleukin (IL)-6, IL-1β, tumor necrosis factor (TNF)-α, and prostaglandins (PGE2) induces uterine contraction, cervical remodeling, and fetal membrane apoptosis, ultimately leading to premature rupture of membranes and labor [1,2]. Additionally, inflammation disrupts placental blood flow through vasoconstriction and intravascular coagulation [5]. Compensatory mechanisms, such as increased iNOS activity and vasodilation, further exacerbate perfusion abnormalities [6,7], leading to placental hypoxia and the generation of reactive oxygen species (ROS), which can negatively impact fetal organ development, particularly pulmonary maturation [6–9]. Pulmonary artery dysfunction due to compromised blood flow is especially concerning as it can result in fetal lung injury and delayed lung maturation.

Despite advancements in neonatal care, pulmonary complications in premature infants remain a significant challenge, necessitating both prenatal and postnatal interventions. Postnatally, respiratory support, including oxygen therapy and mechanical ventilation, increases the risk of complications such as pneumothorax, pneumomediastinum, retinopathy of prematurity, and bronchopulmonary dysplasia. Artificial surfactant therapy can aid lung function, but in extremely preterm neonates—the highest risk group—it is often inadequate, as it may not fully compensate for the profound structural and functional lung immaturity [10]. In late preterm and term neonates, surfactants have been shown to benefit, including reduced mortality, air leaks, and the need for prolonged respiratory support [11]. Prenatally, corticosteroids (e.g., betamethasone or dexamethasone) are routinely administered between 24–34 gestational weeks to accelerate fetal lung maturation, with cohort studies demonstrating significantly lower neonatal mortality—particularly at 24–29 weeks of gestation [12]. However, efficacy remains limited in the most premature neonates (<28 weeks), and systemic risks—such as exacerbated maternal hyperglycemia requiring insulin therapy in gestational and pre-existing diabetes—have also been reported [13]. More recently, emerging therapies such as IGF-1 supplementation and N-acetylcysteine administration have been investigated in clinical studies as potential strategies to improve outcomes in preterm infants [14,15]. While these novel approaches show promise, their therapeutic application is still constrained by limited clinical evidence and methodological challenges, highlighting the need for further studies to establish their safety and efficacy.

Melatonin, a multifunctional antioxidant and anti-inflammatory molecule, offers a promising therapeutic approach for prenatal intervention. Melatonin exhibits potent antioxidant properties and efficiently scavenges free radicals [16]. Maintaining a homeostatic equilibrium between ROS and antioxidants is essential during pregnancy to ensure placental health and stability. Disruption of this balance can lead to complications such as PTB, preeclampsia, and intrauterine growth restriction [17]. Capable of crossing the placental and blood–brain barriers, melatonin modulates inflammation, ROS, and hypoxia—key drivers of PTB-associated lung immaturity [4,18,19]. Also, melatonin modulates immune responses by suppressing the activation of pro-inflammatory transcription factors such as NF-κB [20], thereby downregulating the expression of cytokines, including TNF-α, IL-1β, and IL-6, which are well-known inflammatory mediators that are markedly upregulated during PTB [21]. In addition to intrauterine inflammation, melatonin has shown protective effects in other experimental models, including hyperoxia-induced neonatal lung injury [22] and hypoxic–ischemic brain damage [23], where it reduced oxidative stress, suppressed pro-inflammatory cytokines, and preserved tissue integrity. These findings underscore the dual antioxidant and anti-inflammatory mechanisms of melatonin against both inflammation and oxygen toxicity, supporting its potential to prevent PTB-associated fetal organ damage. However, research on melatonin's role in preventing PTB-associated lung complications and enhancing fetal lung maturation, especially in the context of intrauterine inflammation-induced PTB (IPTB), remains limited. The present study aims to investigate the therapeutic potential of melatonin in promoting fetal lung maturation in a mouse model of IPTB.

2. Materials and Methods

2.1. Animals

All animal care and treatment procedures were approved by the Institutional Animal Care and Use Committee (IACUC) of CHA University (Approval No. IACUC 230152). Pregnant ICR mice (Koatech, Ansan-si, Republic of Korea) were used for this study.

2.2. Establishment of Mouse Models of Intrauterine Inflammation-Induced Preterm Birth

Pregnant ICR mice obtained from Koatech were anesthetized with isoflurane, and a 1.5 cm midline incision was made in the lower abdominal wall after aseptic preparation. At Pregnancy Day 17 (PD17), mice were randomized to receive intrauterine injections of either lipopolysaccharide (LPS) (Sigma-Aldrich, St. Louis, MO, USA) to induce inflammation-induced preterm birth, or phosphate-buffered saline (PBS) as a control. For intrauterine inflammation, LPS (25 μg in 100 μL PBS) was injected into the uterine horn between the first and second gestational sacs. PTB was defined as delivery within 48 h post-injection. Melatonin (10 mg/kg) (Sigma-Aldrich) was administered intraperitoneally 30 min before LPS injection. For postnatal analyses, the number of live pups delivered in each group was recorded at birth, and subsequent evaluations were performed at matched postconceptional ages (PCA = gestational day [GD] at birth + postnatal days [PND]). Thus, pups born at GD20 (PBS) were counted and analyzed at PND1 (PCA21), while pups born at GD18 (LPS) were counted and analyzed at PND3 (PCA21), ensuring developmental equivalence across groups. In this study, PND3 was used as the reference developmental stage based on the LPS group (PCA21).

2.3. RNA Preparation, RT-PCR, and Real-Time RT-PCR

Total RNA was extracted from myometrium, decidua, placenta, and fetal lung tissues using TRIzol reagent (Ambion, Carlsbad, CA, USA), following the manufacturer's instructions. For cDNA synthesis, RNA (1 μg) was subjected to reverse transcription (RT) using M-MLV reverse transcriptase (Promega, Madison, WI, USA) with random primers and oligo dT. The synthesized cDNA was used for polymerase chain reaction (PCR) with gene-specific primers under optimized thermal cycling conditions and annealing temperatures, followed by electrophoresis of the PCR products. DNA fragments were separated by 2% agarose gel electrophoresis in TBE buffer. Real-time RT-PCR was performed using SYBR Green Dye (Bio-Rad, Waltham, MA, USA) to quantify gene expression. A standard curve of cycle thresholds from serial dilutions of a cDNA sample was generated to determine relative transcript levels, normalized to ribosomal protein L7 (*rPL7*) cDNA. All PCRs were performed in duplicate. All raw data values from the real-time RT-PCR experiments are available in the Supplementary Excel File (Supplementary Data S1).

2.4. Western Blot Analysis

Uterine and fetal lung tissues were homogenized using a Polytron homogenizer (Brinkmann, Westbury, NY, USA). Proteins were extracted using PRO-PREP Protein Extraction Solution (iNtRON Biotechnology, Seongnam, Republic of Korea) containing a 1× phosphatase inhibitor (Roche Applied Sciences, Indianapolis, IN, USA). Extracted proteins were separated by sodium dodecyl sulfate-polyacrylamide gel electrophoresis and transferred to a nitrocellulose membrane (Bio-Rad). For Western blot analyses, the membranes were incubated with primary antibodies, followed by horseradish peroxidase-conjugated goat anti-rabbit or anti-mouse secondary antibodies (Invitrogen, Carlsbad, CA, USA). Immunoreactive bands were detected using the enhanced chemiluminescence (ECL) Western blotting substrate kit (Cytiva, Marlborough, MA, USA) and visualized using the ChemiDoc XRS + System (Bio-Rad) with Image Lab software v6.1.

2.5. *In Vivo Imaging of Placental Blood Flow*

To assess placental blood flow in pregnant mice, indocyanine green (ICG; 0.025 mg/100 μL, Cayman Chemical, Ann Arbor, MI, USA) was intravenously injected via the tail vein at PD17. Fluorescence imaging was performed for 30 min under anesthesia using the Pearl Impulse Imager (LI-COR Biosciences, Lincoln, NE, USA). In the LPS-treated group, the maximum placental fluorescence was consistently observed approximately 5 min after injection; therefore, representative images were selected at the time of injection (10 s) and 5 min (300 s). The "time to maximum fluorescence" was defined as the time taken for the placental fluorescence signal to reach its peak and subsequently begin to decline. Fluorescent signals in the placental region were captured, and the intensity of ICG fluorescence was quantified using Image Studio software v5.2 (LI-COR Biosciences, Lincoln, NE, USA).

2.6. *Histochemistry and Immunofluorescence Staining*

Fetal lungs were fixed at PD18 by immersing the entire pup in 4% paraformaldehyde (PFA) (Biosesang, Seongnam, Republic of Korea) prior to the onset of air breathing, whereas PND3 lungs were fixed by immersion in paraformaldehyde following cardiac perfusion with saline to remove blood. Fetuses were then immersed in 30% sucrose (Sigma-Aldrich), embedded in the optimal cutting temperature compound (Leica Microsystems, Wetzlar, Germany), and sectioned at 15 μm using a cryostat (Leica Microsystems). PND3 lungs were processed into paraffin after fixation and subsequently sectioned for histological and immunohistochemical analyses. The sections were stained with hematoxylin and eosin (H&E). For immunofluorescence studies, frozen sections were fixed in 4% PFA for 10 min, washed with PBS, and permeabilized with 0.1% Triton X-100 in PBS for 15 min at 25 °C. Non-specific staining was blocked using Protein Block Serum-Free (Dako, Santa Clara, CA, USA) for 1 h, followed by incubation with the primary antibodies against surfactant protein-B (SP-B) (1:200, ab40876, Abcam, Cambridge, UK) overnight at 4 °C. For fluorescence labeling, sections were incubated with Alexa Fluor 488-conjugated secondary antibodies (1:1000, Invitrogen) for 1 h at 25 °C. Nuclei were counterstained with propidium iodide (1:500, Thermo Fisher Scientific, Waltham, MA, USA). Slides were mounted using Gel Mount Aqueous Mounting Medium (M01, Sigma-Aldrich), and images were captured using a microscope (Carl Zeiss Meditec AG, Jena, Germany) and analyzed with the corresponding software.

2.7. *Transmission Electron Microscopy (TEM)*

Fetal lungs were fixed with 2% glutaraldehyde (Sigma-Aldrich) and 2% PFA in PBS at 4 °C for 8 h, followed by post-fixation in 1% osmium tetroxide. After washing with PBS, samples were dehydrated through a graded ethanol series (60%, 70%, 80%, 90%, 95%, and 100%) and embedded in Epon resin. Thin sections were prepared, counterstained with uranyl acetate and lead citrate, and imaged using a transmission electron microscope (H-7600; Hitachi, Tokyo, Japan).

2.8. *Statistical Analyses*

Statistical analyses were performed using GraphPad Prism (version 8.2, GraphPad Software, San Diego, CA, USA) and SPSS (version 29, IBM Corp., Armonk, NY, USA).

Comparisons among three groups for continuous variables were conducted using the Kruskal–Wallis test, while comparisons between two groups were performed using the Mann–Whitney U test. Categorical variables were analyzed using the chi-square test. A p-value < 0.05 was considered statistically significant.

3. Results

3.1. Melatonin Prevents Preterm Birth and Reduces the Inflammatory Response

To investigate the effects of melatonin on PTB, we used a mouse model of IPTB. Pregnant ICR mice received an intrauterine injection of LPS at PD17 to induce PTB, while the LPS + melatonin group was pretreated with melatonin 30 min before LPS administration (Figure 1A,B). Within 24 h, no cases of PTB were observed in the PBS group (control) (0%, 8/8). In contrast, the LPS group showed a significantly increased PTB rate of 63.6% (7/11), while co-administration of melatonin reduced the rate to 25% (2/8) ($p < 0.05$). By 48 h, PTB remained absent in the PBS group, but occurred in all cases of the LPS group (100%, 11/11) and half of the LPS with melatonin group (50%, 4/8), showing a significant difference between the groups ($p < 0.001$) (Figure 1C). In addition, we quantified the number of surviving pups after delivery (Figure 1D). The results showed a reduction in the LPS group compared with the PBS group, whereas prenatal melatonin treatment significantly increased the number of pups that survived. Melatonin treatment not only reduced the PTB rate but also significantly attenuated the expression of pro-inflammatory genes, including *Ptgs2*, *Il-6*, and *Tnf-α*, in uterine tissues such as myometrium, decidua, and placenta compared to the LPS group (Figure 1E–G).

3.2. Melatonin Restores Reduced LPS-Induced Blood Flow in Inflammation-Induced Preterm Birth in Mice

Adequate placental vascularization is essential for proper fetal development. To assess the impact of melatonin on placental blood flow in IPTB, indocyanine green (ICG) was administered intravenously 6 h after LPS injection at PD17 in pregnant mice, and uterine tissues were imaged using fluorescence detection (Figure 2A). The expression of the vasoconstrictor endothelin-1 (*Edn1*) was significantly increased in the myometrium ($p < 0.001$), decidua ($p < 0.05$), and placenta ($p < 0.05$) of IPTB mice (Figure 2B). Melatonin treatment effectively reduced *Edn1* expression in the uterine tissues ($p < 0.05$). In vivo imaging revealed reduced placental blood flow in IPTB mice (Figure 2C). Dotted lines indicate the location of placentas, showing decreased fluorescence intensity following LPS administration, which was restored by melatonin treatment. Quantitative analysis of ICG fluorescence intensity in the placenta revealed a significant reduction in blood flow in the LPS group compared to both the PBS and melatonin groups at both 10 s ($p < 0.001$) and 300 s ($p < 0.001$) after ICG injection. Melatonin treatment markedly restored placental blood flow at both time points (10 s: $p < 0.001$; 300 s: $p < 0.001$) (Figure 2D). In addition, the time required for placental fluorescence to decline after reaching its highest intensity was defined as the "Time to maximum fluorescence". This time was significantly longer in the LPS group compared with the PBS and melatonin groups ($p < 0.05$), indicating impaired placental hemodynamics. However, in IPTB mice treated with melatonin, the time to maximum fluorescence did not differ significantly from that of the PBS or melatonin groups alone, but it tended to decrease, suggesting a potential partial improvement in placental blood flow.

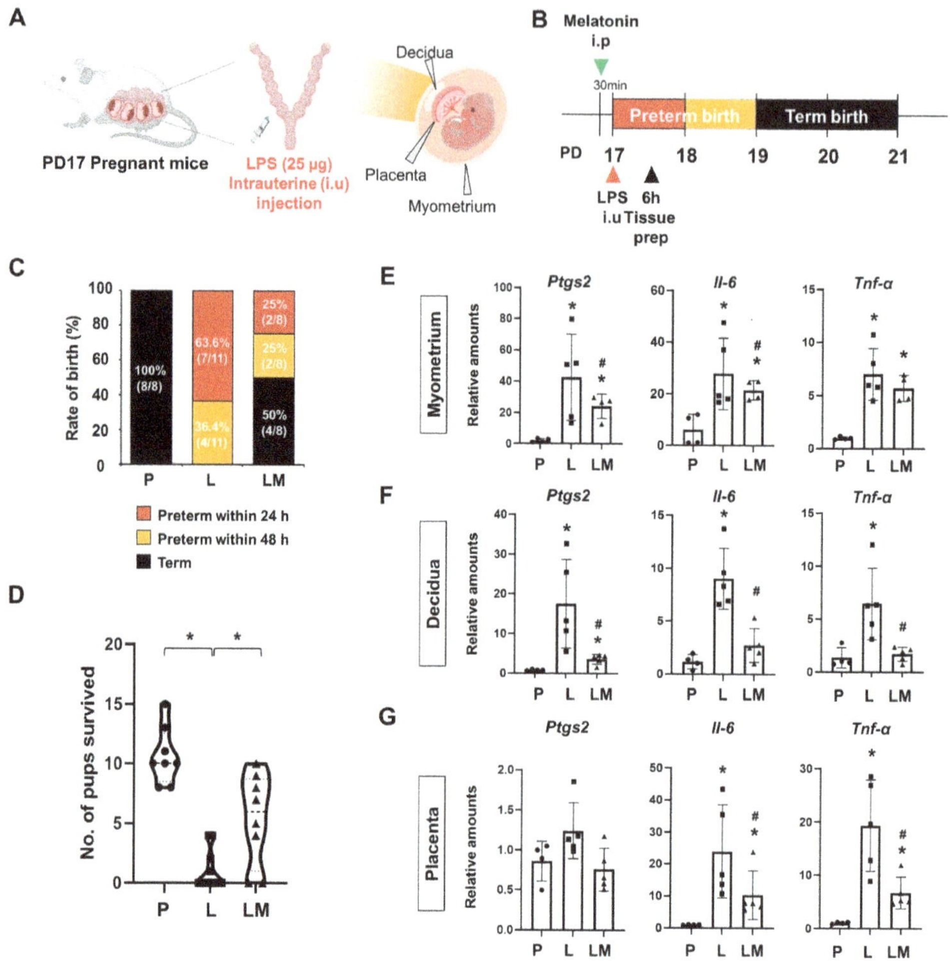

Figure 1. Melatonin prevents preterm birth and reduces inflammatory responses. (**A**) The scheme illustrates the establishment of IPTB mice. The figure was created with BioRender.com. (**B**) The scheme illustrates the time window defined as PTB following IPTB. (**C**) Graph showing the rate of birth after LPS administration with or without prenatal melatonin treatment. Melatonin treatment decreased the rate of PTB. (**D**) The number of surviving pups was quantified at birth. * $p < 0.05$. (**E–G**) Real-time RT-PCR analysis of pro-inflammatory gene expression in the myometrium (**E**), decidua (**F**), and placenta (**G**) following LPS administration, with or without melatonin treatment. *rPL7* was used as an internal loading control. Each plot represents an individual value. * P vs. others, $p < 0.05$; # L vs. LM, $p < 0.05$. Abbreviations: P, PBS; L, LPS; LM, LPS + melatonin.

3.3. Melatonin Improves Fetal Lung Immaturity in Inflammation-Induced Preterm Birth in Mice

To assess the effect of melatonin on fetal lung development in IPTB, the number of lamellar bodies (LBs, secretory organelles present in type II alveolar cells indicative of lung development) was evaluated 24 h after LPS-induced PTB (Figure 3A). Histological analysis showed no significant structural differences in fetal lungs among the groups, regardless of LPS or melatonin treatment (Figure 3B). At PD18, septal thickness was consistent across all groups (no statistically significant difference was observed) (Figure 3C). However,

LPS administration led to a decrease in SP-B expression in the fetal lungs of IPTB mice. Melatonin treatment effectively restored SP-B expression (Figure 3D). Microstructural examination by TEM showed that melatonin administration prevented vacuolation of LBs in the fetal lungs induced by LPS exposure. Additionally, melatonin increased the number of secreted LBs (Figure 3E). Furthermore, LPS treatment tended to reduce the number of LBs in the amniotic fluid ($p < 0.05$), while melatonin treatment significantly restored their numbers ($p < 0.05$) (Figure 3F). These findings suggest that although no immediate structural abnormalities were observed in fetal lungs following LPS exposure, functional impairments, particularly in surfactant production, were evident.

3.4. Melatonin Improves Postnatal Lung Development in Inflammation-Induced Preterm Birth in Mice

Functional impairments observed immediately after LPS exposure also impacted postnatal lung development. At PND3 (PCA21), the body weights of the surviving pups were measured (Figure 4A). Offspring in the LPS group exhibited reduced body weights compared with the PBS group, whereas prenatal melatonin treatment restored the reduction. Histological examination of PND3 lungs by H&E staining revealed that the alveolar septa remained thickened in the LPS group, while this change was ameliorated in the melatonin-treated group ($p < 0.05$) (Figure 4B,C). However, melatonin administration mitigated these abnormalities ($p < 0.01$), suggesting its protective role in postnatal lung maturation. In the LPS group, *Tnf-α* and *Il-1β* were significantly upregulated compared to the PBS group ($p < 0.05$), contributing to impaired lung development at PND3. Melatonin treatment effectively alleviated these inflammatory responses ($p < 0.05$) (Figure 4D,E). Western blot analysis showed that administration of melatonin reduced the expression of inflammatory protein IL-1β compared to the LPS group ($p < 0.05$) (Figure 4F,G). Although the thickness of pulmonary capillary endothelial cells tended to increase in the LPS group, the difference did not reach statistical significance compared to the PBS and LPS with melatonin groups (Figure 4H,I). These findings demonstrate that melatonin not only restores placental blood flow to prevent IPTB but also facilitates lung maturation, mitigating alveolar septal thickness and inflammation in newborn offspring (Figure 5).

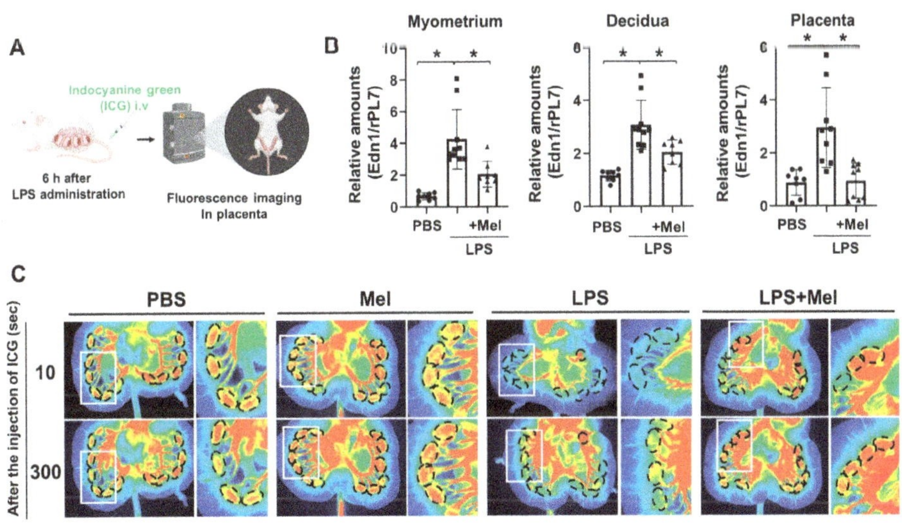

Figure 2. *Cont.*

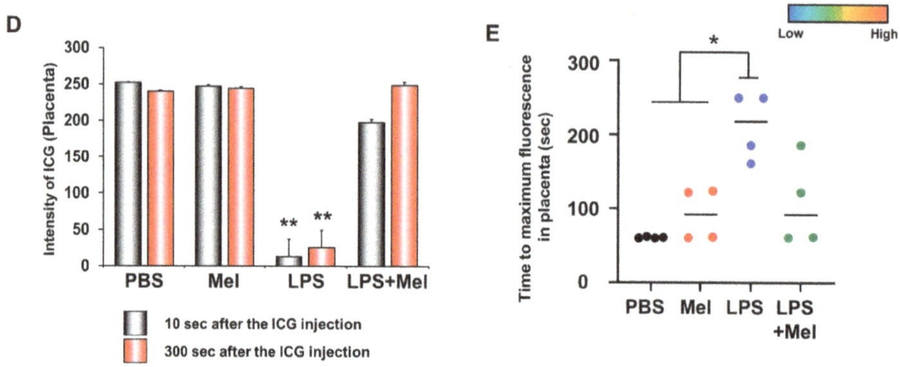

Figure 2. Melatonin restores reduced LPS-induced blood flow in inflammation-induced preterm birth in mice. (**A**) The scheme to explain in vivo fluorescence imaging in placental blood flow. The figure was created with BioRender.com. (**B**) Real-time RT-PCR analyses for relative mRNA expression of *Edn1* in the myometrium, decidua, and placenta. *rPL7* was used as an internal loading control. Each plot represents an individual value. (**C**) In vivo imaging of blood flow in the placenta using the Pearl Impulse Imager. LPS treatment reduced blood flow to the placenta, which was restored by melatonin administration. Regions marked by dotted lines indicate the placental location. The right panel of each group shows a magnified view of the boxed region. (**D**) Graph showing the intensity of blood flow in the placenta at 10 and 300 s after ICG injection. (**E**) Graph showing the time to maximum ICG fluorescence in the placenta. * $p < 0.05$; ** $p < 0.001$. Abbreviations: Mel, melatonin; ICG, Indocyanine green.

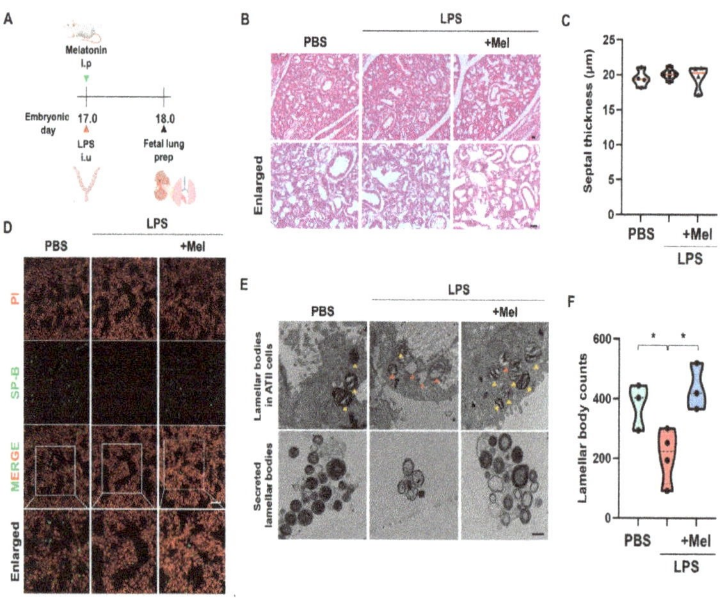

Figure 3. Melatonin improves fetal lung immaturity in inflammation-induced preterm birth in mice. (**A**) The scheme illustrates the fetal lung preparation time point in IPTB mice. (**B**) Histological analysis of fetal lungs at PD18 by H&E staining. (**C**) Quantitative analysis of the septal thickness in PD18 fetal lungs. (**D**) Melatonin increased SP-B expression in fetal lungs exposed to intrauterine inflammation. Scale bar = 20 μm. (**E**) Microstructure of the fetal lung by TEM. Yellow arrowheads indicate normal LBs. Red arrowheads indicate vacuolated and abnormal LBs. Scale bar = 500 nm. (**F**) The LB count in amniotic fluid 24 h after PTB * $p < 0.05$.

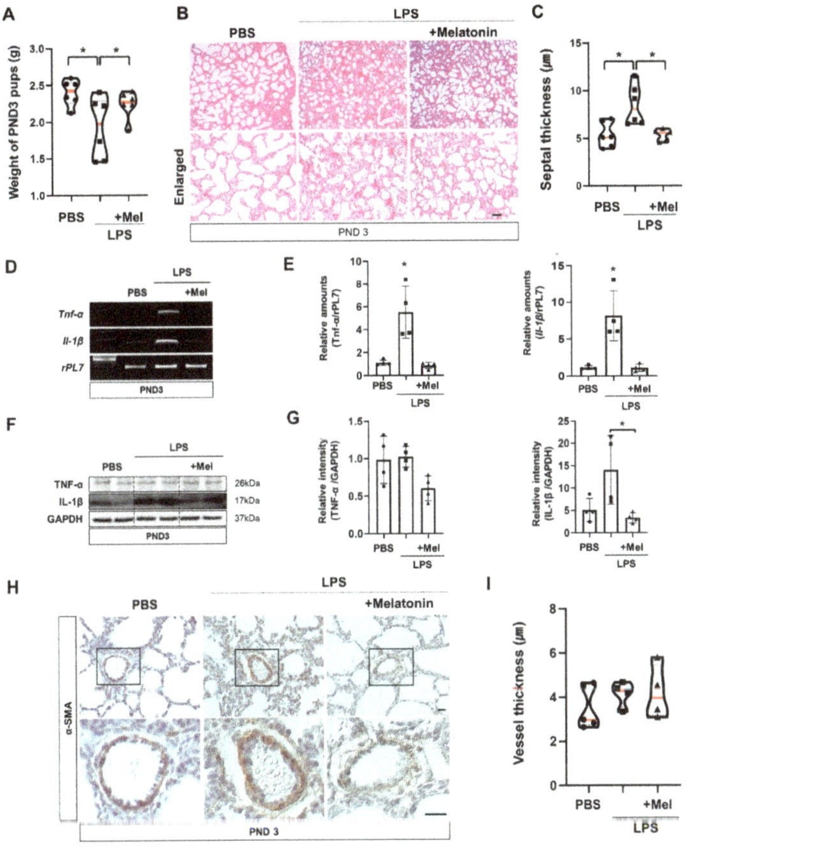

Figure 4. Melatonin improves postnatal lung development in inflammation-induced preterm birth in mice. (**A**) Body weights of surviving pups at PND3. PND3 was defined based on the LPS group to ensure developmental stage equivalence. (**B**) Histological analysis of PND3 lungs in IPTB mice using H&E staining. Scale bar = 20 μm. (**C**) Quantitative analysis of the septal thickness in PND3 lungs. (**D**) RT-PCR analysis of pro-inflammatory genes in PND3 lungs. (**E**) Real-time RT-PCR analysis of pro-inflammatory genes in PND3 lungs. *rPL7* is used as an internal loading control. Each plot represents an individual value. (**F**) Western blot analysis of pro-inflammatory proteins in PND3 lungs. (**G**) The relative intensity of pro-inflammatory genes was quantified using GAPDH as a loading control. Each plot represents an individual value. (**H**) Immunohistochemistry showed the expression of α-SMA in PND3 lungs. Scale bar = 20 μm. (**I**) Quantitative analysis of vessel thickness in PND3 lungs (PBS n = 5, LPS n = 5, LPS + melatonin n = 5). * L vs. others, $p < 0.05$. Abbreviations: PND, Postnatal day; α-SMA, Alpha-smooth muscle actin.

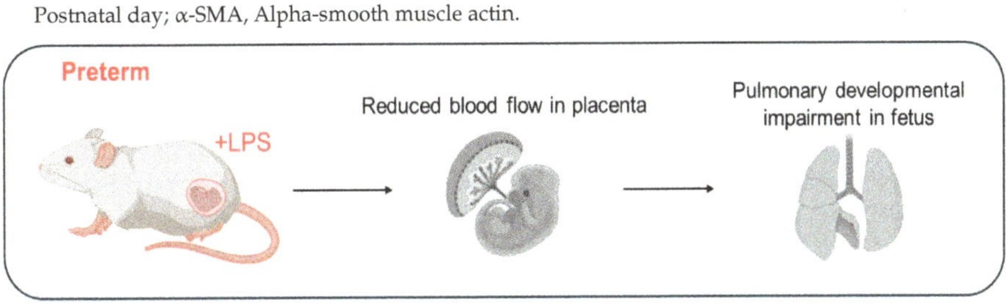

Figure 5. *Cont.*

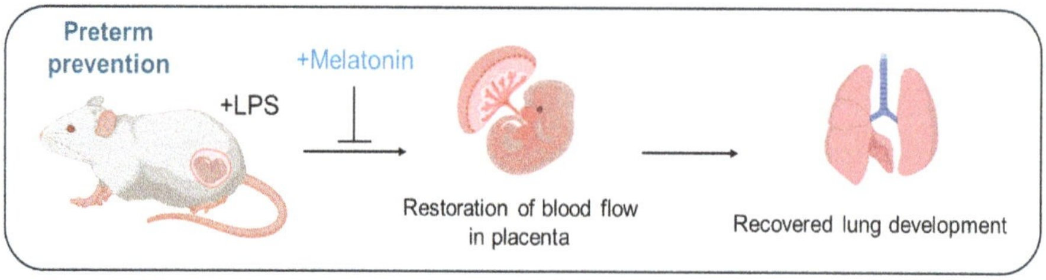

Figure 5. Schematic overview of the role of prenatal melatonin in recovering from preterm birth and enhancing fetal and postnatal lung development in inflammation-induced preterm birth in mice. This figure illustrates the central role of prenatal melatonin administration in regulating PTB and promoting fetal and postnatal lung development in IPTB in mice. The figure was created with BioRender.com.

4. Discussion

PTB, particularly IPTB, poses significant challenges to newborn health and development, impacting crucial organ development such as the lungs. Our study provides compelling evidence for the beneficial effects of melatonin administration in a murine model of IPTB, suggesting its potential to restore fetal lung development. Using IPTB mice, our study showed that inflammation-related genes, *Ptgs2*, *Il-6*, *Tnf-α*, and *Il-1β*, were increased in uterine tissues (Figure 1) and lungs of offspring at PND3 (Figure 4). Similarly, in the chorioamnionitis-induced PTB rat model, LPS administration increased cerebral expression of inflammatory and apoptotic markers, such as IL-1β, TNF-α, RANK, and caspase-8. Similarly, in the intra-amniotic induced PTB mouse model, LPS elevated IL-6 and TNF-α expression in the amniotic fluid [24,25]. LPS exposure has been reported to induce IL-1β expression through activation of the NLRP3 inflammasome, which plays a central role in amplifying inflammatory responses [26]. The increased IL-1β mRNA and protein observed in our study suggest that this pro-inflammatory response may be mediated through NLRP3 inflammasome signaling. Consistent with this, previous studies have demonstrated that melatonin attenuates NRLP3 inflammasome activity via inhibiting nuclear translocation of NF-κB, thereby reducing IL-1β production [27,28]. These findings suggest that the reduction of IL-1β expression by melatonin in our model may be linked to its ability to inhibit NF-κB-driven priming and NLRP3 inflammasome activation. In addition, previous reports using similar PTB models have demonstrated that maternal LPS-induced inflammation not only elevates pro-inflammatory gene expression but also promotes immune cell infiltration into fetal tissues [29,30]. For example, in intraperitoneal LPS-induced PTB mouse models, marked accumulation of neutrophils was observed in fetal lungs and kidneys, coinciding with increased expression of inflammatory cytokines [29]. Also, studies in rhesus macaques have shown that intra-amniotic LPS induces a potent and rapid myeloid cell response in fetal lungs, dominated by infiltrating neutrophils and monocytes/macrophages, with transcriptional profiles indicating exposure to TLR ligands and pro-inflammatory cytokines such as IL-1β and TNF-α [30]. In addition, our unpublished observations in the uterine context revealed that LPS-induced inflammation was accompanied by increased neutrophil and macrophage infiltration, whereas prenatal melatonin treatment attenuated macrophage accumulation. Together, these observations suggest that the elevated cytokine expression in our study may reflect enhanced immune cell recruitment induced by LPS, and it is therefore reasonable to assume that prenatal melatonin treatment could attenuate such neutrophil and macrophage infiltration in the fetal lung of our model.

In our IPTB model, reduced placental blood flow led to uterine vasoconstriction, as indicated by the upregulation of vasoconstrictor markers, such as *Edn1* (Figure 2). Reduced placental blood flow impaired placental function, resulting in a hypoxic environment that initiates a cascade of events characterized by the generation of ROS and tissue inflammation [9]. Melatonin, through its potent antioxidant activity, can neutralize ROS and protect against hypoxia-induced oxidative injury, thereby mitigating the downstream effects of hypoxia in fetal tissues [17,31]. Given that melatonin administration restored placental perfusion and reduced inflammation in our model (Figures 1 and 2), it is plausible that these protective effects were at least partly mediated by melatonin's antioxidant activity.

In the chorioamnionitis-induced PTB rat model, severe hyperemia and edema were observed in various fetal organs, along with hemorrhagic foci in the pleura and lung parenchyma of the LPS group [32]. Particularly inflammatory infiltrates mainly composed of activated T lymphocytes (CD45R0+) were often associated with focal necrosis of adjacent parenchyma, exhibiting severe hypertrophy and disorganization [33]. These findings closely align with our observations of thickened fetal lung septum and inflammatory responses following intrauterine LPS administration. Notably, melatonin treatment effectively mitigated these inflammatory reactions (Figure 4). This protective effect may, in part, stem from melatonin's ability to attenuate hypoxia-induced oxidative stress, thereby preserving vascular integrity and reducing inflammation in the fetal lung. In this study, intrauterine inflammation impaired fetal lung maturation, as demonstrated by reduced SP-B expression and abnormal LB morphology (Figure 3). These changes are consistent with the essential role of surfactant in maintaining alveolar stability and efficient gas exchange [34]. Given the essential roles of SP-B and SP-C in preventing alveolar collapse and supporting fetal lung development [35], the restoration of these proteins by melatonin underscores its potential to mitigate inflammation-induced lung dysfunction.

Melatonin exhibits potent antioxidant properties by neutralizing free radicals and acts as an anti-inflammatory agent by suppressing COX-2 and NF-kB pathways [36,37]. Clinically, melatonin has shown efficacy in improving blood flow and treating conditions such as stroke, depression, glaucoma, hypertension, diabetes, and various cancers [38–40]. In our previous study, melatonin prevented placental blood clot formation in a murine intrauterine inflammation model [41]. Moreover, melatonin has demonstrated anti-fibrotic potential in chemotherapy-induced lung damage [42] and ameliorates neonatal morbidities such as bronchopulmonary dysplasia by reducing oxidative stress and inflammation [41,43,44]. The antioxidant effect of melatonin is primarily mediated through non-receptor mechanisms involving direct scavenging of ROS [45]. The protective effects of melatonin in preventing PTB and improving lung immaturity are likely mediated through receptor-independent antioxidant mechanisms. Furthermore, other studies have demonstrated that melatonin's antioxidant effects are not limited to the fetal lung but also extend to the fetal brain, where they facilitate neurodevelopment in preterm models [17,21]. These observations suggest that the antioxidant properties of melatonin may play a broader role in promoting overall fetal development under inflammatory or oxidative stress conditions. Recent research also highlights melatonin's role in cell differentiation and proliferation [42,46,47], influencing fetal growth and development [48,49]. Our earlier investigation confirmed melatonin's anti-inflammatory effects in preventing PTB in mice and its positive impact on fetal brain and lung maturation [6].

This study provides valuable insights into the impact of melatonin on lung development in premature infants, addressing a critical aspect of PTB. It highlights melatonin's potential as a therapeutic option for enhancing fetal lung maturation, which could improve the prognosis of premature infants. Notably, this study extends beyond previous research focused on prenatal fetuses by demonstrating improved lung maturation in postnatal offspring. Using a mouse model of IPTB enabled a detailed investigation of the molecular pathways and physiological changes associated with fetal lung maturity under controlled experimental conditions.

However, this study has some limitations. Although it identifies changes in gene expression and histological features associated with melatonin treatment, further mechanistic studies are needed to clarify the pathways influencing fetal lung development during PTB. Additionally, the study does not assess long-term effects on respiratory function and neurodevelopment in offspring, highlighting the need for future research in this area. We monitored the offspring up to PND14, primarily recording body weights (unpublished observation). However, in the LPS group, the number of surviving pups was lower than in the other groups (Figure 1D), which reduced litter competition and consequently diminished the body weight differences compared to the other groups. This made it difficult to accurately interpret group differences at this later point. Therefore, we found our analyses on PND3, when the acute effects of intrauterine inflammation and prenatal melatonin treatment were most evident. Given its experimental nature in animals, the direct translation of these findings to clinical practice may be limited. Therefore, further investigations, including clinical trials, are necessary to validate the efficacy and safety of melatonin as a therapeutic intervention for PTB and fetal lung immaturity. Moreover, the relatively small number of biological replicates in some experimental groups could be a limitation of this study. This was largely due to the reduced survival of offspring in the LPS group, which restricted the number of samples available for certain analyses. Although statistical methods were applied appropriately, future studies with larger groups will be required to strengthen these findings. Previous studies have demonstrated that combined administration of melatonin and corticosteroids exerts greater protective effects against acute lung injury than either agent alone, suggesting potential synergistic interactions [50]. Based on these findings, we propose that future studies exploring the combination of melatonin with corticosteroids, which are the current clinical standard for promoting fetal lung maturation in PTB, may help to overcome the residual risks that remain when corticosteroids are used alone in extremely preterm infants. Such an approach could provide an enhanced therapeutic strategy to improve neonatal outcomes in the setting of IPTB.

5. Conclusions

This study demonstrates that melatonin effectively mitigates intrauterine inflammation induced by LPS, which adversely affects the uterine tissues, leading to preterm birth and impaired fetal lung development. The protective effects of prenatal melatonin treatment are likely mediated primarily through anti-inflammatory mechanisms that reduce inflammatory responses. They may also manifest through secondary antioxidant effects, such as the restoration of impaired blood flow. These findings suggest that melatonin holds promise as a therapeutic intervention not only for preventing PTB but also for promoting lung development in premature infants. This underscores the potential of maternal melatonin administration as a viable treatment option for complications associated with PTB. Further investigations, particularly in clinical settings, are warranted to validate the translational potential of melatonin in improving pregnancy outcomes and neonatal health.

Supplementary Materials: The following supporting information can be downloaded at: https://www.mdpi.com/article/10.3390/antiox14091094/s1, Supplementary Data S1 (Excel File): Raw data for Real-time RT-PCR and Western blot band intensity; Supplementary Figure S1: Original uncropped RT-PCR images corresponding to Figure 4D; Figure S2: Original uncropped Western blot images corresponding to Figure 4F.

Author Contributions: S.H.P.; Writing—Original draft, Review and editing, Methodology, Investigation, Formal analysis, Data curation, Visualization. H.Y.C.: Writing—Review and editing, Methodology. Funding acquisition. J.H.J.: Writing—Review and editing, Methodology. H.S.: Writing—Review and editing, Validation, Supervision, Resources, Funding acquisition, Conceptualization. J.Y.L.: Writing—Original draft, review and editing, Validation, Supervision, Methodology, Investigation, Data curation, Funding acquisition, Conceptualization. All authors have read and agreed to the published version of the manuscript.

Funding: This work was supported by the National Research Foundation of Korea (NRF) grants funded by the Korea government (MSIT) (RS-2024-00343078 and 2020R1C1C1012873 to J.Y.L., NRF-2021R1A2C1009821 to H.Y.C., and RS-2025-00517458 and NRF-2019R1A6A1A0303288 to H.S.), by Korean Fund for Regenerative Medicine (KFRM) funded by Ministry of Science and ICT, and Ministry of Health and Welfare, Republic of Korea (RS-2025-02223118 to H.S.), and by a grant of the Korea Health Technology R&D Project through the Korea Health Industry Development Institute (KHIDI), funded by the Ministry of Health & Welfare, Republic of Korea (RS-2025-02215684 to H.S.).

Institutional Review Board Statement: The study was conducted according to the guidelines of the Declaration of Helsinki and approved by the Institutional Animal Care and Use Committee of CHA University (approval no. IACUC 230152). The date of the ethics approval is 10 October 2023.

Data Availability Statement: The data that support the findings of this study are available from the corresponding author upon reasonable request.

Acknowledgments: Figures were created with Biorender.com.

Conflicts of Interest: The authors declare that they have no known competing financial interests or personal relationships that could have appeared to influence the work reported in this paper.

References

1. Vivekanand, K.; Jyotsana, P.; Akshunna, K. Preterm Birth: An Overview. *Cureus* **2022**, *14*, e33006. [CrossRef]
2. Vrachnis, N.; Vitoratos, N.; Iliodromiti, Z.; Sifakis, S.; Deligeoroglou, E.; Creatsas, G. Intrauterine inflammation and preterm delivery. *Ann. N. Y. Acad. Sci.* **2010**, *1205*, 118–122. [CrossRef]
3. Fraser, J.; Walls, M.; McGuire, W. Respiratory complications of preterm birth. *BMJ* **2004**, *329*, 962–965. [CrossRef]
4. Cannavo, L.; Perrone, S.; Viola, V.; Marseglia, L.; Di Rosa, G.; Gitto, E. Oxidative Stress and Respiratory Diseases in Preterm Newborns. *Int. J. Mol. Sci.* **2021**, *22*, 12504. [CrossRef] [PubMed]
5. Stanifer, J.W. Virchow's triad: Kussmaul, Quincke and von Recklinghausen. *J. Med. Biogr.* **2016**, *24*, 89–100. [CrossRef]
6. Lee, J.Y.; Na, Q.; Shin, N.E.; Shin, H.E.; Kang, Y.; Chudnovets, A.; Lei, J.; Song, H.; Burd, I. Melatonin for prevention of fetal lung injury associated with intrauterine inflammation and for improvement of lung maturation. *J. Pineal Res.* **2020**, *69*, e12687. [CrossRef]
7. Sutherland, M.R.; Bertagnolli, M.; Lukaszewski, M.-A.; Huyard, F.; Yzydorczyk, C.; Luu, T.M.; Nuyt, A.M. Preterm birth and hypertension risk: The oxidative stress paradigm. *Hypertension* **2014**, *63*, 12–18. [CrossRef] [PubMed]
8. Preston, M.; Hall, M.; Shennan, A.; Story, L. The role of placental insufficiency in spontaneous preterm birth: A literature review. *Eur. J. Obstet. Gynecol. Reprod. Biol.* **2024**, *295*, 136–142. [CrossRef] [PubMed]
9. Tafani, M.; Sansone, L.; Limana, F.; Arcangeli, T.; De Santis, E.; Polese, M.; Fini, M.; Russo, M.A. The Interplay of Reactive Oxygen Species, Hypoxia, Inflammation, and Sirtuins in Cancer Initiation and Progression. *Oxidative Med. Cell. Longev.* **2016**, *2016*, 3907147. [CrossRef]
10. Hentschel, R.; Bohlin, K.; van Kaam, A.; Fuchs, H.; Danhaive, O. Surfactant replacement therapy: From biological basis to current clinical practice. *Pediatr. Res.* **2020**, *88*, 176–183. [CrossRef]
11. Ramaswamy, V.V.; Abiramalatha, T.; Bandyopadhyay, T.; Boyle, E.; Roehr, C.C. Surfactant therapy in late preterm and term neonates with respiratory distress syndrome: A systematic review and meta-analysis. *Arch. Dis. Child. Fetal Neonatal Ed.* **2022**, *107*, 393–397. [CrossRef]
12. Committee on Obstetric, P. Committee Opinion No. 713: Antenatal Corticosteroid Therapy for Fetal Maturation. *Obstet. Gynecol.* **2017**, *130*, e102–e109. [CrossRef] [PubMed]

13. Lofberg, L.; Serenius, F.; Hellstrom-Westas, L.; Olhager, E.; Ley, D.; Farooqi, A.; Stephansson, O.; Abrahamsson, T. Postnatal betamethasone treatment in extremely preterm infants and risk of neurodevelopmental impairment: A cohort study. *Arch. Dis. Child. Fetal Neonatal Ed.* **2025**, *110*, 382–387. [CrossRef]
14. Hellstrom, A.; Ley, D.; Hallberg, B.; Lofqvist, C.; Hansen-Pupp, I.; Ramenghi, L.A.; Borg, J.; Smith, L.E.H.; Hard, A.L. IGF-1 as a Drug for Preterm Infants: A Step-Wise Clinical Development. *Curr. Pharm. Des.* **2017**, *23*, 5964–5970. [CrossRef] [PubMed]
15. De Rose, D.U.; Landolfo, F.; Pugnaloni, F.; Giliberti, P.; Santisi, A.; Columbo, C.; Martini, L.; Ronchetti, M.P.; Schingo, P.M.; Salvatori, G.; et al. Use of N-Acetylcysteine in Preterm Neonates with Enteral Feeding Intolerance and Intestinal Obstruction: A Case Series and Review of the Literature. *Children* **2024**, *11*, 873. [CrossRef]
16. Reiter, R.J.; Mayo, J.C.; Tan, D.X.; Sainz, R.M.; Alatorre-Jimenez, M.; Qin, L. Melatonin as an antioxidant: Under promises but over delivers. *J. Pineal Res.* **2016**, *61*, 253–278. [CrossRef]
17. Joseph, T.T.; Schuch, V.; Hossack, D.J.; Chakraborty, R.; Johnson, E.L. Melatonin: The placental antioxidant and anti-inflammatory. *Front. Immunol.* **2024**, *15*, 1339304. [CrossRef] [PubMed]
18. Claustrat, B.; Leston, J. Melatonin: Physiological effects in humans. *Neurochirurgie* **2015**, *61*, 77–84. [CrossRef]
19. Tarocco, A.; Caroccia, N.; Morciano, G.; Wieckowski, M.R.; Ancora, G.; Garani, G.; Pinton, P. Melatonin as a master regulator of cell death and inflammation: Molecular mechanisms and clinical implications for newborn care. *Cell Death Dis.* **2019**, *10*, 317. [CrossRef]
20. Shi, D.; Xiao, X.; Wang, J.; Liu, L.; Chen, W.; Fu, L.; Xie, F.; Huang, W.; Deng, W. Melatonin suppresses proinflammatory mediators in lipopolysaccharide-stimulated CRL1999 cells via targeting MAPK, NF-kappaB, c/EBPbeta, and p300 signaling. *J. Pineal Res.* **2012**, *53*, 154–165. [CrossRef]
21. Lee, J.Y.; Song, H.; Dash, O.; Park, M.; Shin, N.E.; McLane, M.W.; Lei, J.; Hwang, J.Y.; Burd, I. Administration of melatonin for prevention of preterm birth and fetal brain injury associated with premature birth in a mouse model. *Am. J. Reprod. Immunol.* **2019**, *82*, e13151. [CrossRef]
22. He, F.; Wang, Q.F.; Li, L.; Yu, C.; Liu, C.Z.; Wei, W.C.; Chen, L.P.; Li, H.Y. Melatonin Protects Against Hyperoxia-Induced Apoptosis in Alveolar Epithelial type II Cells by Activating the MT2/PI3K/AKT/ETS1 Signaling Pathway. *Lung* **2023**, *201*, 225–234. [CrossRef] [PubMed]
23. Gou, Z.; Su, X.; Hu, X.; Zhou, Y.; Huang, L.; Fan, Y.; Li, J.; Lu, L. Melatonin improves hypoxic-ischemic brain damage through the Akt/Nrf2/Gpx4 signaling pathway. *Brain Res. Bull.* **2020**, *163*, 40–48. [CrossRef]
24. Gomez-Lopez, N.; Romero, R.; Arenas-Hernandez, M.; Panaitescu, B.; Garcia-Flores, V.; Mial, T.N.; Sahi, A.; Hassan, S.S. Intra-amniotic administration of lipopolysaccharide induces spontaneous preterm labor and birth in the absence of a body temperature change. *J. Matern.-Fetal Neonatal Med.* **2018**, *31*, 439–446. [CrossRef] [PubMed]
25. Motomura, K.; Romero, R.; Galaz, J.; Tao, L.; Garcia-Flores, V.; Xu, Y.; Done, B.; Arenas-Hernandez, M.; Miller, D.; Gutierrez-Contreras, P. Fetal and maternal NLRP3 signaling is required for preterm labor and birth. *JCI Insight* **2022**, *7*, e158238. [CrossRef]
26. Chen, Z.; Shan, Y.; You, X.; Gu, H.; Xu, C.; Long, J.; Ni, X. NLRP3 inflammasome is involved in uterine activation for labor at term and preterm. *Reproduction* **2021**, *162*, 449–460. [CrossRef] [PubMed]
27. Zhu, H.; Zhang, L.; Xiao, F.; Wu, L.; Guo, Y.; Zhang, Z.; Xiao, Y.; Sun, G.; Yang, Q.; Guo, H. Melatonin-Driven NLRP3 Inflammation Inhibition Via Regulation of NF-kappaB Nucleocytoplasmic Transport: Implications for Postoperative Cognitive Dysfunction. *Inflammation* **2023**, *46*, 1471–1492. [CrossRef]
28. Arioz, B.I.; Tarakcioglu, E.; Olcum, M.; Genc, S. The Role of Melatonin on NLRP3 Inflammasome Activation in Diseases. *Antioxidants* **2021**, *10*, 1020. [CrossRef]
29. Hudalla, H.; Karenberg, K.; Kuon, R.J.; Poschl, J.; Tschada, R.; Frommhold, D. LPS-induced maternal inflammation promotes fetal leukocyte recruitment and prenatal organ infiltration in mice. *Pediatr. Res.* **2018**, *84*, 757–764. [CrossRef]
30. Jackson, C.M.; Demmert, M.; Mukherjee, S.; Isaacs, T.; Thompson, R.; Chastain, C.; Gray, J.; Senthamaraikannan, P.; Presicce, P.; Chetal, K.; et al. A potent myeloid response is rapidly activated in the lungs of premature Rhesus macaques exposed to intra-uterine inflammation. *Mucosal Immunol.* **2022**, *15*, 730–744. [CrossRef]
31. Zhu, H.L.; Shi, X.T.; Xu, X.F.; Zhou, G.X.; Xiong, Y.W.; Yi, S.J.; Liu, W.B.; Dai, L.M.; Cao, X.L.; Xu, D.X.; et al. Melatonin protects against environmental stress-induced fetal growth restriction via suppressing ROS-mediated GCN2/ATF4/BNIP3-dependent mitophagy in placental trophoblasts. *Redox Biol.* **2021**, *40*, 101854. [CrossRef] [PubMed]
32. Yavuz, A.; Sezik, M.; Eris Yalcin, S.; Asci, H.; Ozmen, O. Immunohistochemical analysis of protective effects of maternal fingolimod on the placenta and fetal lung and brain in chorioamnionitis-induced preterm birth rat model. *Immunopharmacol. Immunotoxicol.* **2020**, *42*, 564–571. [CrossRef]
33. Bayes-Genis, A. Hypertrophy and inflammation. Too much for one heart. *Eur. Heart J.* **2007**, *28*, 661–663. [CrossRef] [PubMed]
34. Cañadas, O.; Olmeda, B.; Alonso, A.; Pérez-Gil, J. Lipid–protein and protein–protein interactions in the pulmonary surfactant system and their role in lung homeostasis. *Int. J. Mol. Sci.* **2020**, *21*, 3708. [CrossRef]

35. Beers, M.F.; Shuman, H.; Liley, H.G.; Floros, J.; Gonzales, L.W.; Yue, N.; Ballard, P.L. Surfactant protein B in human fetal lung: Developmental and glucocorticoid regulation. *Pediatr. Res.* **1995**, *38*, 668–675. [CrossRef]
36. Choi, E.-Y.; Jin, J.-Y.; Choi, J.-I.; Choi, I.S.; Kim, S.-J. DHA suppresses Prevotella intermedia lipopolysaccharide-induced production of proinflammatory mediators in murine macrophages. *Br. J. Nutr.* **2014**, *111*, 1221–1230. [CrossRef]
37. Deng, W.-G.; Tang, S.-T.; Tseng, H.-P.; Wu, K.K. Melatonin suppresses macrophage cyclooxygenase-2 and inducible nitric oxide synthase expression by inhibiting p52 acetylation and binding. *Blood* **2006**, *108*, 518–524. [CrossRef]
38. Bonnefont-Rousselot, D.; Collin, F. Melatonin: Action as antioxidant and potential applications in human disease and aging. *Toxicology* **2010**, *278*, 55–67. [CrossRef]
39. Yeh, C.-M.; Su, S.-C.; Lin, C.-W.; Yang, W.-E.; Chien, M.-H.; Reiter, R.J.; Yang, S.-F. Melatonin as a potential inhibitory agent in head and neck cancer. *Oncotarget* **2017**, *8*, 90545–90556. [CrossRef]
40. Karaaslan, C.; Suzen, S. Antioxidant properties of melatonin and its potential action in diseases. *Curr. Top. Med. Chem.* **2015**, *15*, 894–903. [CrossRef] [PubMed]
41. Lee, J.Y.; Li, S.; Shin, N.E.; Na, Q.; Dong, J.; Jia, B.; Jones-Beatty, K.; McLane, M.W.; Ozen, M.; Lei, J. Melatonin for prevention of placental malperfusion and fetal compromise associated with intrauterine inflammation-induced oxidative stress in a mouse model. *J. Pineal Res.* **2019**, *67*, e12591. [CrossRef]
42. Chu, J.; Tu, Y.; Chen, J.; Tan, D.; Liu, X.; Pi, R. Effects of melatonin and its analogues on neural stem cells. *Mol. Cell. Endocrinol.* **2016**, *420*, 169–179. [CrossRef]
43. Gharehbaghi, M.M.; Yeganedoust, S.; Shaseb, E.; Fekri, M. Evaluation of melatonin efficacy in prevention of bronchopulmonary dysplasia in preterm newborn infants. *Turk. J. Pediatr.* **2022**, *64*, 79–84. [CrossRef]
44. Pan, L.; Fu, J.H.; Xue, X.D.; Xu, W.; Zhou, P.; Wei, B. Melatonin protects against oxidative damage in a neonatal rat model of bronchopulmonary dysplasia. *World J. Pediatr.* **2009**, *5*, 216–221. [CrossRef]
45. Tamura, H.; Takasaki, A.; Taketani, T.; Tanabe, M.; Lee, L.; Tamura, I.; Maekawa, R.; Aasada, H.; Yamagata, Y.; Sugino, N. Melatonin and female reproduction. *J. Obstet. Gynaecol. Res.* **2014**, *40*, 1–11. [CrossRef] [PubMed]
46. Domínguez Rubio, A.P.; Sordelli, M.S.; Salazar, A.I.; Aisemberg, J.; Bariani, M.V.; Cella, M.; Rosenstein, R.E.; Franchi, A.M. Melatonin prevents experimental preterm labor and increases offspring survival. *J. Pineal Res.* **2014**, *56*, 154–162. [CrossRef]
47. Li, X.; Chen, X.; Zhou, W.; Ji, S.; Li, X.; Li, G.; Liu, G.; Wang, F.; Hao, A. Effect of melatonin on neuronal differentiation requires CBP/p300-mediated acetylation of histone H3 lysine 14. *Neuroscience* **2017**, *364*, 45–59. [CrossRef] [PubMed]
48. Chen, Y.H.; Xu, D.X.; Wang, J.P.; Wang, H.; Wei, L.Z.; Sun, M.F.; Wei, W. Melatonin protects against lipopolysaccharide-induced intra-uterine fetal death and growth retardation in mice. *J. Pineal Res.* **2006**, *40*, 40–47. [CrossRef]
49. Polglase, G.R.; Barbuto, J.; Allison, B.J.; Yawno, T.; Sutherland, A.E.; Malhotra, A.; Schulze, K.E.; Wallace, E.M.; Jenkin, G.; Ricardo, S.D. Effects of antenatal melatonin therapy on lung structure in growth-restricted newborn lambs. *J. Appl. Physiol.* **2017**, *123*, 1195–1203. [CrossRef] [PubMed]
50. Crisafulli, C.; Mazzon, E.; Muia, C.; Bella, P.; Esposito, E.; Meli, R.; Cuzzocrea, S. Effects of combination of melatonin and dexamethasone on acute lung injury in a mice model of carrageenan-induced pleurisy. *J. Pineal Res.* **2006**, *41*, 228–237. [CrossRef]

antioxidants

MDPI

Article

Composition and Antioxidant Status of Vegan Milk—Pilot Study

Agnieszka Chrustek [1], Agnieszka Dombrowska-Pali [2,*] and Dorota Olszewska-Słonina [1]

1 Department of Pathobiochemistry and Clinical Chemistry, Faculty of Pharmacy, Collegium Medicum in Bydgoszcz, Nicolaus Copernicus University, M. Curie-Skłodowskiej 9, 85-094 Bydgoszcz, Poland; a.chrustek@cm.umk.pl (A.C.); dorolsze@cm.umk.pl (D.O.-S.)

2 Department of Perinatology, Gynecology and Gynecologic Oncology, Faculty of Health Sciences, Collegium Medicum in Bydgoszcz, Nicolaus Copernicus University, Łukasiewicza 1, 85-821 Bydgoszcz, Poland

* Correspondence: agnieszka.pali@cm.umk.pl

Abstract: Background: More and more women are following a vegan and vegetarian diet. For some, the use of a vegan diet during lactation is controversial. Purpose: The aim of the study was to comparatively analyze the concentration of selected hormones, micro- and macronutrients, vitamins, and the basic composition and antioxidant status of the milk of vegan women, compared to the milk of omnivorous women. Methods: The study included 17 breastfeeding vegan women and 27 omnivorous women. The basic composition of human milk was analyzed using the MIRIS HMATM analyzer (Uppsala, Sweden) The levels of hormones and vitamins were determined by the enzyme-linked immunosorbent method. In order to determine the antioxidant activity and micro- and macroelements, spectrophotometric methods were used. Results: The vegan group was characterized by a lower average age, lower BMI, and lower WHR index compared to the control group. The milk of vegan women showed significantly higher cortisol concentrations and lower iron, vitamin B6, and antioxidant status than the milk of omnivorous women. Conclusions: A vegan diet helps maintain a healthy body weight and is more popular among younger women, under 30 years of age. Higher levels of milk cortisol in vegan women may indicate a high level of anxiety and stress experienced by breastfeeding women, which may have negative consequences not only for breastfeeding mothers but also for the development of their children. Lack of appropriate supplementation in women who do not consume meat and animal products may cause a deficiency of micro- and macroelements in breast milk.

Keywords: human milk; vegans; antioxidants; nutrition

1. Introduction

Proper nutrition during pregnancy and lactation is crucial for the health of both the mother and the child. A balanced diet rich in essential nutrients supports the development of the fetus and then ensures adequate milk production. Human milk is usually sufficient to cover the caloric requirements of the child and thus support its growth and development. It is widely believed that the mother's diet affects the composition of human milk, especially the fatty acid profile, which is well documented [1–3]. However, the remaining variability of nutrients and macro- and microelements in human milk attributable to diet remains largely unknown [4]. It should be noted that restrictive dietary restrictions of lactating women may affect the nutrient content of human milk as well as reduce the volume of milk produced [5]. In the context of optimal nutrient intake, the use of a vegan diet by breastfeeding women is controversial. The vegan diet is the most restrictive form of vegetarianism. Vegans eliminate all animal products from their diet. In recent years, it has been observed that plant-based diets are becoming increasingly popular in Europe [6]. It is estimated that the prevalence of vegans in Europe is from 1 to 10% [7]. In turn, in 2024, according to a report conducted by CEOWORLD Consumer Insights magazine, Poland ranked 19th in the world in terms of the population of vegetarians and vegans, 8.4% and 1.8%, respectively [8]. The more restrictive the vegan diet, the greater the risk of dietary deficiencies. Hence, doubts have arisen as to the adequacy of nutrient intake by vegan women and the impact of this diet on the composition of breast milk [9].

According to the American Academy of Nutrition and Dietetics, the composition of vegan milk is similar to that of non-dieting mothers, except for the concentration of fatty acids. They recommend this diet, properly balanced, with well-chosen supplementation [10]. The Swiss Federal Food Commission and the German Nutrition Society have expressed an opposing position, claiming that a vegan diet cannot be recommended for pregnant and breastfeeding women nor for infants, children, and the elderly [11,12]. According to the guidelines of the European Milk Bank Association (EMBA), breastfeeding vegan women can be recruited as donors to the Human Milk Bank, provided that they regularly supplement vitamin B12 [13]. In turn, national guidelines for individual countries vary; for example, in Switzerland, Germany, and Australia, vegan mothers cannot donate their milk to other children [14]. In Poland, a vegan diet is a permanent exclusion criterion for donating human milk [15]. Given the scientific evidence for the varied intake of some nutrients during a vegan diet, which may translate into differences in the composition of human milk, the aim of this study was to conduct a comparative analysis between the milk of vegan women and the milk of women consuming all animal products.

Human milk consists, among others, of carbohydrates, fats, proteins, bioactive factors (immunoglobulins, hormones, growth factors), vitamins, micro- and macronutrients, as well as the microbiome [16–22].

Carbohydrates are utilized in energy processes but also for the proper maturation of the nervous system and the maintenance of a healthy bacterial flora in the gastrointestinal tract of infants [16]. Human milk oligosaccharides (HMOs) exhibit anti-adhesive, anti-infective, and probiotic properties [17]. Meanwhile, proteins in human milk not only provide nutrition but also serve bioactive functions. They act as carriers of other nutrients (e.g., lactoferrin, alpha-lactalbumin, β-casein), regulate intestinal development (e.g., growth factors, lactoferrin), and nutrient absorption (e.g., bile salt-stimulated lipase, amylase), and demonstrate immunological and antimicrobial activity (e.g., lactoferrin, secretory IgA, cytokines, lysozyme) [18,23]. Fats constitute the primary energy source (approximately 50% of calories, equivalent to 25 g per day up to 6 months of age) and are an important source of essential nutrients such as polyunsaturated fatty acids (PUFAs), fat-soluble vitamins, complex lipids, and bioactive compounds [18]. The lipids in human milk positively influence child development. Long-chain polyunsaturated fatty acids (LCPUFAs) stimulate the functional development of the retina and brain in infants, participate in inflammatory and immune processes, which affects the optimal maturation of the immune system, and may additionally influence adipocyte differentiation and contribute to reducing the risk of obesity [19].

Bioactive factors in human milk include immunoglobulins, hormones, growth factors, glycans, nucleotides, blood cells, cytokines, stem cells, lactoferrin, antioxidant compounds such as vitamins A, E, C, β-carotene, and glutathione, as well as antioxidant enzymes (e.g., glutathione peroxidase, catalase, superoxide dismutase) [20,21,24–26].

Human milk contains numerous hormones, including thyroid hormones (thyroxine, triiodothyronine), thyrotropin (TSH), thyrotropin-releasing hormone (TRH), parathormone, cortisol, melatonin, progesterone, estrogens, prolactin, insulin, adipose tissue hormones (adiponectin, resistin, leptin, ghrelin), and growth hormone [16,20,25,26]. Cortisol (a stress hormone, glucocorticoid) is produced by the adrenal cortex, influences metabolism, and possesses anti-inflammatory properties. It is a crucial component of human milk due to its role in regulating the digestive system and influencing the neurological development of the child [20].

Adipose tissue hormones have a positive effect on early satiety control in infants and the regulation of energy balance during childhood and adulthood, protecting against later obesity. These hormones are likely secreted and synthesized by the mammary gland and pass from the serum into human milk. Adipose tissue hormones present in human milk include leptin, adiponectin, resistin, ghrelin, obestatin, irisin, apelin, and visfatin [25,26]. Adiponectin has anti-inflammatory and anti-atherogenic properties and regulates metabolism [25], while leptin improves insulin sensitivity, regulates energy balance, and appetite [26].

Human milk is rich in all vitamins, micro- and macroelements necessary for the proper development of the child. Macronutrients include Na, Cl, K, Mg, Ca, phosphates, and sulfates, while microelements include Fe, Cu, Mn, Zn, Co, Mo, Se, Cr, F, and I [24,27–29]. Vitamin B6 supports the body's immune resistance, participates in gluconeogenesis and antibody formation, and aids the child's nervous system [27]. Vitamin D is involved in bone formation and plays a crucial role in calcium–potassium balance. Its deficiency causes rickets in children [30].

Due to the limited scientific literature on the impact of veganism on the composition and antioxidant status of human milk, the aim of our study was to assess the basic composition, hormones (adiponectin, leptin, cortisol), vitamins (vitamin D, vitamin B6), micro- and macroelements (iron, phosphorus, magnesium, calcium), and the antioxidant status of vegan mothers' milk.

2. Materials and Methods

The research was approved by the Bioethics Committee of the Nicolaus Copernicus University in Toruń at the Ludwik Rydygier Collegium Medicum in Bydgoszcz (consent no. KB437/2018).

All women participating in the study read information about it, completed the questionnaire, and expressed their conscious opinion, written consent to participate in it. The questionnaire included questions about the age of women, residence, professional activity, weight and height of a breastfeeding woman, duration of pregnancy, fertility (primiparous/multiparous), type of delivery, diet, and taking medications and supplements.

2.1. Characteristics of the Vegan and Control Groups

The study group initially included 23 vegan women and 27 women who consumed all animal products. In the group of vegan women recruited for the study, only 5 women consciously supplemented vitamin B12. The remaining 17 women only used supplementation due to breastfeeding when recommended by medical personnel. Finally, the study included 17 women who had excluded all animal products such as meat, milk, honey, dairy products, and eggs from their diet for at least 3 years and did not supplement vitamin B12.

Both the study group (n = 17) and the control group (n = 27) excluded women with diagnosed diseases: thyroid, circulatory system, digestive system, respiratory system, cancer, and allergies. The study also did not include women addicted to nicotine or women taking any medications. The entire study group came from one region of the Kujawsko-pomorskie Voivodeship (Poland). The women resided in urban areas and held higher education degrees. They were primiparous, having given birth via vaginal delivery, and were in the 3rd to 6th month of lactation. The mothers exclusively breastfed their infants without supplementation with formula or solid foods (Figure 1).

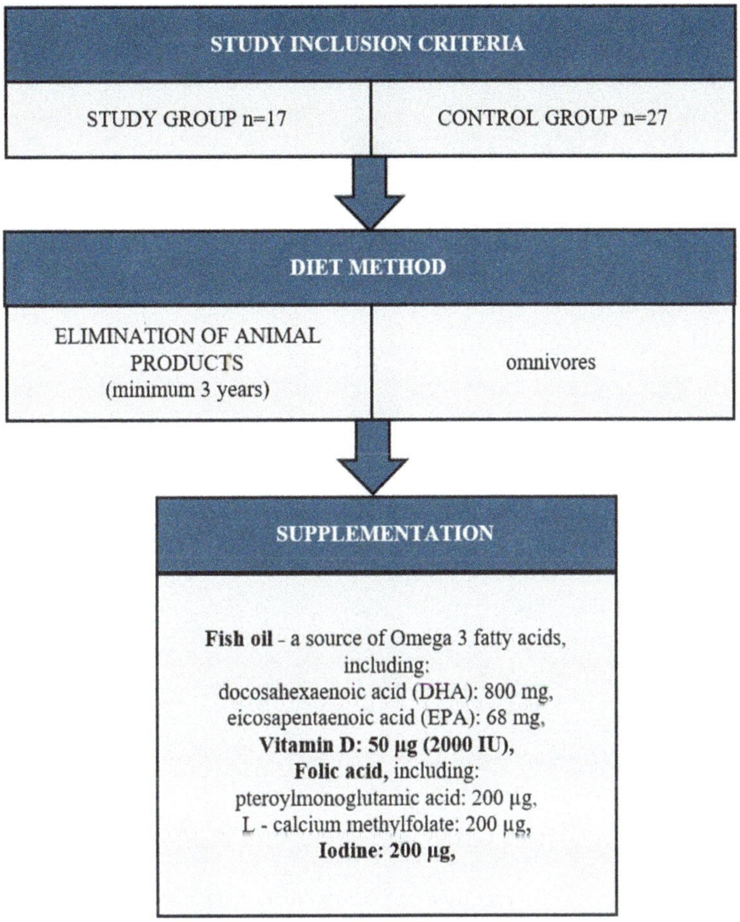

Figure 1. Study group inclusion criteria.

The material for the study was human milk from breastfeeding women. The study participants volunteered to participate in the research project via social media in response to the information about the study. Human milk came from a daily collection (40 mL). Each woman expressed milk using a breast pump according to the daily milk collection protocol four times a day, from four time intervals: 06:00–12:00, 12:00–18:00, 18:00–24:00, and 24:00–6:00. The desired amount of expressed milk from each time interval was 10 mL, with 5 mL before putting the baby to the breast and 5 mL after the end of feeding. Each portion of milk was poured into one collective bottle. The material was provided by breastfeeding women within 24 h of collection, then aliquoted into 2 mL Eppendorf tubes (MedLab, Raszyn, Poland) and frozen at −20 °C and then at −80 °C. The material was stored in these conditions until the parameters described below were determined, but no longer than 3 months.

2.2. Determination of the Basic Composition of Breast Milk

The basic composition of breast milk includes the content of fats [mg/100 mL], total protein [mg/100 mL], crude protein [mg/100 mL], carbohydrates [mg/100 mL], total solids [mg/100 mL], and energy value [kcal/100 mL]. The determination was performed using the MIRIS HMAT analyzer (MIRIS AB, (Uppsala, Sweden)) in accordance with the manufacturer's procedure. Miris HMA™ software (2022) processes the measurement data via internal calibration. Detection limit: for fat content—0.06 g/100 mL, crude protein—0.06 g/100 mL, total protein—0.05 g/100 mL, and for carbohydrates—0.04 g/100 mL.

Breast milk samples were heated at 40 °C in a thermostatic bath (BIONOVO, Legnica, Poland) before analysis and then homogenized with MIRIS Sonicator (Białystok, Poland) [1.5 s/mL]. Each sample was analyzed in three replicates.

2.3. Determination of Leptin Concentration

In order to determine leptin, a ready-made, commercial enzyme-linked immunosorbent assay, Human Leptin ELISA (BioVender, Karásek, Czech Republic) was used in accordance with the manufacturer's procedure. This method consisted of 60 min of incubation of milk samples and standards (100 µL each) on a 96-well plate coated with polyclonal antibodies against human leptin. The next step was the application of 100 µL of polyclonal antibody conjugated to horseradish peroxidase and incubation for 60 min. The resulting conjugate reacted with 100 µL of TMB, then the reaction was stopped by the addition of an acidic solution (100 µL of sulfuric acid VI). The absorbance was measured using the MultiSkan Go (Vantaa, Finland) reader at a wavelength of 450 nm relative to the reference wave (630 nm). Detection limit: 0.2 ng/mL. Each sample was analyzed in three replicates.

2.4. Determination of Adiponectin Concentration

In order to determine adiponectin in human milk, a commercial enzyme-linked immunosorbent assay (Human Adiponectin/Acrp30, DuoSet ELISA, R&D Systems, Minneapolis, MN, USA) was used.

Prior to analysis, a 96-well plate was coated with antibodies and incubated overnight at room temperature, then rinsed, and a reagent containing 1% BSA (bovine serum albumin) in PBS was added and incubated for 1 h. After this time, 100 µL of human milk and standard samples were applied and incubated for 2 h, then the plate was rinsed and a detection antibody (100 µL) was added, incubated for another 2 h, washed, and 100 µL Streptavidin-HRP was added and incubated for 20 min. In the next step, the plate was rinsed, 50 µL of "Substrate Solution" (containing TMB) was added and incubated for 20 min at room temperature to protect it from light, finally "Stop Solution" (containing acid) was added to stop the reaction, the plate was gently shaken and the absorbance was measured at a wavelength of 450 nm using the MultiSkan Go reader. Detection limit: 10 pg/mL. Each sample was analyzed in three replicates.

2.5. Determination of Cortisol Concentration

A commercial enzyme-linked immunosorbent assay, Cortisol Elisa (DiaMetra, Via Pozzuolo, Italy), was used to determine cortisol in human milk according to the manufacturer's procedure. Human milk samples and standards (20 µL each) were applied to a 96-well plate coated with antibodies, followed by the addition of 200 µL of antigen-conjugated horseradish peroxidase solution, and incubated at 37 °C for 60 min. Subsequently, 100 µL of TMB was added to obtain a colored solution, incubated for 15 min at room temperature in the dark, and the reaction was stopped by the application of sulfuric acid VI (100 µL). Absorbance was read at a wavelength of 450 relative to the reference wave (620–630 nm) using the MultiSkan Go reader. Detection limit: 0.12 ng/mL. Each sample was analyzed in three replicates.

2.6. TAS Marking

A ready-to-use, commercially sourced enzyme-linked immunosorbent assay (TAS, München, Germany) was used to determine the total antioxidant capacity in accordance with the manufacturer's procedure. A 10 μL sample of human milk, a calibrator, and a light were applied to the 96-well plate. A 100 μL reagent containing a peroxide solution was added and prepared and then incubated for 10 min at 37 °C. Reagent with an enzyme was added to some wells, and a reagent without an enzyme was added to the rest. Then, 50 μL of "Stop Solution" (containing sulfuric acid) was added to stop the reaction. The absorbance at a wavelength of 450 nm was read.

The total antioxidant capacity was calculated using the formula:

$$TAS = 392 - (392 - \text{calibrator concentration}) \times \left(\frac{\text{change in absorbance of the sample}}{\text{change in absorbance of the calibrator}}\right) [\mu\text{mol/L}]$$

Reference Range:
<250 μmol/L—low antioxidant capacity
250–300 μmol/L—medium antioxidant capacity
>300 μmol/L—high antioxidant capacity
Each sample was analyzed in three replicates.

2.7. Determination of the Antioxidant Activity of Human Milk Using the DPPH• (2,2-Diphenyl-1-picrylhydrazyl) Radical

The method of Atanassov et al. was used in the study. (2011) with minor modifications [31].

Prior to the experiment, a 100 μM solution of DPPH• (2,2-diphenyl-1-picrylhydrazyl, Sigma Aldrich, Darmstadt, Germany) was prepared by dissolving 4 mg of the standard substance in 100 mL of methanol (POCH, Gliwice, Poland).

Briefly, 1 mL of prepared DPPH solution was applied to Eppendorf tubes (Warsaw, Poland), 250 μL of human milk was added, vortexed, and then incubated for 60 min in the dark at room temperature. Before reading, the samples were centrifuged for 2 min at 1500× *g*, at room temperature. The absorbance was measured at a wavelength of 517 nm against the reference sample, which was methanol. The control sample was a 100 μM methanol DPPH solution, which was measured at the beginning and end of the experiment. The solutions for obtaining the curve were prepared in a similar way, replacing 250 μL of human milk with Trolox solutions (Morąg, Poland).

Each sample was analyzed in three replicates.

Preparation of the calibration curve

Five solutions of Trolox (TE, (+−)-hydroxy-2,5,7,8-tetramethylchromatane-2-carboxylic acid, Sigma Aldrich, Darmstadt, Germany) at concentrations of 10–80 mg/1000 mL were prepared to determine the ability to deactivate free radicals in human milk. The calibration curve shows the dependence of the absorbance value of Trolox on its concentration. The results were expressed in terms of Trolox equivalent (μmol TE/1000 mL, mg TE/100 mL). In addition, the percentage ability of human milk to reduce the DPPH radical was calculated:

$$\%\text{inhibition} = (\frac{\text{A} - \text{Ab}}{\text{A}}) \times 100$$

A—control absorbance;
Ab—average absorbance value of the tested human milk;
Detection limit: 0.1 μg TE/mL.

2.8. Determination of the Ability of Human Milk to Reduce Fe (III) Ions

In the experiment, the FRAP method (iron ion reduction ability) according to Benzie and Strain (1999) was used with modifications [32].

Labeling in human milk rolls

Before the experiment, the following were prepared:

(a) 300 mM acetate buffer, pH 3.6;
(b) 10 mM TPTZ (2,4,6-Tri(2-pyridyl)-s-triazine, Sigma-Aldrich, Darmstadt, Germany) in 40 mM HCl;
(c) 20 mM $FeCl_3 \times 6H_2O$. (Chempur, Piekary Śląskie, Poland).

The first step was to prepare the FRAP reagent by combining solutions a, b, and c in a ratio of 10:1:1. For the purpose of determination, a blank sample was prepared, which was a FRAP reagent, and 100 µL of human milk samples and calibration curve standards. The next step was to apply 100 µL of sample and 3 mL of FRAP reagent, vortexing, followed by immediate reading of sample absorbance at 0 min at 593 nm and incubation at 37 °C in a water bath for 4 min. The last stage was vortexing and subsequent measurement of absorbance at the same wavelength.

Each sample was analyzed in three replicates.

Preparation of the calibration curve

Six solutions of ascorbic acid (Chempur, Piekary Śląskie, Poland) with concentrations of 100–1000 µM were prepared in order to determine the ability to reduce Fe(III) ions in human milk. The standard curve shows the dependence of the absorbance value of ascorbic acid on its concentration. FRAP concentrations are expressed as the equivalent of ascorbic acid (µM).

The FRAP value is calculated from the formula:

$$\mathrm{FRAP} = \frac{\text{change in sample absorbance from 0 min to 4 min}}{\text{change in absorbance of the standard from 0 min to 4 min}} \times \text{standard concentration} \times 2\ [\mu\mathrm{M}]$$

Ascorbic acid has a constant stoichiometric coefficient of 2.0 in the FRAP test.

Detection limit: 2 µM.

2.9. Determination of the Content of the Total Sum of Polyphenols

Determination of human milk samples

The method of Vazquez et al. was used for the determination (2015) with modifications [33]. The milk samples were gradually thawed in the refrigerator and vortexed. The first stage of the assay was the deproteinization of the samples by mixing 1 mL of a sample of human milk with 2 mL of 50% methanol, 100 µL of Carrez I and Carrez II reagent, and 1 mL of acetonitrile The samples were then vortexed and topped up with 50% methanol to a volume of 5 mL. The samples were incubated at room temperature for 25 min, and finally centrifuged at 4500× *g* at 4 °C for 15 min.

The total sum of polyphenols was determined in the aqueous layer of the sample by the FC (Folin–Ciocâlteu, Chempur, Poland) method. Briefly,100 µL of methanol extract was mixed with 1 mL of previously prepared FC reagent (FC:H20; 1:1, *v*/*v*) and with 3 mL of 20% Na_2CO_3. After vortexing, the mixture was incubated for 30 min in the dark at room temperature. The absorbance was then measured at a wavelength of 765 nm using a UV/VIS spectrophotometer (Biosense, Wrocław, Poland). The solutions necessary to obtain the calibration curve were prepared by replacing 100 µL of methanolic extract with gallic acid solutions (Merck, Darmstadt, Germany) of known concentration. The reference sample was 100 µL of methanol.

Each sample was analyzed in three replicates.

Preparation of the calibration curve

Six aqueous standard solutions of gallic acid (GAE) were prepared with concentrations: 0.0–0.10 mg/mL. The total polyphenol content was calculated on the basis of a calibration curve describing the dependence of absorbance values on the concentration of gallic acid. Results are presented as the equivalent of gallic acid in 1000 mL of human milk (mg GAE/L).

Detection limit: 5 mg GAE/L.

2.10. Determination of Paraoxonase 1 (PON1) Concentration

In order to determine the concentration of paraoxonase 1, a ready-made, commercial enzyme-linked immunosorbent assay (Human Paraoxonase 1 ELISA, BioVender, Karásek, Czech Republic) was used. Standards and samples of human milk (100 µL each) were incubated on a 96-well plate coated with polyclonal antibodies, then after 60 min of incubation (37 °C) and washing, 100 µL of biotin-labeled antibody was applied and incubated for another 60 min at 37 °C. The next stage was rinsing and an application of 100 µL of Streptavidin-HRP and another 30 min incubation (37 °C), then the plate was rinsed for the last time, and 100 µL of TMB and 10 min incubation were added. The last step was to stop the reaction by adding 100 µL of sulfuric acid (VI) each. The absorbances were read at a wavelength of 450 nm relative to the reference wave (620–630 nm). Detection limit: 0.1 ng/mL. Each sample was analyzed in three replicates.

2.11. Determination of Iron Concentration

A set of reagents from BioMaxima (Lublin, Poland) was used for the determination of iron (Table 1).

Table 1. Determination of iron concentration in human milk.

	Reagent Test	Human Milk Samples	Pattern
R1 reagent	750 µL	750 µL	750 µL
Standard/Sample	-	75 µL	75 µL
Distilled water	75 µL	-	-
The samples were thoroughly mixed, incubated for 5 min at 37 °C, centrifuged for several seconds at the highest speed, and then the absorbance A1 was read against the reagent test at a wavelength of 590 nm.			
R2 Reagent	75 µL	75 µL	75 µL
The samples were mixed, incubated at 37 °C for 5 min, after which time the samples were centrifuged for several minutes at the highest speed, and then the A2 absorbance was read against the reagent test at a wavelength of 590 nm.			

Making the marking:

A ready-made standard from BioMaxima (Lublin, Poland) (200 µg/dL) was used.

The concentration of iron in the human milk samples was calculated using the formula:

$$Iron\ concentration = \frac{A2(PB) - A1(PB)}{A2(PW) - A1(PW)} \times standard\ concentration\ [\mu g/dL]$$

A1(PB)—absorbance of 1 test sample;

A2(PB)—absorbance of 2 test sample;

A1(PW)—absorbance of standard 1;

A2(PW)—absorbance 2 of the standard;

Each sample was analyzed in three replicates.

2.12. Determination of Magnesium, Calcium, and Phosphorus Concentration

In order to determine magnesium, calcium, and phosphorus, the BioMaxima kit (Lublin, Poland) was used. Each sample was analyzed in three replicates (Table 2).

Table 2. Determination of magnesium, calcium, and phosphorus concentrations in human milk.

	Reagent Test	Human Milk Samples	Pattern
R1 reagent	750 µL	750 µL	750 µL
Standard/Sample	-	75 µL	75 µL
Distilled water	75 µL	-	-
The samples were thoroughly mixed, incubated for 5 min at 37 °C, centrifuged for several seconds at the highest speed, and then the absorbance A1 was read against the reagent test at a wavelength of 590 nm.			
R2 Reagent	75 µL	75 µL	75 µL
The samples were mixed, incubated at 37 °C for 5 min, after which time the samples were centrifuged for several minutes at the highest speed, and then the A2 absorbance was read against the reagent test at a wavelength of 590 nm.			

Making the marking:

$$\text{Magnesium concentration} = [\text{mg/dL}]\frac{A\ (PB)}{A(PW)} \times \textit{standard concentration}$$

A(PB)—absorbance of the test sample;
A(PW)—standard absorbance.

2.13. Determination of Vitamin D Concentration

To determine the concentration of vitamin 25(OH)D3 (HVD3) in human milk, a commercial enzyme-linked immunosorbent assay from Cloud-Clone Corp (Katy, TX, USA) was used. The test was performed in accordance with the rules of the procedure given by the test manufacturer. Briefly, 50 µL of standards and human milk samples were pipetted onto a 96-well plate, and then 50 µL of reagent A was added to each well and incubated for 1 h at 37 °C. Subsequently, the plate was rinsed 3 times, and 100 µL of reagent B was added at 37 °C. The penultimate stages were rinsing the plate and adding 90 µL of "Substrate Solution" (incubation 15 min, 37 °C). Finally, 50 µL of "Stop Solution" was added to each well, and the absorbance at a wavelength of 450 nm was read. Detection limit: 4.83 pg/mL. Each sample was analyzed in three replicates.

2.14. Determination of Vitamin B6

In order to determine the concentration of vitamin B6 in human milk, an enzyme-linked immunosorbent assay from Cloud-Clone Corp. (USA) was used. The principle of the method and the performance of the test are identical to the above-described test for the determination of vitamin D.

Detection limit: 1.00 ng/mL. Each sample was analyzed in three replicates.

2.15. Statistical Analysis

In order to carry out the statistical analysis, the Statistica 13.1 software package from StatSoft® (Kraków, Poland) was used.

The normality of the schedule was verified by the Shapiro–Wilk test. No normality of the distribution of the analyzed quantitative variables was found.

A nonparametric Mann–Whitney U test was used to assess statistical significance in two groups of independent variables. The variability of the parameters is presented in the form of a median, minimum, and maximum (Min–Max).

In order to analyze many factors per variable in the studied groups, multiple linear regression was used. The variability of the parameters is presented in the form of a regression coefficient, error of the standard, regression coefficient, and the level of statistical significance $p < 0.05$.

To assess the correlation between the parameters studied, the Spearman correlation test was used. The results at the level of $p < 0.05$ were considered statistically significant.

The correlations that occur are represented by a correlation matrix.

3. Results

The group of vegan women had a lower mean age, lower BMI, and lower WHR compared to the control group (Table 3).

Table 3. Characteristics of the study groups.

	Vegans n = 17	Control Group n = 27	p
Age (years) $\bar{x} \pm SD$	29.23 ± 4.29	30.74 ± 2.68	**0.049**
BMI (kg/m^2) (Me; Min–Max)	21.26; 17.31–27.14	23.05; 17.93–29.30	0.066
HBD (week) (Me, Min–Max)	40.00; 35.00–42.00	40.00; 35.00–41.00	0.425
WHR (Me, Min–Max)	0.78; 0.71–0.85	0.81; 0.72–1.06	**0.039**

n—size; SD—standard deviation; BMI—body mass index [kg/m$\bar{x}^2$]; WHR—ratio of waist circumference to hip circumference; HBD—week of pregnancy; Me—median; Min–Max—minimum and maximum values; p—level of statistical significance (Mann–Whitney U test).

3.1. Basic Composition of Human Milk

No significant differences were observed in the parameters of the basic composition of human milk between the group of vegans and the control group (Table 4).

Table 4. Comparison of the basic composition of human milk in vegan and non-dieting women.

Variable	Study Group: Vegans n = 17 Me (Min–Max)	Control Group: Women Not Following Diets n = 27 Me (Min–Max)	p
Fat [g/100 mL]	3.40 (1.30–6.90)	3.30 (0.80–7.300)	0.847
Total protein [g/100 mL]	1.20 (0.40–1.60)	1.20 (0.70–1.50)	0.615
Carbohydrates [g/100 mL]	7.90 (6.60–8.30)	7.80 (6.20–8.30)	0.734
Total solids [g/100 mL]	12.60 (10.10–16.30)	12.20 (10–16.00)	0.847
Energy value [g/100 mL]	68 (47–101)	65 (44–100)	0.426
Crude protein [g/100 mL]	1.00 (0.30–1.30)	0.90 (0.50–1.20)	0.419

Me—median; Min–Max—minimum and maximum values; p—level of statistical significance (Mann–Whitney U test); n—number.

3.2. Hormones in Human Milk

No significant differences were observed in the concentration of adipose tissue hormones (adiponectin, leptin) in human milk between the vegan group and the control group (Table 3). A median increase in cortisol concentration of 61.10% (p = 0.022) in the milk of vegan women was shown compared to the control group (Table 5).

Table 5. Hormone content in the milk of vegans and women who do not follow a diet.

Variable	Study Group: Vegans n = 17 Me (Min–Max)	Control Group: Women Not Following Diets n = 27 Me (Min–Max)	p
Adiponectin [pg/mL]	3233 (540–5230)	2900 (335–7891)	0.735
Leptin [ng/mL]	1.52 (1.04–10.46)	1.65 (1.10–9.06)	0.628
Cortisol [ng/mL]	9.36 (3.97–48.02)	5.81 (1.48–16.29)	**0.022**

Me—median; Min-Max—minimum and maximum values; p—level of statistical significance (Mann–Whitney U test); n—number.

3.3. Antioxidant Status of Human Milk

The antioxidant status of vegan women shows significant differences compared to the milk of women who do not follow a diet. In the group of vegans, the median total antioxidant status of human milk was lower by 32.82% (p < 0.001) and the ability of human milk to reduce iron ions by 27.50% (p = 0.049), as well as the concentration of paraoxonase 1 by 29.03% (p = 0.021) and polyphenols by 11.18% (p = 0.039), compared to the control group (Table 6).

Table 6. Antioxidant status of human milk in vegan and non-dieting women.

Variable	Study Group: Vegans n = 17 Me (Min–Max)	Control Group: Women Not Following Diets n = 27 Me (Min–Max)	p
TAS [µM]	200.13 (156.60–272.99)	298.03 (170.99–403.96)	**<0.001**
DPPH [% Inhibition]	47.35 (4.37–76.30)	53.24 (24.95–74.91)	0.419
DPPH [µM TE/L]	208.50 (68.60–302.76)	227.67 (135.59–298.21)	0.419
FRAP [µM]	482.97 (65.10–887.21)	666.16 (155.94–2272.51)	**0.049**
PON1 [ng/mL]	1.10 (0.46–2.12)	1.55 (0.46–4.46)	**0.021**
Polyphenols [mg GAE/L]	10.33 (7.40–23.98)	11.63 (7.08–102.064)	**0.039**

TAS—total antioxidant status; FRAP—iron ion reduction ability; DPPH—2,2-diphenyl-1-picrylhydrazyl radical reduction method; TE—trolox; GAE—gallic acid; PON1—paraoxonase 1; Me—median; Min-Max—minimum and maximum values; p—level of statistical significance (Mann–Whitney U test), n—abundance.

In the group of vegans, high positive correlations were found between BMI/DPPH (r = 0.556, p = 0.020) and WHR/polyphenols (r = 0.659; p = 0.007), and high negative correlations between age/FRAP (r = −0.553; p = 0.021) and cortisol/TAS (r = −0.557; p = 0.020) (Tables S1–S3).

3.4. Content of Vitamins, Micro-, and Macroelements in Human Milk Samples

The content of iron and vitamin B6 in human milk differs significantly in the group of vegans (p = 0.037; p = 0.024) compared to the control group. The median concentration of iron in human milk is lower by 41.88% in the study group compared to the control group (p = 0.037), while vitamin B6 is lower by 3.49%. The remaining micro- and macroelements examined did not show any significant statistical differences (Table 7).

Table 7. The content of selected vitamins, micro-, and macronutrients in the milk of vegans and women who do not follow a diet.

Variable	Study Group: Vegans n = 17 Me (Min–Max)	Control Group: Women Not Following Diets n = 27 Me (Min–Max)	p
Iron [µg/dL]	8.52 (1.38–21.63)	14.66 (1.98–65.01)	**p = 0.037**
Phosphorus [mg/dL]	6.26 (2.25–12.64)	5.16 (2.57–7.45)	p = 0.231
Magnesium [mg/dL]	4.02 (2.74–8.73)	4.02 (2.51–5.94)	p = 0.774
Calcium [mg/dL]	14.06 (8.29–32.82)	13.90 (6.79–21.88)	p = 0.703
Vitamin D [pg/mL]	138.60 (62.05–804.87)	162.10 (44.38–354.40)	p = 0.398
Vitamin B6 [ng/mL]	273.20 (1.05–296.80)	283.10 (155.50–350.89)	**p = 0.024**

p—level of statistical significance (Mann–Whitney U test).

In the group of vegans, a high negative correlation was observed between magnesium and leptin (r = −0.565; p = 0.018) and calcium-polyphenols (r = −0.514; p = 0.035). In addition, high positive correlations were found between vitamin B6 and adiponectin (r = 0.747; p = 0.001) and vitamin D and calcium (r = 0.641; p = 0.007), (Tables S1–S3).

3.5. Analysis of the Influence of Different Predictors on the Tested Parameters

No significant statistical difference was found when examining multiple linear regressions between the examined parameters (DPPH, FRAP, cortisol, fat, total protein, carbohydrates, total solids, caloric value) and predictors (HBD, BMI, WHR, age, stage of lactation, type of delivery, place of residence) in the study groups.

4. Discussion

Breast milk is sufficient for the development of a child in most cases due to its rich composition and adaptation to the needs of the child. Permanent dietary restrictions during lactation may result in the exhaustion of the breastfeeding woman's body and negatively affect both the volume of breast milk and the content of certain nutrients in breast milk. Of great importance for the development of a child in the first months are proteins, especially amino acids, polyunsaturated fatty acids, vitamin D, B vitamins, iron, and iodine [34]. A vegan diet is often accused of not providing the right amount of nutrients [12].

Our studies did not show any significant differences in parameters such as BMI, HBD, but a positive correlation was observed between HBD and WHR ($r = 0.601$; $p < 0.001$). These results indicate that younger women are adopting veganism and that diet and HBD can affect the figure of a breastfeeding woman.

There is little scientific research on the effect of a vegan diet on the composition of human milk. According to literature data, a vegan and vegetarian diet may affect the composition of breast milk, mainly the fatty acid composition and vitamin concentration [5,9,35].

Milk from vegetarian women contains reduced levels of saturated long-chain fatty acids and higher levels of polyunsaturated fatty acids [5]. Sanders et al. (1978) [35] showed significantly lower amounts of C16:0, C16:1, and C18:0, and higher C18:2 and C18:3 in the milk of vegan women compared to women not following a diet. A higher percentage of ALA (2.09%) was shown in milk from vegan women compared to milk from vegetarians (1.55%) and from women not following a diet (1.19%). There was no statistically significant difference in the concentration of DHA in the study groups ($p = 0.543$) [21]. Patel and Lovelady (1998) observed a higher amount of short-chain fatty acids in the milk of vegan women compared to the milk of vegetarians and women who did not follow a diet, similarly for linoleic acid and α-linolenic acid [36]. The amount of long-chain fatty acids in human milk was higher in the milk of omnivorous women than in the milk of vegetarians and vegans. Compared to other dietary groups, the share of DHA in the milk of vegan women was lower, and in addition, a lower LA/ALA ratio and a higher omega-6/omega-3 ratio were observed in the milk of vegan women than in the other groups [35]. The results of our own studies did not show significant differences in the basic composition of human milk between the group of vegans and the group of omnivorous women.

Further studies show a lower concentration of taurine in the milk of vegan women (35 mg/dL) compared to the milk of omnivorous women (53 mg/dL) [37,38]. Due to the various guidelines regarding the vegan diet during breastfeeding, mainly caused by reports of insufficient demand for vitamins, micro-, and macroelements during the above diet, it was decided to extend our own research by analyzing selected compounds in human milk. Our own studies showed that the concentration of iron in the milk of vegan women is 41.88% lower compared to the control group, while the content of phosphorus, magnesium, and calcium does not differ significantly. There is no reference to the results in the scientific literature, but there are a few studies comparing the iron status in the blood plasma of vegans to people who do not follow a diet. Scientists have shown that the iron concentration between the two groups does not differ significantly, but it was noted that vegans have lower plasma ferritin concentrations compared to the control group [39]. Low levels of ferritin, an iron-binding protein, may be related to insufficient iron storage, which in turn leads to lower concentrations in breast milk.

Concerned about the content of vitamins, micro-, and macroelements, researchers have studied the effect of diet on this topic in a rather narrow scope. It was shown that the concentration of vitamin B12 in milk from vegetarian women was significantly lower ($p = 0.006$) [36]. Similar results were obtained by Specker et al. (1990) [40]. Pawlak et al. (2018) [41] analyzed the concentration of a given vitamin in human milk from vegans, vegetarians, and women who did not follow a diet, and no significant differences were observed. The results of our own studies showed that the milk of vegans contains a lower concentration of vitamin B6 compared to the control group ($p = 0.024$), while no significant differences were observed with regard to vitamin D ($p = 0.398$). There is a lack of comparative literature regarding the content of vitamin B6 in human milk in the groups discussed, but we can refer to the content of this substance in blood plasma. Atamasirikul et al. (2008) [42] confirm the effect of a vegan diet on the content of vitamin B6 in the human body, showing a lower median concentration of the substance in the blood plasma of vegans (37.40 nmol/L) compared to people not following a diet (47.40 nmol/L). Schüpbach et al. (2017) [43] had the opposite opinion, showing no effect of diet on the concentration of vitamin B6 in the body of vegans (27 nmol/L), but they showed lower values in vegetarians (16 nmol/L) compared to the control group (22 nmol/L).

These data draw particular attention to the precise preparation of the meal in vegans and possible deficiencies in supplementation.

To date, the scientific literature does not contain studies comparing the content of hormones in human milk, as well as the antioxidant status of human food between the groups described above. The results of our own studies showed an increase in the median concentration of cortisol in milk from vegan women (9.36 ng/mL) compared to the control group (5.81 ng/mL). Its increased concentration may indicate increased stress factors affecting vegans, but it should also be remembered that cortisol in appropriate concentrations stimulates the digestive system in children. In this study, a decrease in TAS in daytime samples and FRAP, polyphenol concentration, and PON1 in nighttime samples of milk from vegan women was also observed. These results indicate lower protection against free radicals and, consequently, in the future against free radical diseases, such as cancer, cardiovascular diseases, diabetes, and neurodegenerative diseases. These results are surprising considering the fact that vegans consume foods rich in antioxidants. On the other hand, analyzing our own results showing deficiencies in the content of vitamin B6, which inhibits oxidative stress [44,45], suggests a significant effect of the content of a given vitamin on antioxidant status.

5. Conclusions

A plant-based diet is becoming increasingly popular. People who do not consume meat and animal products have a lower body mass index (BMI) and are slimmer than people who eat meat.

Our research indicates that the elimination of meat and animal products alone does not have a significant effect on the nutritional composition of breast milk. Vegan mothers are able to produce nutritionally valuable milk for their infants. This is particularly significant for vegan women during lactation, as their dietary choices should not serve as a permanent exclusion criterion for their participation as donors in human milk banks—a practice that currently occurs, among other places, in Poland. However, the results of this study draw attention to the need to introduce recommendations for supplementation in breastfeeding women who follow a vegan diet. The milk of women who follow a vegan diet, in the absence of appropriately selected supplementation, has lower values of micro- and macroelements and a lower antioxidant status, which may affect the growth and development of the child. Higher levels of milk cortisol were also observed in the breast milk of vegan women, indicating that these women experienced more long-term stress than omnivores. Therefore, special attention should be paid to the mood of a postpartum woman who follows a vegan diet, as the higher levels of milk cortisol in our study probably indicate that the elimination diet itself is a risk factor for long-term stress, which may consequently affect mental health.

Supplementary Materials: The following supporting information can be downloaded at: https://www.mdpi.com/article/10.3390/antiox14050505/s1.

Author Contributions: Conceptualization: A.C. and A.D.-P.; methodology: A.D.-P. and A.C.; validation: D.O.-S. formal analysis: A.C. investigation and data curation, A.C.; writing—original draft preparation, A.C. and A.D.-P. writing—review and editing, D.O.-S., visualization, A.D.-P. and D.O.-S. All authors have read and agreed to the published version of the manuscript.

Funding: This research received no external funding.

Institutional Review Board Statement: The research was approved by the Bioethics Committee of the Nicolaus Copernicus University in Toruń at the Ludwik Rydygier Collegium Medicum in Bydgoszcz (consent no. KB437/2018).

Informed Consent Statement: The study participants expressed their informed and voluntary consent to participate in the research.

Data Availability Statement: Data will be made available upon request. Data are stored at the Department of Pathobiochemistry and Clinical Chemistry, Collegium Medicum in Bydgoszcz (Poland). Person responsible for providing data: Agnieszka Chrustek (mail: a.chrustek@cm.umk.pl).

Conflicts of Interest: The authors declare no conflicts of interest.

Abbreviations

ALA	alpha-lipoic acid
BMI	Body Mass Index
DHA	docosahexaenoic acid
DPPH	2,2-diphenyl-1-picrylhydrazyl radical reduction method
FRAP	iron ion reduction ability
GAE	gallic acid
HBD	hebdomas graviditatis
LA	lipoic acid
Me	median
Min–Max	minimum and maximum values
n	abundance
p	level of statistical significance
PON1	paraoxonase 1
TAS	total antioxidant status
TE	trolox
WHR	waist-to-hip ratio

References

1. Nasser, R.; Stephen, A.M.; Goh, Y.K.; Clandinin, M.T. The effect of a controlled manipulation of maternal dietary fat intake on medium and long chain fatty acids in human breast milk in Saskatoon, Canada. *Int. Breastfeed. J.* **2010**, *5*, 3. [CrossRef] [PubMed] [PubMed Central]
2. Innis, S.M. Maternal Nutrition, Genetics, and Human Milk Lipids. *Curr. Nutr. Rep.* **2013**, *2*, 151–158. [CrossRef]
3. Huang, G.; Li, N.; Wu, X.; Zheng, N.; Zhao, S.; Zhang, Y.; Wang, J. Nutrition, production, and processing of virgin omega-3 polyunsaturated fatty acids in dairy: An integrative review. *Heliyon* **2024**, *10*, e39810. [CrossRef] [PubMed] [PubMed Central]
4. Bravi, F.; Wiens, F.; Decarli, A.; Dal Pont, A.; Agostoni, C.; Ferraroni, M. Impact of maternal nutrition on breast-milk composition: A systematic review. *Am. J. Clin. Nutr.* **2016**, *104*, 646–662. [CrossRef] [PubMed]
5. Karcz, K.; Królak-Olejnik, B. Vegan or vegetarian diet and breast milk composition—A systematic review. *Crit. Rev. Food Sci. Nutr.* **2021**, *61*, 1081–1098. [CrossRef] [PubMed]
6. Gandino, S.; Bzikowska-Jura, A.; Karcz, K.; Cassidy, T.; Wesolowska, A.; Królak-Olejnik, B.; Klotz, D.; Arslanoglu, S.; Picaud, J.C.; Boquien, C.Y.; et al. Vegan/vegetarian diet and human milk donation: An EMBA survey across European milk banks. *Matern. Child Nutr.* **2024**, *20*, e13564. [CrossRef] [PubMed] [PubMed Central]
7. Allès, B.; Baudry, J.; Méjean, C.; Touvier, M.; Péneau, S.; Hercberg, S.; Kesse-Guyot, E. Comparison of Sociodemographic and Nutritional Characteristics between Self-Reported Vegetarians, Vegans, and Meat-Eaters from the NutriNet-Santé Study. *Nutrients* **2017**, *9*, 1023. [CrossRef] [PubMed] [PubMed Central]

8. Wilson, D. Revealed: Countries with the Most Vegetarians in the World. 2024. Available online: https://ceoworld.biz/2024/01/21/revealed-countries-with-the-most-vegetarians-in-the-world-2024/ (accessed on 22 February 2025).
9. Perrin, M.T.; Pawlak, R.; Dean, L.L.; Christis, A.; Friend, L. A cross-sectional study of fatty acids and brain-derived neurotrophic factor (BDNF) in human milk from lactating women following vegan, vegetarian, and omnivore diets. *Eur. J. Nutr.* **2019**, *58*, 2401–2410. [CrossRef] [PubMed]
10. Melina, V.; Craig, W.; Levin, S. Position of the Academy of Nutrition and Dietetics: Vegetarian Diets. *J. Acad. Nutr. Diet.* **2016**, *116*, 1970–1980. [CrossRef] [PubMed]
11. Federal Commission for Nutrition (FCN). *Vegan Diets: Review of Nutritional Benefits and Risks. Expert Report of the FCN*; Federal Food Safety and Veterinary Office: Bern, Switzerland, 2018; Volume 28–34, pp. 61–62.
12. Richter, M.; Boeing, H.; Grünewald-Funk, D.; Heseker, H.; Kroke, A.; Leschik-Bonnet, E.; Oberritter, H.; Strohm, D.; Watzl, B. The German Nutrition Society (DGE). Vegan Diet Position of the German Nutrition Society (DGE). *Ernahr. Umsch. Int.* **2016**, *4*, 92–102. [CrossRef]
13. Weaver, G.; Bertino, E.; Gebauer, C.; Grovslien, A.; Mileusnic-Milenovic, R.; Arslanoglu, S.; Barnett, D.; Boquien, C.Y.; Buffin, R.; Gaya, A.; et al. Recommendations for the Establishment and Operation of Human Milk Banks in Europe: A Consensus Statement From the European Milk Bank Association (EMBA). *Front. Pediatr.* **2019**, *7*, 53. [CrossRef] [PubMed] [PubMed Central]
14. EFCNI. The European Foundation for the Care of Newborn Infants Position Paper. Recommendations for Promoting Human Milk Banks in Germany, Austria, and Switzerland. Available online: https://www.efcni.org/activities/projects/milk-banks/ (accessed on 26 February 2025).
15. Wesołowska, A.; Pawlus, B.; Bernatowicz-Łojko, U.; Borszewska-Kornacka, M.K. Operating of human milk banks in Poland—Practical recommendations. *Postępy Neonatol.* **2018**, *1*, 59–64. [CrossRef]
16. Guo, M. Chemical composition of human milk. In *Human Milk Biochemistry and Infant Formula Manufacturing Technology*; Woodhead Publishing Series in Food Science, Technology and Nutrition; Woodhead Publishing: Cambridge, UK, 2014; pp. 19–32. ISBN 9781845697242. [CrossRef]
17. György, P.; Norris, R.F.; Rose, C.S. Bifidus factor I. A variant of Lactobacillus bifidus requiring a special growth factor. *Arch. Biochem. Biophys.* **1954**, *48*, 193–201. [CrossRef] [PubMed]
18. Mosca, F.; Giannì, M.L. Human milk: Composition and health benefits. *Pediatr. Medica Chir.* **2017**, *39*, 47–52. [CrossRef] [PubMed]
19. Demmelmair, H.; Koletzko, B. Importance of fatty acids in the perinatal period. *World Rev. Nutr. Diet.* **2015**, *112*, 31–47. [CrossRef] [PubMed]
20. Hart, S.; Boylan, L.M.; Border, B.; Carroll, S.R.; Mcgunegle, D.; Lampe, R.M. Breast milk levels of cortisol and secretory Immunoglobulin A (SIgA) differ with maternal mood and infant neuro-behavioral functioning. *Infant Behav. Dev.* **2004**, *27*, 101–106. [CrossRef]
21. Gila-Diaz, A.; Arribas, S.M.; Algara, A.; Martín-Cabrejas, M.A.; de Pablo, A.L.; de Pipaón, M.S.; Ramiro-Cortijo, D. A review of bioactive factors in human breastmilk: A focus on prematurity. *Nutrients* **2019**, *11*, 1307. [CrossRef] [PubMed] [PubMed Central]
22. Boix-Amoros, A.; Collado, M.; Mira, A. Relationship between milk microbiota, bacterial load, macronutrients, and human cells during lactation. *Front. Microbiol.* **2016**, *7*, 492. [CrossRef] [PubMed] [PubMed Central]
23. Lis, J.; Orczyk-Pawiłowicz, M.; Kątnik-Prastowska, I. Proteins of human milk involved in immunological processes. *Postep. Hig. Med. Dosw.* **2013**, *67*, 529–547. (In Polish) [CrossRef] [PubMed]
24. Picciano, M.F. Nutrient composition of human milk. *Breastfeeding* **2001**, *48*, 53–67. [CrossRef] [PubMed]
25. Savino, F.; Liguori, S.A.; Fissore, M.F.; Ogger, R. Breast Milk Hormonesand Their Protective Effecton Obesity. *Int. J. Pediatr. Endocrinol.* **2009**, *2009*, 327505. [CrossRef] [PubMed] [PubMed Central]
26. Badillo-Suárez, P.A.; Rodríguez-Cruz, M.; Nieves-Morale, X. Impact of metabolic hormones secreted in human breast milkon nutritional programming in childhood obesity. *J. Mammary Gland. Biol. Neoplasia* **2017**, *22*, 171–191. [CrossRef] [PubMed]
27. Ford, J.E.; Zechalko, A.; Murphy, J.; Brooke, O.G. Comparison of the B vitamin composition of milk from mothers of preterm and term babies. *Arch. Dis. Child.* **1983**, *58*, 367–372. [CrossRef] [PubMed] [PubMed Central]
28. Koreti, S.; Prasad, N. Micronutrient content of breast milk. *J. Evol. Med. Dent. Sci.* **2014**, *3*, 1633–1638. [CrossRef]
29. Wack, R.P.; Lien, E.L.; Taft, D.; Roscelli, J.D. Electrolyte composition of human breast milk beyond the early postpartum period. *Nutrition* **1997**, *13*, 774–777. [CrossRef] [PubMed]
30. Bae, Y.J.; Kratzsch, J. Vitamin D and calcium in the human breast milk. *Best Pract. Res. Clin. Endocrinol. Metab.* **2018**, *32*, 39–45. [CrossRef] [PubMed]
31. Atanassova, M.; Georgieva, S.; Ivancheva, K. Total Phenolic And Total Flavonoid Contents, Antioxidant Capacity And Biological Contaminants. *J. Univ. Chem. Technol. Metall.* **2011**, *46*, 81–88.

32. Benzie, I.F.; Strain, J.J. Ferric reducing/antioxidant power assay: Direct measure of total antioxidant activity of biological fluids and modified version for simultaneous measurement of total antioxidant power and ascorbic acid concentration. *Methods Enzymol.* **1999**, *299*, 15–27. [CrossRef] [PubMed]
33. Vazquez, M.I.; Catalan-Dibene, J.; Zlotnik, A. B cells responses and cytokine production are regulated by their immune microenvironment. *Cytokine* **2015**, *74*, 318–326. [CrossRef] [PubMed] [PubMed Central]
34. Kleinman, R.E. *Pediatric Nutrition Handbook*, 6th ed.; American Academy of Pediatrics: Elk Grove Village, IL, USA, 2009.
35. Sanders, T.A.; Ellis, F.R.; Dickerson, J.W. Studies of vegans: The fatty acid composition of plasma choline phosphoglycerides, erythrocytes, adipose tissue, and breast milk, and some indicators of susceptibility to ischemic heart disease in vegans and omnivore controls. *Am. J. Clin. Nutr.* **1978**, *31*, 805–813. [CrossRef] [PubMed]
36. Patel, K.D.; Lovelady, C.A. Vitamin B12 status of East Indian vegetarian lactating women living in the United States. *Nutr. Res.* **1998**, *18*, 1839–1846. [CrossRef]
37. Sanders, T.A.; Reddy, S. The influence of a vegetarian diet on the fatty acid composition of human milk and the essential fatty acid status of the infant. *J. Pediatr.* **1992**, *120*, S71–S77. [CrossRef] [PubMed]
38. Rana, S.K.; Sanders, T.A. Taurine concentrations in the diet, plasma, urine and breast milk of vegans compared with omnivores. *Br. J. Nutr.* **1986**, *56*, 17–27. [CrossRef] [PubMed]
39. Alexy, U.; Fischer, M.; Weder, S.; Längler, A.; Michalsen, A.; Sputtek, A.; Keller, M. Nutrient Intake and Status of German Children and Adolescents Consuming Vegetarian, Vegan or Omnivore Diets: Results of the VeChi Youth Study. *Nutrients* **2021**, *13*, 1707. [CrossRef] [PubMed] [PubMed Central]
40. Specker, B.L.; Black, A.; Allen, L.; Morrow, F. Vitamin B-12: Low milk concentrations are related to low serum concentrations in vegetarian women and to methylmalonic aciduria in their infants. *Am. J. Clin. Nutr.* **1990**, *52*, 1073–1076. [CrossRef] [PubMed]
41. Pawlak, R.; Vos, P.; Shahab-Ferdows, S.; Hampel, D.; Allen, L.H.; Perrin, M.T. Vitamin B-12 content in breast milk of vegan, vegetarian, and nonvegetarian lactating women in the United States. *Am. J. Clin. Nutr.* **2018**, *108*, 525–531. [CrossRef] [PubMed]
42. Atamasirikul, K.; Kajanachumpol, S.; Wilairat, P.; Tantibhedhyangkul, P. Clinical usefulness of a HPLC method for simultaneous quantitation of plasma homocysteine and cysteine. *J. Med. Assoc. Thail.* **2008**, *91*, 338–344. [PubMed]
43. Schüpbach, R.; Wegmüller, R.; Berguerand, C.; Bui, M.; Herter-Aeberli, I. Micronutrient status and intake in omnivores, vegetarians and vegans in Switzerland. *Eur. J. Nutr.* **2017**, *56*, 283–293. [CrossRef] [PubMed]
44. Hashim, A.; Wang, L.; Juneja, K.; Ye, Y.; Zhao, Y.; Ming, L.J. Vitamin B6s inhibit oxidative stress caused by Alzheimer's disease-related Cu(II)-β-amyloid complexes-cooperative action of phospho-moiety. *Bioorg. Med. Chem. Lett.* **2011**, *21*, 6430–6432. [CrossRef] [PubMed]
45. Yokochi, N.; Morita, T.; Yagi, T. Inhibition of diphenolase activity of tyrosinase by vitamin b(6) compounds. *J. Agric. Food Chem.* **2003**, *51*, 2733–2736. [CrossRef] [PubMed]

antioxidants

MDPI

Review

Oxidant Stress, Hyperoxia/Hypoxia and Neonatal Respiratory Disorders

Ourania Kaltsogianni [1,2], Theodore Dassios [1,2] and Anne Greenough [1,2,*]

1 Women and Children's Health, School of Life Course and Population Sciences, Faculty of Life Sciences and Medicine, King's College London, London SE5 9RJ, UK; ourania.kaltsogianni@nhs.net (O.K.); theodore.dassios@kcl.ac.uk (T.D.)

2 Neonatal Intensive Care Centre, King's College Hospital NHS Foundation Trust, London SE5 9RS, UK

* Correspondence: anne.greenough@kcl.ac.uk

Abstract: Neonates, especially those born prematurely, have low antioxidant capacity and are highly exposed to oxidant stress during the perinatal period. Oxidant stress damage has been associated with several diseases of prematurity, including respiratory distress syndrome (RDS), bronchopulmonary dysplasia (BPD), and pulmonary hypertension. In addition, preterm infants are frequently exposed to hypoxia or hyperoxia, which further increases oxidant stress and morbidity. This narrative review describes the relationship between oxidant stress, hyperoxia/hypoxia, and neonatal respiratory disorders. Preterm infants with respiratory distress syndrome and BPD have higher levels of oxidative stress biomarkers in plasma and in tracheal aspirates and reduced activity of antioxidant enzymes. Early, prolonged, and frequent intermittent hypoxaemic episodes are related to BPD development. Exposure to hyperoxia is linked to longer duration of respiratory support and higher BPD rates. Preclinical data showed that intermittent hypoxia and hyperoxia are associated with pulmonary hypertension (PH) and that hyperoxia can negatively affect the response to pulmonary vasodilators. Antioxidant treatments are not routinely implemented into clinical care due to their modest effect on clinical outcomes, associated complications, and limited clinical data. Optimisation of oxygen delivery and monitoring with closed-loop automated oxygen control systems could potentially reduce oxidant stress in the neonatal environment.

Keywords: oxidative stress; neonate; hyperoxia; hypoxia

1. Introduction

Reactive oxygen species (ROS) are a group of molecules that are formed from molecular oxygen through reduction–oxidation or electronic excitation. These molecules can be classified as free radicals and non-radicals. ROS are generated endogenously or as a result of environmental exposure, and several mechanisms exist to control their production and availability. Under physiological conditions, ROS function as second messengers that regulate cellular processes through different mechanisms, including changes to enzyme activity, gene transcription, and membrane integrity [1]. The imbalance between increased ROS production and deficient intracellular antioxidant systems leads to the generation of free radicals and subsequent cell damage or disrupted redox signalling, a condition frequently described as oxidant stress [2]. More recently, the term "oxidative distress" has been used to describe this imbalance to account for the presence of oxidant stress and for the importance of redox signalling under physiological conditions [3]. A key metabolite in oxidant stress is hydrogen peroxide (H_2O_2), which is derived from molecular oxygen by NADPH oxidases, superoxide dismutases (SOD), and the mitochondrial electron transport chain in mammalian cells with aerobic metabolism [3]. Biological signalling is mainly mediated by oxidation of sulfur in target proteins and affects several cellular responses, including cell proliferation, migration, differentiation, and angiogenesis [4]. Antioxidant enzymes (catalases, glutathione peroxidases, and peroxiredoxins) are responsible for the removal of H_2O_2 [5]. Higher levels of H_2O_2 lead to nonspecific protein oxidation and altered signalling, which cause damage to all cellular components and subsequently lead to inflammation, growth arrest, and cell death [1].

Free radicals are reactive oxygen, nitrogen, or sulfur species that contain one or more unpaired electrons [6]. Free radicals are a product of metabolic redox reactions, mainly in the respiratory chain, as a consequence of exogenous and endogenous processes such as hypoxia, ischemia, ischemia–reperfusion, inflammation, hyperoxia, neutrophil and macrophage activation, and mitochondrial dysfunction [7]. Free radicals are unstable molecules that react with cellular components and affect cellular and tissue functions, leading to organ damage. Antioxidants are produced endogenously or potentially can be administered exogenously and protect from free radicals by either neutralising or removing them and by repairing the free-radical-induced cellular damage (Table 1) [8,9].

Table 1. Major oxidants and antioxidants.

Oxidants	
Reactive oxygen species	
	Superoxide ($O_2^{\cdot -}$)
	Hydrogen peroxide (H_2O_2)
	Hydroxyl radical ($HO^{\cdot}$)
	Hypochlorite (HOCl)
Reactive nitrogen species	
	Nitric oxide (NO)
	Nitric dioxide ($NO_2^{\cdot}$)
	Nitric trioxide ($NO_3^{\cdot}$)
	Nitrate (NO_3^-)
	Peroxynitrite ($ONOO^{\cdot}$)
Reactive sulfur species	
Radicals	Thiyl radical ($RS^{\cdot}$)
	Glutathionyl radical ($GSSG^{\cdot}$)
	Reactive sulfane species ($RSR^{\cdot}$)
Non-radicals	Reactive sulfane species (RSR)
	Reducing sulfur species (hydrogen sulfide (H_2S), reduced glutathione (GSH))
	Reactive sulfur substances (sulfur dioxide (SO_2), sulfur trioxide (SO_3))
	Sulfur secondary metabolites (Allicin)
Antioxidants	
Enzymatic	Primary enzymes: superoxide dismutase, catalase, glutathione peroxidase

Table 1. *Cont.*

	Secondary enzymes: glutathione reductase, glucose-6-phosphate-dehydrogenase
Non- enzymatic	Flavonoids
	Cofactors: coenzyme Q10
	Minerals: zinc, selenium
	Organosulfur compounds: glutathione
	Vitamins and derivatives: A (retinol), C (Ascorbic acid), E (tocopherol), K
	Carotenoids: β-carotene, lycopene, lutein, zeaxanthin
	Nitrogen non-protein compounds: uric acid
	Phenolic acids: hydroxycinnamic acids, hydroxybenzoic acids

Pregnancy complications and conditions, including pre-eclampsia, diabetes, undernutrition, smoking, infection or inflammation, may trigger an acute increase in free radical formation and, thus, create an adverse intrauterine environment that affects foetal development [10]. Foetal oxidant stress has also been demonstrated in pregnancies complicated by intrauterine growth restriction (IUGR), as this condition is often linked with impaired placental blood flow and intrauterine hypoxia that causes the generation of free radicals [11]. It has also been hypothesised that foetal oxidant stress may be the underlying mechanism that links placental dysfunction to foetal programming that predisposes to adult diseases such as metabolic syndrome, obesity, diabetes, hyperinsulinemia, hypertension, and cardiovascular disease [10]. An impaired oxidant/ antioxidant status, as demonstrated by significantly higher isoprostane and prostaglandin F2 alpha levels, was found in children born small or large for gestation when compared with those who had an appropriate for gestation birth weight [12]. In addition, animal studies link foetal oxidant stress to renal and cardiovascular alterations, which contribute to hypertension, endothelial dysfunction, ventricular hypertrophy, and susceptibility to develop arrhythmias [13].

Newborn infants are susceptible to significant oxidant stress during the transition from the hypoxic intrauterine environment to the relatively high oxygen exposure after birth. This transition, in combination with the use of supplementary oxygen during resuscitation, assisted ventilation, administration of surfactant, total parenteral nutrition, and blood transfusions, further enhances the production of free radicals and exacerbates oxidant stress [5].

Newborn infants have decreased antioxidant capacity and are highly exposed to oxidant stress. Antioxidant defence mechanisms progressively increase during gestation, and many important vitamins and minerals with antioxidant roles are delivered to the foetus during the third trimester of pregnancy [14]. Premature birth interrupts the synthesis and passage of maternal antioxidants and, consequently, increases the susceptibility of preterm infants to oxidant stress damage, which has been associated with several diseases of prematurity [14].

Neonatal patients, and especially those born prematurely, experience frequent fluctuations of their transcutaneous oxygen saturation (SpO_2) levels, either due to respiratory pauses or apnoea and ineffective breathing efforts in spontaneously breathing infants [15]. Other physiologic parameters that contribute to desaturations are the low functional residual capacity (FRC) of preterm infants, which results from atelectasis and high chest wall compliance, and low blood oxygen capacity, including blood volume and haemoglobin content [16]. In ventilated infants, episodes of intermittent hypoxia (IH) occur due to ineffective respiratory support and loss of lung volume [17]. In a cohort study, extremely preterm infants experienced up to 600 intermittent hypoxic episodes within one week, and these were associated with a higher risk of developing retinopathy of prematurity (ROP) requiring laser treatment [18]. In addition, newborn infants receiving supplemental oxygen are frequently exposed to hyperoxia and its associated complications. Studies on the achievement of oxygen saturation targets in extremely preterm and very low birth weight (VLBW) infants showed that they spent 20% to 73% of the time in hyperoxia [19,20].

Hypoxia, hyperoxia, and fluctuations between these conditions increase oxidant stress and generate reactive oxygen species that result in alterations in cellular proliferation and apoptosis (Figure 1) [21]. Hypoxaemic events during early postnatal life have been associated with multiple poor outcomes, including growth restriction [22], sleep-disordered breathing [23], ROP [24], neurodevelopmental impairment [24], and mortality [24]. Hyperoxia has been related to lung, retina, central nervous system (CNS), and red blood cell injuries, as well as generalised tissue damage in the neonatal period [25]. Recurrent IH and hyperoxia exposure may result in long-term changes in respiratory mechanics with increased airway resistance and decreased compliance and may contribute to wheezing in former preterm infants [26].

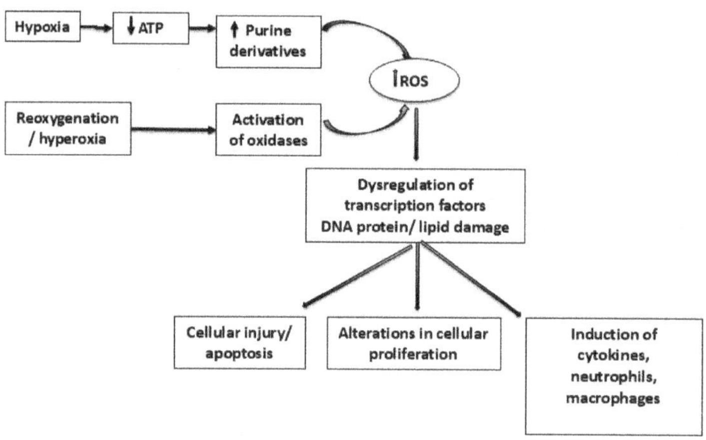

Figure 1. Diagram demonstrating the generation of reactive oxygen species (ROS) relative to hypoxia and hyperoxia in neonatal respiratory disorders. Hypoxia leads to adenosine triphosphate (ATP) exhaustion and subsequent increased formation of purine derivatives. Reoxygenation and excess oxygen/hyperoxia activate oxidases that metabolise purine derivatives and generate reactive oxygen species (ROS) (e.g., superoxide anion) that cause oxidative stress and tissue damage. ATP: adenosine triphosphate; DNA: deoxyribonucleic acid; ROS: reactive oxygen species.

Early exposure to oxidative stress has been associated with altered lung development processes and susceptibility to neonatal respiratory disorders, including respiratory distress syndrome (RDS), bronchopulmonary dysplasia (BPD), and persistent pulmonary hypertension (PH) [27]. This narrative review describes the relationship between oxidant stress, hyperoxia/ hypoxia, and neonatal respiratory disorders.

2. Respiratory Distress Syndrome

Respiratory distress syndrome (RDS) is the most common respiratory disorder in preterm infants. Among other factors, oxidative stress plays a significant role in the pathophysiology of RDS. Preterm infants at birth frequently experience periods of hypoxia and reoxygenation that lead to an increase in oxidant stress. ROS increase proinflammatory cytokines, activate neutrophils, and lead to an acute inflammatory state with subsequent alveolar injury and pulmonary damage [28]. Bodha et al. found lower levels of reduced glutathione (GSH) in the tracheal aspirates of preterm infants with RDS compared with controls that were electively intubated for surgery. In addition, oxidised glutathione (GSSG) and the ratio of GSSG to GSH were significantly increased and correlated with disease severity [29]. Dizdar et al. showed an association of lower serum total antioxidant capacity (TAC) levels with longer duration of respiratory support in preterm infants with RDS. Surfactant administration significantly increased TAC levels and improved oxidant–antioxidant balance [30]. A pilot study demonstrated a significant increase in the plasma levels of interleukin (IL)-6, IL-8, and tumour necrosis factor (TNF)-α in preterm ventilated infants with high oxygen requirements ($FiO_2 > 0.3$) compared with infants requiring lower inspired oxygen concentration ($FiO_2 < 0.3$) [31]. In addition, in a subsequent case–control study, preterm infants with RDS (n = 31) had significantly higher plasma levels of oxidative stress markers (malondialdehyde (MDA) and hydrogen peroxide (H_2O_2)) and reduced activity of antioxidant enzymes (catalase (CAT) and superoxide dismutase (SOD)) when compared with healthy preterm newborns (n = 36) [32]. However, more infants in the RDS group were born by caesarean section, which may also increase oxidative stress [33].

3. Bronchopulmonary Dysplasia

Despite advances in neonatal care, bronchopulmonary dysplasia (BPD) is one of the most common serious morbidities of prematurity and is associated with adverse respiratory and neurodevelopmental outcomes [34]. BPD is a multifactorial disease, and its pathogenesis entails an ongoing process of injury and recovery of an immature lung [35]. Intermittent hypoxaemic episodes have the potential to trigger a proinflammatory cascade [36] and oxidative stress [37], which have both been implicated in the pathogenesis of BPD. Chronic oxygen toxicity damages the pulmonary epithelium, inactivates surfactant, and causes intra-alveolar oedema and interstitial wall thickening with fibrosis, leading to pulmonary atelectasis. This results in disturbed alveolarisation, which is a pathological characteristic of BPD [25]. Inflammatory markers, including neutrophils, interleukin 8 (IL-8), and elastase, were increased in the bronchoalveolar lavage fluid of infants with BPD, supporting the role of inflammation and oxidative stress in the development of the disease [38]. In addition, IL-6, as a marker of airway inflammation, and 8-hydroxy-20-deoxyguanosine (8-OHdG), which indicates oxidative DNA damage, were higher in tracheal aspirates of preterm ventilated infants with BPD on postnatal days 1 and 28 compared with preterm infants without BPD [39]. Among 104 preterm infants born at less than 30 weeks of gestation, higher carboxyhaemoglobin (COHb) levels from blood gas analysis in the first seven days of life were significantly associated with the risk of developing BPD [40]. The level of COHb is a marker of endogenous production of carbon monoxide as a consequence of oxidative stress [41].

Fairchild et al. studied episodes of bradycardia or hypoxaemia ($SpO_2 < 80\%$ for ≥ 10 s) in 502 VLBW infants, 37% of whom developed BPD, in the first four weeks after birth. The frequency and duration of hypoxaemic episodes overall and during the fourth week were significantly associated with the development of BPD [42]. In a retrospective cohort of 137 extremely preterm infants, a diagnosis of BPD at 36 weeks postmenstrual age was associated with increased daily frequency of IH episodes, longer duration, and a higher IH nadir from day seven to day 26 of life. Increased numbers of IH events occurring 1–20 min apart were also related to this outcome. Infants with BPD had a lower mean SpO_2

and higher mean airway pressure and inspired oxygen concentration requirements [43]. Similarly, a post hoc analysis of the Canadian Oxygen Trial that included 1018 extremely preterm infants demonstrated significant differences in hypoxaemic exposure among infants who developed BPD compared to those who did not, and these differences increased in magnitude over the first four weeks of life. Both the frequency of prolonged intermittent hypoxic episodes (SpO_2 < 80% and duration of ≥60 s) and the proportion of time per day with SpO_2 < 80% between birth and 36 weeks postmenstrual age (PMA) were significantly associated with the risk of severe BPD [44]. In the same cohort of extremely preterm infants, intermittent hypoxic episodes were linked to increased risk of late death or childhood disability at 18 months corrected age [24]. BPD increases the risk of adverse neurodevelopment [45]. Therefore, early and prolonged intermittent hypoxic episodes may independently contribute to an increased risk of poor neurodevelopmental outcomes and BPD.

Oxygen toxicity has been well recognised as one of the factors contributing to the development of BPD. Animal models suggest that even brief exposure to hyperoxia can cause long-term morphologic and functional changes in the lung [46,47], resulting in a phenotype comparable with BPD [47]. Those data are in agreement with clinical studies showing that restricted use of supplemental oxygen or lower SpO_2 targets is associated with a reduction in lung inflammation and BPD rates [48]. In a prospective observational study in 502 VLBW infants, the implementation of lower SpO_2 targets (85 to 93% versus 92 to 100%) reduced the incidence of BPD and the use of postnatal corticosteroids; the number needed to treat was six [49].

Vento et al. randomised 78 extremely preterm infants to a lower (0.3) or higher (0.9) fraction of the inspired oxygen concentration during resuscitation at birth. Patients in the lower FiO_2 group required significantly fewer days of supplemental oxygen, mechanical ventilation, and continuous positive airway pressure support and had lower BPD rates (15.4% versus 31.7%). In addition, infants who received higher FiO_2 were characterised by increased expression of proinflammatory cytokines and oxidative stress biomarkers in the first week after birth, and those indices correlated significantly with BPD development [50]. Birenbaum et al. showed a 50% reduction in the incidence of BPD among 145 VLBW infants with the use of lower oxygen concentrations during resuscitation and lower oxygen saturation targets (88–92%). However, the study was limited by its retrospective design and several changes implemented into the neonatal care during the study period that prevented the authors from being able to accurately attribute the relative influence of each factor on the outcome of interest [51].

4. Pulmonary Hypertension

During the intrauterine period, the foetus is subjected to a relative hypoxaemia, which results in a physiological state of pulmonary hypertension (PH). At birth, lung aeration leads to a rise in oxygen tension that contributes to rapid pulmonary vasodilation. Failure of the pulmonary vasculature to relax results in increased pulmonary arterial pressure and vascular resistance with right-to-left shunting of deoxygenated blood from the pulmonary to the systemic circulation and the development of persistent pulmonary hypertension of the newborn (PPHN) [52]. Oxygen is often used in the treatment of PPHN as a potent pulmonary vasodilator. Preclinical studies have shown that even short-lasting alveolar hyperoxia can be harmful for newborns with PPHN, as it seems to increase pulmonary vascular contractility [53] and impairs the subsequent response to pulmonary vasodilators, including inhaled nitric oxide (iNO) [54]. In a lamb model of PPHN, a gradual increase in the FiO_2 up to 0.5 significantly reduced the pulmonary vascular resistance, but no additional decrease was observed with a further increase in the oxygen concentration to 100% [54].

The alveolar and lung vascular changes in infants with BPD may also result in an increase in pulmonary vascular resistance and subsequently in the development of PH [55]. Animal data suggest that intermittent hypoxia is implicated in the pathogenesis of PH. In a rat model of BPD, IH during recovery from hyperoxia-induced lung injury prevented recovery of alveologenesis and increased pulmonary vascular resistance, the Fulton index (which is the weight ratio of the right ventricle to the sum of the left ventricle and septum [56]), and arterial wall thickness [57]. Exposure of neonatal mice to IH from day seven of postnatal life and for three weeks resulted in altered lung endothelial cell function, mitochondrial DNA lesions, and impaired lung angiogenesis. When superimposed on hyperoxia, IH induced a severe lung vascular phenotype that is seen in preterm infants with PH [58]. In a case–control study that included 80 extremely preterm infants with BPD, prolonged severe hypoxaemic episodes (SpO_2 < 70% and <80%) were associated with PH and increased mortality [59].

5. Antioxidant Treatments

Many antioxidant therapeutic strategies have been studied to determine if they would reduce oxidative stress in newborn infants. Vitamins A and E have been extensively investigated as potential antioxidant therapies for the prevention of BPD with regard to whether they could eliminate lipid peroxidation induced by ROS. A Cochrane review and meta-analysis included 10 randomised controlled trials (RCTs) and 1460 infants comparing vitamin A with a control and showed a marginal reduction (risk reduction (RR): 0.87, 95% confidence interval (CI): 0.77 to 0.99) in the incidence of BPD at 36 weeks PMA and no significant differences in neurodevelopmental outcomes at 18–22 months corrected age [60]. Vitamin A supplementation has not been implemented into standard clinical practice due to its modest effect and the need for intramuscular administration [61]. A more recent systematic review and meta-analysis showed that enteral compared with parenteral vitamin A is equally effective in reducing BPD rates at 36 weeks postmenstrual age, but the benefit was limited to infants with low baseline intake (<1500 IUIU/kg/day) [61]. Vitamin E was not found to play a significant role in the prevention of BPD (RR = 0.659, 95% CI = 0.243–1.786; $p = 0.412$) [62]. Vitamin E supplementation may have a neuroprotective effect [63] and reduce severe ROP, but high intravenous doses in VLBW infants have been associated with increased risk of sepsis [64].

Vitamin C (ascorbic acid: AA) is a water-soluble antioxidant in cells and plasma that is actively transported across the placenta [65]. In addition, in vitro studies demonstrate a prooxidant activity via the Fenton reaction [66]. Evidence regarding the association between AA concentration and respiratory morbidity in preterm infants is conflicting. In a prospective observational study that included 144 infants of any gestation, higher AA concentrations on day two of life were associated with a higher risk of developing BPD [67]. A double-blind RCT that included 119 VLBW infants did not show any significant association between AA levels throughout the first 28 days of life and the incidence of BPD at 36 weeks postmenstrual age. Higher AA supplementation was associated with a trend towards reduced need for supplemental oxygen at 36 weeks postmenstrual age [68]. A more recent systematic review and meta-analysis included two studies (160 VLBW infants) and demonstrated a significant reduction in the risk for BPD with vitamin C supplementation but with very low certainty evidence [69]. In another systematic review and meta-analysis, vitamin C supplementation in pregnancies exposed to smoking was found to reduce the incidence of wheeze in offspring at 12 months and five years. No significant association was found between vitamin C and E supplementation and the risk for RDS [70].

Superoxide dismutase (SOD), an antioxidant enzyme, converts the superoxide radical that is generated through metabolism or other cell reactions to hydrogen peroxide (H_2O_2) and molecular oxygen (O_2). Antioxidant enzymes, including catalase, glutathione peroxidase, and peroxiredoxins, break down H_2O_2 to water and molecular oxygen [5]. SOD is decreased in the pulmonary vascular endothelium of preterm infants with BPD, leading to reduced vasodilation induced by nitric oxide [71]. Studies in lamb models of PPHN demonstrated that intratracheal administration of recombinant human SOD (r-h SOD) reduced oxidant stress [72,73] and improved nitric oxide-mediated vasodilation [72]. In an RCT in preterm infants with RDS, intratracheal administration of r-h SOD every 48 h during mechanical ventilation (MV) for up to one month of age did not significantly affect the rates of death or BPD at 36 weeks postmenstrual age. At follow-up, however, infants treated with r-h SOD had fewer wheezing episodes and respiratory illnesses requiring hospitalisation compared with controls when assessed at one year of age [74].

Breast milk consists of many antioxidants, including vitamins, SOD, glutathion, melatonin, and short-chain fatty acids [75]. Therefore, in addition to its other advantages with regard to nutrition and immunity, it may contribute to reducing oxidative stress for newborn infants. Human milk following preterm birth has been found to have higher antioxidant capacity when compared to term milk and may help reduce the incidence of oxidative stress-related diseases of prematurity [76]. A systematic review and meta-analysis included 15 studies (n = 4984 infants) evaluating the effect of mother's own milk on the incidence of BPD. They found that breast milk as an exclusive diet was associated with a significant reduction in the risk of BPD (RR: 0.74, 95% CI: 0.57–0.96, five studies) when compared with preterm formula, regardless of gestation at birth [77]. A recent review concluded that the antioxidant capacity of breast milk may be the primary mechanism through which it contributes to reduced BPD rates [75]. These results further emphasise the well-established benefits of breast milk for preterm infants.

Novel antioxidant therapies aim at the induction of endogenous antioxidant systems, as, following systemic administration, they may not reach the target cells where oxidative stress causes injury [78]. One example is melatonin, a neurohormone that induces several endogenous antioxidant systems. In an RCT that included 110 infants < 32 weeks of gestation, intravenous administration of melatonin in divided doses and within six hours of life led to significant reductions in inflammatory cytokines and ventilator and oxygen requirements compared with controls. Furthermore, none of the infants treated with melatonin were oxygen dependent at 28 days of life, whereas there were six infants (7.5%) with oxygen dependency in the control group [79]. These promising results need to be examined in a larger RCT, preferably with long-term follow-up.

A recent review highlighted the limitations of therapeutic interventions in targeting oxidative stress [80]. Oxidative stress is usually a contributor to the disease rather than its primary cause, and this limits the effectiveness of antioxidant treatments. In addition, agents that appear to work in vitro may not provide the same results in vivo due to difficulty in achieving adequate concentrations [80]. Therapeutic strategies should selectively target disease-related mechanisms to avoid disrupting some important signalling processes mediated by oxidants [1]. Preventing oxidant production or inhibiting oxidant-induced signalling and increasing the synthesis of endogenous antioxidant enzymes should be the focus of emerging antioxidant therapies [80].

6. Closed-Loop Automated Oxygen Control Systems

Closed-loop automated feedback-controlled (CLAC) systems of oxygen delivery monitor oxygen saturation values in real time to calculate and make an adjustment to the inspired oxygen concentration, without any human intervention [81]. Several algorithms have been developed to provide automated oxygen control in preterm infants. All of them use SpO_2 levels measured by pulse oximetry as an input, but they process this information in different ways in order to adjust the inspired oxygen concentration. CLAC has been extensively studied in very preterm or low birth weight infants. The available studies demonstrated that CLAC monitoring increases the time spent within the target SpO_2 range, improves the stability of partial arterial oxygen pressure (PaO_2), and reduces the time spent in hypoxia and hyperoxia, with fewer manual adjustments to the inspired oxygen concentration [81–86]. Results appear to be consistent for all the control algorithms [86,87], across different SpO_2 target ranges [87], and in infants receiving either mechanical ventilation or non-invasive respiratory support [85]. In addition, CLAC facilitates earlier weaning of the FiO_2 compared with manual oxygen control [88,89]. In a randomised crossover study in ventilated preterm infants, hourly median FiO_2 values were lower during the automated period (median 24 h FiO_2: 0.32 versus 0.37, $p < 0.001$) and infants spent longer time with $FiO_2 < 0.25$ (median % time with $FiO_2 < 0.25$: 8% versus 0%, $p = 0.018$) [89]. Reynolds et al. [88] demonstrated that preterm infants receiving high-flow nasal cannula oxygen spent more time in air during automated oxygen control when compared with manual control [88]. However, there is limited evidence on the effect of CLAC on longer-term respiratory and other outcomes [81,85,86]. The majority of studies had a randomised crossover design and were not sufficiently powered to detect significant differences in long-term outcomes. A recent Cochrane review [85] included 18 randomised trials, which compared CLAC with manual oxygen control or different systems of automated oxygen control in preterm infants and concluded that none of the studies assessed time on oxygen therapy, BPD, severe ROP, necrotising enterocolitis, or mortality [85].

The improvement in the achievement of oxygen saturation targets and the reduction in the duration of supplemental oxygen treatment could ameliorate the risks related to hypoxia and hyperoxia and improve neonatal respiratory morbidity. We have recently reported that, in an RCT in preterm ventilated infants, CLAC monitoring was associated with significant reductions in the duration of MV and supplemental oxygen treatment, incidences of BPD and moderate or severe BPD or death, and the need for home oxygen [90]. These promising results should be further explored in large multicentre studies before the routine implementation of CLAC in clinical practice.

7. Conclusions

Newborn infants, especially those born prematurely, are exposed to oxidative stress, which increases their susceptibility to respiratory disorders and other free-radical-related diseases of prematurity. Oxygen is the most used drug in neonatal intensive care, and inappropriate dosing may lead to hypoxia or hyperoxia that further increases oxidative damage. Strategies to limit oxidant stress need to be developed.

Despite promising early data, no specific antioxidant treatments are currently used in the neonatal setting. Major barriers to effective treatments include the lack of targets of the antioxidant factors, their limited bioavailability to the tissue in which oxidative stress is ongoing, and the lack of knowledge regarding appropriate dosing, as well as the timing of therapy, as there is evidence of oxidative stress occurring even in the perinatal period. Further, the genetic variability of the subjects may affect their susceptibility to oxidative stress and response to treatment [91]. Genetic differences between antioxidant systems may influence individual responses to hypoxia and hyperoxia. In a cohort of 284 VLBW infants with BPD, genetic variants of a transcription factor that induces antioxidant enzymes (nuclear factor erythroid-2 related factor-2 (NFE2L2)) were significantly associated with the risk and severity of BPD after adjusting for epidemiological confounders [92]. Therefore, genomic sequencing may allow for the identification of individual differences in endogenous antioxidant systems and the development of tailored therapeutic interventions.

The complex interplay between oxidant stress and antioxidants has recently been reviewed [1,80]. These reviews highlight that the results of clinical trials in this field are rather disappointing and emphasise that a greater understanding of the mechanisms of action of antioxidants is necessary.

Future studies should explore the biomarkers that would help identify patients at high risk of oxidative injury at an early stage and develop novel, tailored therapies.

Importantly, the implementation of strategies that will promote an optimal perinatal environment with minimal risk of oxidative stress appears necessary. As oxygen can cause oxidative damage, the optimisation of oxygen delivery and monitoring should also be the focus of future research. CLAC systems appear promising in reducing the risks related to oxygen treatment. Future research should further explore the effects of CLAC on clinical outcomes and evaluate its implementation in clinical practice.

Author Contributions: O.K. wrote the first draft of this manuscript; all authors, O.K., T.D. and A.G. have contributed to and approved the final manuscript. All authors have read and agreed to the published version of the manuscript.

Funding: This research received no external funding.

Institutional Review Board Statement: This research did not require ethical approval.

Data Availability Statement: No new data were created or analyzed in this study. Data sharing is not applicable to this article.

Conflicts of Interest: The authors have no conflicts of interest or financial statements to disclose. Inspiration Healthcare loaned the equipment used in our studies on closed-loop automated oxygen control but was not involved in the study design, data collection, or data analysis. Ourania Kaltsogianni is the King's Medical Research Trust Clinical Lecturer and is supported by the King's College Hospital Charity. The charity was not involved in the study design, data collection, or analysis; in the writing of the manuscript; or in the decision to publish this review.

Abbreviations

The following abbreviations are used in this manuscript:

AA	Ascorbic acid
BPD	Bronchopulmonary dysplasia
CAT	Catalase
CLAC	Closed-loop automated oxygen control
CNS	Central nervous system
COHb	Carboxyhaemoglobin
FiO_2	Fraction of inspired oxygen
FRC	Functional residual capacity
GSH	Reduced glutathione
GSSG	Oxidised glutathione
H_2O_2	Hydrogen peroxide
IH	Intermittent hypoxia
IL	Interleukin
iNO	Inhaled nitric oxide

IUGR	Intrauterine growth restriction
MDA	Malondialdehyde
MV	Mechanical ventilation
NFE2L2	Nuclear factor erythroid-2 related factor-2
8-OHdG	8-hydroxy-20-deoxyguanosine
PH	Pulmonary hypertension
PMA	Postmenstrual age
PaO_2	Partial arterial oxygen pressure
PPHN	Persistent pulmonary hypertension of the newborn
RCT	Randomised controlled trial
RDS	Respiratory distress syndrome
ROP	Retinopathy of prematurity
ROS	Reactive oxygen species
SOD	Superoxide dismutase
SpO_2	Peripheral oxygen saturation
TAC	Total antioxidant capacity
TNF	Tumour necrosis factor
VLBW	Very low birth weight

References

1. Sies, H.; Jones, D.P. Reactive oxygen species (ROS) as pleiotropic physiological signalling agents. *Nat. Rev. Mol. Cell Biol.* **2020**, *21*, 363–383. [CrossRef]
2. Rahal, A.; Kumar, A.; Singh, V.; Yadav, B.; Tiwari, R.; Chakraborty, S.; Dhama, K. Oxidative stress, prooxidants, and antioxidants: The interplay. *BioMed Res. Int.* **2014**, *2014*, 761264. [CrossRef]
3. Sies, H. Hydrogen peroxide as a central redox signaling molecule in physiological oxidative stress: Oxidative eustress. *Redox Biol.* **2017**, *11*, 613–619. [CrossRef]
4. Zeida, A.; Babbush, R.; Lebrero, M.C.; Trujillo, M.; Radi, R.; Estrin, D.A. Molecular basis of the mechanism of thiol oxidation by hydrogen peroxide in aqueous solution: Challenging the SN2 paradigm. *Chem. Res. Toxicol.* **2012**, *25*, 741–746. [CrossRef] [PubMed]
5. Lembo, C.; Buonocore, G.; Perrone, S. Oxidative Stress in Preterm Newborns. *Antioxidants* **2021**, *10*, 1672. [CrossRef] [PubMed]
6. Aranda-Rivera, A.K.; Cruz-Gregorio, A.; Arancibia-Hernández, Y.L.; Hernández-Cruz, E.Y.; Pedraza-Chaverri, J. RONS and Oxidative Stress: An Overview of Basic Concepts. *Oxygen* **2022**, *2*, 437–478. [CrossRef]
7. Davies, K.J. Oxidative stress, antioxidant defenses, and damage removal, repair, and replacement systems. *IUBMB Life* **2000**, *50*, 279–289. [CrossRef] [PubMed]
8. Pham-Huy, L.A.; He, H.; Pham-Huy, C. Free radicals, antioxidants in disease and health. *Int. J. Biomed. Sci.* **2008**, *4*, 89–96. [CrossRef]
9. Ali, S.S.; Ahsan, H.; Zia, M.K.; Siddiqui, T.; Khan, F.H. Understanding oxidants and antioxidants: Classical team with new players. *J. Food Biochem.* **2020**, *44*, e13145. [CrossRef]
10. Perrone, S.; Santacroce, A.; Picardi, A.; Buonocore, G. Fetal programming and early identification of newborns at high risk of free radical-mediated diseases. *World J. Clin. Pediatr.* **2016**, *5*, 172–181. [CrossRef]
11. Longini, M.; Perrone, S.; Kenanidis, A.; Vezzosi, P.; Marzocchi, B.; Petraglia, F.; Centini, G.; Buonocore, G. Isoprostanes in amniotic fluid: A predictive marker for fetal growth restriction in pregnancy. *Free Radic. Biol. Med.* **2005**, *38*, 1537–1541. [CrossRef]
12. Chiavaroli, V.; Giannini, C.; D'Adamo, E.; de Giorgis, T.; Chiarelli, F.; Mohn, A. Insulin resistance and oxidative stress in children born small and large for gestational age. *Pediatrics* **2009**, *124*, 695–702. [CrossRef] [PubMed]
13. Rodriguez-Rodriguez, P.; Ramiro-Cortijo, D.; Reyes-Hernandez, C.G.; Lopez de Pablo, A.L.; Gonzalez, M.C.; Arribas, S.M. Implication of Oxidative Stress in Fetal Programming of Cardiovascular Disease. *Front. Physiol.* **2018**, *9*, 602. [CrossRef]
14. Matyas, M.; Hasmasanu, M.G.; Zaharie, G. Antioxidant Capacity of Preterm Neonates Assessed by Hydrogen Donor Value. *Medicina* **2019**, *55*, 720. [CrossRef] [PubMed]
15. Di Fiore, J.M.; Arko, M.K.; Miller, M.J.; Krauss, A.; Betkerur, A.; Zadell, A.; Kenney, S.R.; Martin, R.J. Cardiorespiratory events in preterm infants referred for apnea monitoring studies. *Pediatrics* **2001**, *108*, 1304–1308. [CrossRef]
16. Di Fiore, J.M.; MacFarlane, P.M.; Martin, R.J. Intermittent Hypoxemia in Preterm Infants. *Clin. Perinatol.* **2019**, *46*, 553–565. [CrossRef]

17. Bolivar, J.M.; Gerhardt, T.; Gonzalez, A.; Hummler, H.; Claure, N.; Everett, R.; Bancalari, E. Mechanisms for episodes of hypoxemia in preterm infants undergoing mechanical ventilation. *J. Pediatr.* **1995**, *127*, 767–773. [CrossRef]
18. Di Fiore, J.M.; Bloom, J.N.; Orge, F.; Schutt, A.; Schluchter, M.; Cheruvu, V.K.; Walsh, M.; Finer, N.; Martin, R.J. A higher incidence of intermittent hypoxemic episodes is associated with severe retinopathy of prematurity. *J. Pediatr.* **2010**, *157*, 69–73. [CrossRef] [PubMed]
19. Ford, S.P.; Leick-Rude, M.K.; Meinert, K.A.; Anderson, B.; Sheehan, M.B.; Haney, B.M.; Leeks, S.R.; Simon, S.D.; Jackson, J.K. Overcoming barriers to oxygen saturation targeting. *Pediatrics* **2006**, *118* (Suppl. 2), S177–S186. [CrossRef]
20. Hagadorn, J.I.; Furey, A.M.; Nghiem, T.H.; Schmid, C.H.; Phelps, D.L.; Pillers, D.A.; Cole, C.H.; Group, A.V.S. Achieved versus intended pulse oximeter saturation in infants born less than 28 weeks' gestation: The AVIOx study. *Pediatrics* **2006**, *118*, 1574–1582. [CrossRef]
21. Mohamed, T.; Abdul-Hafez, A.; Gewolb, I.H.; Uhal, B.D. Oxygen injury in neonates: Which is worse? hyperoxia, hypoxia, or alternating hyperoxia/hypoxia. *J. Lung Pulm. Respir. Res.* **2020**, *7*, 4–13.
22. Warburton, A.; Monga, R.; Sampath, V.; Kumar, N. Continuous pulse oximetry and respiratory rate trends predict short-term respiratory and growth outcomes in premature infants. *Pediatr. Res.* **2019**, *85*, 494–501. [CrossRef] [PubMed]
23. Rosen, C.L.; Larkin, E.K.; Kirchner, H.L.; Emancipator, J.L.; Bivins, S.F.; Surovec, S.A.; Martin, R.J.; Redline, S. Prevalence and risk factors for sleep-disordered breathing in 8- to 11-year-old children: Association with race and prematurity. *J. Pediatr.* **2003**, *142*, 383–389. [CrossRef] [PubMed]
24. Poets, C.F.; Roberts, R.S.; Schmidt, B.; Whyte, R.K.; Asztalos, E.V.; Bader, D.; Bairam, A.; Moddemann, D.; Peliowski, A.; Rabi, Y.; et al. Association Between Intermittent Hypoxemia or Bradycardia and Late Death or Disability in Extremely Preterm Infants. *JAMA* **2015**, *314*, 595–603. [CrossRef]
25. Perrone, S.; Bracciali, C.; Di Virgilio, N.; Buonocore, G. Oxygen Use in Neonatal Care: A Two-edged Sword. *Front. Pediatr.* **2016**, *4*, 143. [CrossRef]
26. Dylag, A.M.; Mayer, C.A.; Raffay, T.M.; Martin, R.J.; Jafri, A.; MacFarlane, P.M. Long-term effects of recurrent intermittent hypoxia and hyperoxia on respiratory system mechanics in neonatal mice. *Pediatr. Res.* **2017**, *81*, 565–571. [CrossRef]
27. Cannavo, L.; Perrone, S.; Viola, V.; Marseglia, L.; Di Rosa, G.; Gitto, E. Oxidative Stress and Respiratory Diseases in Preterm Newborns. *Int. J. Mol. Sci.* **2021**, *22*, 12504. [CrossRef]
28. Marseglia, L.; D'Angelo, G.; Granese, R.; Falsaperla, R.; Reiter, R.J.; Corsello, G.; Gitto, E. Role of oxidative stress in neonatal respiratory distress syndrome. *Free Radic. Biol. Med.* **2019**, *142*, 132–137. [CrossRef]
29. Boda, D.; Nemeth, I.; Pinter, S. Surface tension, glutathione content and redox ratio of the tracheal aspirate fluid of premature infants with IRDS. *Biol. Neonate* **1998**, *74*, 281–288. [CrossRef]
30. Dizdar, E.A.; Uras, N.; Oguz, S.; Erdeve, O.; Sari, F.N.; Aydemir, C.; Dilmen, U. Total antioxidant capacity and total oxidant status after surfactant treatment in preterm infants with respiratory distress syndrome. *Ann. Clin. Biochem.* **2011**, *48 Pt 5*, 462–467. [CrossRef] [PubMed]
31. Carvalho, C.G.; Procianoy, R.S.; Neto, E.C.; Silveira, R.C. Preterm Neonates with Respiratory Distress Syndrome: Ventilator-Induced Lung Injury and Oxidative Stress. *J. Immunol. Res.* **2018**, *2018*, 6963754. [CrossRef] [PubMed]
32. Hamid, E.R.A.; Ali, W.H.; Azmy, A.; Ahmed, H.H.; Sherif, L.S.; Saleh, M.T. Oxidative Stress and Anti-Oxidant Markers in Premature Infants with Respiratory Distress Syndrome. *Open Access Maced. J. Med. Sci.* **2019**, *7*, 2858–2863. [CrossRef]
33. Mutlu, B.; Aksoy, N.; Cakir, H.; Celik, H.; Erel, O. The effects of the mode of delivery on oxidative-antioxidative balance. *J. Matern. Fetal Neonatal Med.* **2011**, *24*, 1367–1370. [CrossRef]
34. Naples, R.; Ramaiah, S.; Rankin, J.; Berrington, J.; Harigopal, S. Life-threatening bronchopulmonary dysplasia: A British Paediatric Surveillance Unit Study. *Arch. Dis. Child. Fetal Neonatal Ed.* **2022**, *107*, 13–19. [CrossRef] [PubMed]
35. Jobe, A.H. Mechanisms of Lung Injury and Bronchopulmonary Dysplasia. *Am. J. Perinatol.* **2016**, *33*, 1076–1078. [CrossRef] [PubMed]
36. Darnall, R.A.; Chen, X.; Nemani, K.V.; Sirieix, C.M.; Gimi, B.; Knoblach, S.; McEntire, B.L.; Hunt, C.E. Early postnatal exposure to intermittent hypoxia in rodents is proinflammatory, impairs white matter integrity, and alters brain metabolism. *Pediatr. Res.* **2017**, *82*, 164–172. [CrossRef]
37. Di Fiore, J.M.; Vento, M. Intermittent hypoxemia and oxidative stress in preterm infants. *Respir. Physiol. Neurobiol.* **2019**, *266*, 121–129. [CrossRef]
38. Kotecha, S.; Chan, B.; Azam, N.; Silverman, M.; Shaw, R.J. Increase in interleukin-8 and soluble intercellular adhesion molecule-1 in bronchoalveolar lavage fluid from premature infants who develop chronic lung disease. *Arch. Dis. Child. Fetal Neonatal Ed.* **1995**, *72*, F90–F96. [CrossRef]

39. Hsiao, C.C.; Chang, J.C.; Tsao, L.Y.; Yang, R.C.; Chen, H.N.; Lee, C.H.; Lin, C.Y.; Tsai, Y.G. Correlates of Elevated Interleukin-6 and 8-Hydroxy-2′-Deoxyguanosine Levels in Tracheal Aspirates from Very Low Birth Weight Infants Who Develop Bronchopulmonary Dysplasia. *Pediatr. Neonatol.* **2017**, *58*, 63–69. [CrossRef]
40. Bednarczuk, N.; Williams, E.E.; Greenough, A.; Dassios, T. Carboxyhaemoglobin levels and free-radical-related diseases in prematurely born infants. *Early Hum. Dev.* **2022**, *164*, 105523. [CrossRef]
41. Dani, C.; Remaschi, G.; Monti, N.; Pratesi, S. Carboxyhemoglobin as biomarker of late-onset sepsis in preterm infants. *Eur. J. Pediatr.* **2023**, *182*, 4523–4528. [CrossRef] [PubMed]
42. Fairchild, K.D.; Nagraj, V.P.; Sullivan, B.A.; Moorman, J.R.; Lake, D.E. Oxygen desaturations in the early neonatal period predict development of bronchopulmonary dysplasia. *Pediatr. Res.* **2019**, *85*, 987–993. [CrossRef] [PubMed]
43. Raffay, T.M.; Dylag, A.M.; Sattar, A.; Abu Jawdeh, E.G.; Cao, S.; Pax, B.M.; Loparo, K.A.; Martin, R.J.; Di Fiore, J.M. Neonatal intermittent hypoxemia events are associated with diagnosis of bronchopulmonary dysplasia at 36 weeks postmenstrual age. *Pediatr. Res.* **2019**, *85*, 318–323. [CrossRef]
44. Jensen, E.A.; Whyte, R.K.; Schmidt, B.; Bassler, D.; Vain, N.E.; Roberts, R.S.; Canadian Oxygen Trial, I. Association between Intermittent Hypoxemia and Severe Bronchopulmonary Dysplasia in Preterm Infants. *Am. J. Respir. Crit. Care Med.* **2021**, *204*, 1192–1199. [CrossRef]
45. DeMauro, S.B. Neurodevelopmental outcomes of infants with bronchopulmonary dysplasia. *Pediatr. Pulmonol.* **2021**, *56*, 3509–3517. [CrossRef]
46. Ambalavanan, N.; Morty, R.E. Searching for better animal models of BPD: A perspective. *Am. J. Physiol. Lung Cell Mol. Physiol.* **2016**, *311*, L924–L927. [CrossRef]
47. Yee, M.; White, R.J.; Awad, H.A.; Bates, W.A.; McGrath-Morrow, S.A.; O'Reilly, M.A. Neonatal hyperoxia causes pulmonary vascular disease and shortens life span in aging mice. *Am. J. Pathol.* **2011**, *178*, 2601–2610. [CrossRef]
48. Davidson, L.M.; Berkelhamer, S.K. Bronchopulmonary Dysplasia: Chronic Lung Disease of Infancy and Long-Term Pulmonary Outcomes. *J. Clin. Med.* **2017**, *6*, 4. [CrossRef]
49. Deulofeut, R.; Critz, A.; Adams-Chapman, I.; Sola, A. Avoiding hyperoxia in infants < or = 1250 g is associated with improved short- and long-term outcomes. *J. Perinatol.* **2006**, *26*, 700–705.
50. Vento, M.; Moro, M.; Escrig, R.; Arruza, L.; Villar, G.; Izquierdo, I.; Roberts, L.J., 2nd; Arduini, A.; Escobar, J.J.; Sastre, J.; et al. Preterm resuscitation with low oxygen causes less oxidative stress, inflammation, and chronic lung disease. *Pediatrics* **2009**, *124*, e439–e449. [CrossRef] [PubMed]
51. Birenbaum, H.J.; Dentry, A.; Cirelli, J.; Helou, S.; Pane, M.A.; Starr, K.; Melick, C.F.; Updegraff, L.; Arnold, C.; Tamayo, A.; et al. Reduction in the incidence of chronic lung disease in very low birth weight infants: Results of a quality improvement process in a tertiary level neonatal intensive care unit. *Pediatrics* **2009**, *123*, 44–50. [CrossRef]
52. Lakshminrusimha, S.; Konduri, G.G.; Steinhorn, R.H. Considerations in the management of hypoxemic respiratory failure and persistent pulmonary hypertension in term and late preterm neonates. *J. Perinatol.* **2016**, *36* (Suppl. 2), S12–S19. [CrossRef]
53. Lakshminrusimha, S.; Russell, J.A.; Steinhorn, R.H.; Ryan, R.M.; Gugino, S.F.; Morin, F.C., 3rd; Swartz, D.D.; Kumar, V.H. Pulmonary arterial contractility in neonatal lambs increases with 100% oxygen resuscitation. *Pediatr. Res.* **2006**, *59*, 137–141. [CrossRef]
54. Lakshminrusimha, S.; Swartz, D.D.; Gugino, S.F.; Ma, C.X.; Wynn, K.A.; Ryan, R.M.; Russell, J.A.; Steinhorn, R.H. Oxygen concentration and pulmonary hemodynamics in newborn lambs with pulmonary hypertension. *Pediatr. Res.* **2009**, *66*, 539–544. [CrossRef]
55. Sheth, S.; Goto, L.; Bhandari, V.; Abraham, B.; Mowes, A. Factors associated with development of early and late pulmonary hypertension in preterm infants with bronchopulmonary dysplasia. *J. Perinatol.* **2020**, *40*, 138–148. [CrossRef] [PubMed]
56. Park, J.B.; Suh, M.; Park, J.Y.; Park, J.K.; Kim, Y.I.; Kim, H.; Cho, Y.S.; Kang, H.; Kim, K.; Choi, J.H.; et al. Assessment of Inflammation in Pulmonary Artery Hypertension by (68)Ga-Mannosylated Human Serum Albumin. *Am. J. Respir. Crit. Care Med.* **2020**, *201*, 95–106. [CrossRef] [PubMed]
57. Mankouski, A.; Kantores, C.; Wong, M.J.; Ivanovska, J.; Jain, A.; Benner, E.J.; Mason, S.N.; Tanswell, A.K.; Auten, R.L.; Jankov, R.P. Intermittent hypoxia during recovery from neonatal hyperoxic lung injury causes long-term impairment of alveolar development: A new rat model of BPD. *Am. J. Physiol. Lung Cell Mol. Physiol.* **2017**, *312*, L208–L216. [CrossRef]
58. Damianos, A.; Kulandavelu, S.; Chen, P.; Nwajei, P.; Batlahally, S.; Sharma, M.; Alvarez-Cubela, S.; Dominguez-Bendala, J.; Zambrano, R.; Huang, J.; et al. Neonatal intermittent hypoxia persistently impairs lung vascular development and induces long-term lung mitochondrial DNA damage. *J. Appl. Physiol.* **2022**, *133*, 1031–1041. [CrossRef] [PubMed]
59. Gentle, S.J.; Travers, C.P.; Nakhmani, A.; Indic, P.; Carlo, W.A.; Ambalavanan, N. Intermittent Hypoxemia and Bronchopulmonary Dysplasia with Pulmonary Hypertension in Preterm Infants. *Am. J. Respir. Crit. Care Med.* **2023**, *207*, 899–907. [CrossRef]

60. Darlow, B.A.; Graham, P.J. Vitamin A supplementation for preventing morbidity and mortality in very low birthweight infants. *Cochrane Database Syst. Rev.* **2000**, *4*, CD000501.
61. Rakshasbhuvankar, A.A.; Pillow, J.J.; Simmer, K.N.; Patole, S.K. Vitamin A supplementation in very-preterm or very-low-birth-weight infants to prevent morbidity and mortality: A systematic review and meta-analysis of randomized trials. *Am. J. Clin. Nutr.* **2021**, *114*, 2084–2096. [CrossRef]
62. Farahbakhsh, N.; Sharma, D.; Fatahi, S.; Fathi, M.; Vakili, K.; Deravi, N.; Tutunchian, Z.; Ahsan, E.; Yaghoobpoor, S.; Tabatabaii, S.A. Effects of Vitamin D and E Supplementation on Prevention of Bronchopulmonary Dysplasia (BPD) in Premature Neonates: A Systematic Review and Meta-Analysis. *Curr. Pediatr. Rev.* **2024**, *21*, 362–373. [CrossRef]
63. Gouhie, F.A.; Cezar, A.C.N.; Silva, D.A.; Parreira, B.R.; Machado, L.F.; de Freitas, L.C.F.; da Silva Lucena Patriota, P.R. Vitamin E for intraventricular hemorrhage prevention in preterm neonates: A systematic review and meta-analysis. *Neuroprotection* **2025**, *3*, 165–171. [CrossRef]
64. Brion, L.P.; Bell, E.F.; Raghuveer, T.S. Vitamin E supplementation for prevention of morbidity and mortality in preterm infants. *Cochrane Database Syst. Rev.* **2003**, *4*, CD003665. [CrossRef] [PubMed]
65. Streeter, M.L.; Rosso, P. Transport mechanisms for ascorbic acid in the human placenta. *Am. J. Clin. Nutr.* **1981**, *34*, 1706–1711. [CrossRef]
66. Carr, A.; Frei, B. Does vitamin C act as a pro-oxidant under physiological conditions? *FASEB J.* **1999**, *13*, 1007–1024. [CrossRef]
67. Silvers, K.M.; Gibson, A.T.; Russell, J.M.; Powers, H.J. Antioxidant activity, packed cell transfusions, and outcome in premature infants. *Arch. Dis. Child. Fetal Neonatal Ed.* **1998**, *78*, F214–F219. [CrossRef]
68. Darlow, B.A.; Buss, H.; McGill, F.; Fletcher, L.; Graham, P.; Winterbourn, C.C. Vitamin C supplementation in very preterm infants: A randomised controlled trial. *Arch. Dis. Child. Fetal Neonatal Ed.* **2005**, *90*, F117–F122. [CrossRef] [PubMed]
69. Bronnert, A.; Bloomfield, P.M.; Paramo, L.D.; Lin, L.; Bloomfield, F.H.; Cormack, B.E. The effect of vitamin supplementation on neurodevelopmental and clinical outcomes in very low birth weight and very preterm infants: A systematic review and meta-analysis. *PLoS ONE* **2025**, *20*, e0327628. [CrossRef] [PubMed]
70. Murphy, V.E.; Jensen, M.E.; Harvey, S.; Beyene, T.; Gregson, J.; Islam, F.; Huang, W.; Aistrope, K.; Collison, A. Vitamin A, C and/or E Intake During Pregnancy and Offspring Respiratory Health: A Systematic Review and Meta-Analysis. *J. Hum. Nutr. Diet.* **2025**, *38*, e70086. [CrossRef]
71. Roberts, K.; Stepanovich, G.; Bhatt-Mehta, V.; Donn, S.M. New Pharmacologic Approaches to Bronchopulmonary Dysplasia. *J. Exp. Pharmacol.* **2021**, *13*, 377–396. [CrossRef]
72. Farrow, K.N.; Lakshminrusimha, S.; Czech, L.; Groh, B.S.; Gugino, S.F.; Davis, J.M.; Russell, J.A.; Steinhorn, R.H. SOD and inhaled nitric oxide normalize phosphodiesterase 5 expression and activity in neonatal lambs with persistent pulmonary hypertension. *Am. J. Physiol. Lung Cell Mol. Physiol.* **2010**, *299*, L109–L116. [CrossRef]
73. Farrow, K.N.; Lakshminrusimha, S.; Reda, W.J.; Wedgwood, S.; Czech, L.; Gugino, S.F.; Davis, J.M.; Russell, J.A.; Steinhorn, R.H. Superoxide dismutase restores eNOS expression and function in resistance pulmonary arteries from neonatal lambs with persistent pulmonary hypertension. *Am. J. Physiol. Lung Cell Mol. Physiol.* **2008**, *295*, L979–L987. [CrossRef] [PubMed]
74. Davis, J.M.; Parad, R.B.; Michele, T.; Allred, E.; Price, A.; Rosenfeld, W.; North American Recombinant Human CuZn SODSG. Pulmonary outcome at 1 year corrected age in premature infants treated at birth with recombinant human CuZn superoxide dismutase. *Pediatrics* **2003**, *111*, 469–476. [CrossRef] [PubMed]
75. Yang, X.; Jiang, S.; Deng, X.; Luo, Z.; Chen, A.; Yu, R. Effects of Antioxidants in Human Milk on Bronchopulmonary Dysplasia Prevention and Treatment: A Review. *Front. Nutr.* **2022**, *9*, 924036. [CrossRef]
76. Chrustek, A.; Dombrowska-Pali, A.; Olszewska-Slonina, D. Analysis of the composition and antioxidant status of breast milk in women giving birth prematurely and on time. *PLoS ONE* **2021**, *16*, e0255252. [CrossRef]
77. Villamor-Martinez, E.; Pierro, M.; Cavallaro, G.; Mosca, F.; Villamor, E. Mother's Own Milk and Bronchopulmonary Dysplasia: A Systematic Review and Meta-Analysis. *Front. Pediatr.* **2019**, *7*, 224. [CrossRef] [PubMed]
78. Mathias, M.; Chang, J.L.; Perez, M.; Saugstad, O. Supplemental Oxygen in the Newborn: Historical Perspective and Current Trends. *Antioxidants* **2021**, *10*, 1879. [CrossRef]
79. Gitto, E.; Reiter, R.J.; Sabatino, G.; Buonocore, G.; Romeo, C.; Gitto, P.; Bugge, C.; Trimarchi, G.; Barberi, I. Correlation among cytokines, bronchopulmonary dysplasia and modality of ventilation in preterm newborns: Improvement with melatonin treatment. *J. Pineal Res.* **2005**, *39*, 287–293. [CrossRef]
80. Forman, H.J.; Zhang, H. Targeting oxidative stress in disease: Promise and limitations of antioxidant therapy. *Nat. Rev. Drug Discov.* **2021**, *20*, 689–709. [CrossRef]
81. Dani, C. Automated control of inspired oxygen (FiO_2) in preterm infants: Literature review. *Pediatr. Pulmonol.* **2019**, *54*, 358–363. [CrossRef]

82. Sturrock, S.; Ambulkar, H.; Williams, E.E.; Sweeney, S.; Bednarczuk, N.F.; Dassios, T.; Greenough, A. A randomised crossover trial of closed loop automated oxygen control in preterm, ventilated infants. *Acta Paediatr.* **2021**, *110*, 833–837. [CrossRef]
83. Abdo, M.; Hanbal, A.; Asla, M.M.; Ishqair, A.; Alfar, M.; Elnaiem, W.; Ragab, K.M.; Nourelden, A.Z.; Zaazouee, M.S. Automated versus manual oxygen control in preterm infants receiving respiratory support: A systematic review and meta-analysis. *J. Matern. Fetal Neonatal Med.* **2022**, *35*, 6069–6076. [CrossRef] [PubMed]
84. Mitra, S.; Singh, B.; El-Naggar, W.; McMillan, D.D. Automated versus manual control of inspired oxygen to target oxygen saturation in preterm infants: A systematic review and meta-analysis. *J. Perinatol.* **2018**, *38*, 351–360. [CrossRef] [PubMed]
85. Stafford, I.G.; Lai, N.M.; Tan, K. Automated oxygen delivery for preterm infants with respiratory dysfunction. *Cochrane Database Syst. Rev.* **2023**, *11*, CD013294. [CrossRef]
86. Sturrock, S.; Williams, E.; Dassios, T.; Greenough, A. Closed loop automated oxygen control in neonates—A review. *Acta Paediatr.* **2020**, *109*, 914–922. [CrossRef]
87. van Kaam, A.H.; Hummler, H.D.; Wilinska, M.; Swietlinski, J.; Lal, M.K.; te Pas, A.B.; Lista, G.; Gupta, S.; Fajardo, C.A.; Onland, W.; et al. Automated versus Manual Oxygen Control with Different Saturation Targets and Modes of Respiratory Support in Preterm Infants. *J. Pediatr.* **2015**, *167*, 545–550. [CrossRef] [PubMed]
88. Reynolds, P.R.; Miller, T.L.; Volakis, L.I.; Holland, N.; Dungan, G.C.; Roehr, C.C.; Ives, K. Randomised cross-over study of automated oxygen control for preterm infants receiving nasal high flow. *Arch. Dis. Child. Fetal Neonatal Ed.* **2019**, *104*, F366–F371. [CrossRef]
89. Claure, N.; Bancalari, E.; D'Ugard, C.; Nelin, L.; Stein, M.; Ramanathan, R.; Hernandez, R.; Donn, S.M.; Becker, M.; Bachman, T. Multicenter crossover study of automated control of inspired oxygen in ventilated preterm infants. *Pediatrics* **2011**, *127*, e76–e83. [CrossRef]
90. Kaltsogianni, O.; Dassios, T.; Jenkinson, A.; Jeffreys, E.; Ikeda, K.; Sugino, M.; Greenough, A. Closed-loop automated oxygen control in preterm ventilated infants: A randomised controlled trial. *Arch. Dis. Child. Fetal Neonatal Ed.* 2025, *in press*.
91. Ofman, G.; Tipple, T.E. Antioxidants & bronchopulmonary dysplasia: Beating the system or beating a dead horse? *Free Radic. Biol. Med.* **2019**, *142*, 138–145. [PubMed]
92. Sampath, V.; Garland, J.S.; Helbling, D.; Dimmock, D.; Mulrooney, N.P.; Simpson, P.M.; Murray, J.C.; Dagle, J.M. Antioxidant response genes sequence variants and BPD susceptibility in VLBW infants. *Pediatr. Res.* **2015**, *77*, 477–483. [CrossRef] [PubMed]

antioxidants

MDPI

Review

Does the Maternal Gut Microbiome Influence the Outcome of Perinatal Asphyxia?

Vlad-Petru Morozan [1,2,*], Mara I. Ionescu [1,3,*], Carmen M. D. Zahiu [1], Ana Maria Catrina [4], Andreea Racoviță [5], Ana-Teodora Chirilă [1], Ioana-Alexandra Dogaru [1], Cristian Ciotei [1], Gratiela Gradisteanu Pircalabioru [6,7,8] and Ana-Maria Zăgrean [1]

1 Division of Physiology—Neuroscience, Carol Davila University of Medicine and Pharmacy, 02001 Bucharest, Romania; carmen.zahiu@umfcd.ro (C.M.D.Z.); ana-teodora.chirila0721@stud.umfcd.ro (A.-T.C.); ioana-alexandra.dogaru0125@rez.umfcd.ro (I.-A.D.); cristian.ciotei0720@stud.umfcd.ro (C.C.); ana-maria.zagrean@umfcd.ro (A.-M.Z.)

2 Department of Plastic and Reconstructive Microsurgery, Central Military Emergency University Hospital 'Dr. Carol Davila', 010825 Bucharest, Romania

3 Department of Pediatrics II, Marie Curie Emergency Children's Hospital, 75534 Bucharest, Romania

4 Cantacuzino National Military Medical Institute for Research and Development, 050096 Cernica, Romania; catrina.ana-maria@cantacuzino.ro

5 Department of Dermatology, Elias University Hospital, 011461 Bucharest, Romania; andreea-stefania.racovita@rez.umfcd.ro

6 Research Institute of the University of Bucharest, 050663 Bucharest, Romania; gratiela.gradisteanu@icub.unibuc.ro

7 eBio-Hub Centre of Excellence in Bioengineering, National University of Science and Technology Politehnica Bucharest, 061344 Bucharest, Romania

8 Department of Botany and Microbiology, University of Bucharest, 060101 Bucharest, Romania

* Correspondence: vlad-petru.morozan@drd.umfcd.ro (V.-P.M.); mara-ioana.ionescu@umfcd.ro (M.I.I.)

Abstract: This review explores the maternal gut microbiome's role in shaping neonatal neurodevelopmental outcomes following perinatal asphyxia (PA), a leading cause of infant mortality and disability with limited therapeutic options beyond hypothermia. We synthesized current evidence on microbiome-mediated neuroprotective mechanisms against hypoxic-ischemic brain injury. The maternal microbiome influences fetal development through bioactive metabolites (short-chain fatty acids, indole derivatives) that cross the placental barrier, bacterial antigen regulation, and infant microbiome colonization. These pathways activate multiple protective mechanisms: anti-inflammatory signaling via NF-κB suppression and regulatory T cell expansion; antioxidant defenses through Nrf2 activation; neural repair via BDNF upregulation and neurogenesis; and oxytocin system modulation. Animal models demonstrate that maternal dysbiosis from high-fat diet or antibiotics exacerbates PA-induced brain damage, increasing inflammatory markers and hippocampal injury. Conversely, probiotic supplementation, dietary fiber, and specific interventions (omega-3, resveratrol) reduce neuroinflammation and oxidative injury. Human studies link maternal dysbiosis-associated conditions (obesity, gestational diabetes) with adverse pregnancy outcomes, though direct clinical evidence for PA severity remains limited. Understanding the maternal microbiome-fetal brain axis opens therapeutic avenues, including prenatal probiotics, dietary modifications, and targeted metabolite supplementation to prevent or mitigate PA-related neurological sequelae, potentially complementing existing neuroprotective strategies.

Keywords: maternal microbiome; gut-brain axis; perinatal asphyxia; oxidative stress; neuroinflammation; neuroprotection; Nrf2 pathway; short-chain fatty acids; ferroptosis; blood-brain barrier

1. Introduction

Perinatal Asphyxia (PA) is a medical condition in newborns caused by oxygen deprivation and hypercapnia that lasts sufficiently long to cause physical damage, particularly to the nervous system. By accounting for about half of neonatal encephalopathies in term and near-term infants, PA inflicts significant neurodevelopmental disability and ranks as the second-leading contributor to neonatal mortality [1–3].

Despite its high prevalence, therapeutic interventions for PA remain limited, with most offering only partial benefit [4]. While hypothermia remains the sole established treatment in clinical practice, its modest effectiveness (one out of seven treated infants avoids death or major disability) and narrow therapeutic window following birth [5,6] call for novel strategies to mitigate the consequences of hypoxic-ischemic events. Emerging evidence suggests that the maternal microbiome may be a novel and beneficial tool to address these therapeutic gaps [7–9].

The maternal microbiome influences foetal brain development, particularly through influencing inflammation and oxidative stress, two key drivers of hypoxic-ischemic injury in PA [7,9]. A balanced maternal microbiota creates a potent anti-inflammatory and antioxidative environment that regulates maternal immune activation and modulates microglial activity in the foetus, guiding adequate foetal neurodevelopment [10–13]. This protective milieu can buffer the developing brain against the detrimental effects of oxygen deprivation, while enhancing its neural repair capacity [10,11].

Conversely, maternal microbiota impairment, known as dysbiosis, has been linked to heightened pro-inflammatory states and disrupted metabolic signalling, exacerbating the severity of hypoxic-ischemic damage [14]. Given these findings, this narrative review aims to examine the mechanistic links and current evidence regarding the role of the maternal microbiota in PA outcomes.

2. Perinatal Asphyxia: Pathophysiology and Outcomes

To understand how the maternal microbiota might influence PA outcomes, it is essential to consider the pathophysiological cascade underlying this condition.

Classically, PA undergoes three distinct phases interweaved with a latent period [15] The primary phase represents the injury itself, where low oxygen and high carbon dioxide levels trigger the chemo-reflexes, which further induce bradycardia and intense peripheral vasoconstriction to maintain perfusion of central organs [16,17]. Hypoxemia increases inflammation and oxidative stress, causes cell oedema and amplifies cell injury, increasing cellular death [18,19]. These neuronal lesions cause waves of anoxic depolarisation, which provide positive feedback in this loop and further aggravate the injuries [20,21].

Following moderate or severe hypoxic injury, a latent and apparently quiescent phase appears [22]. The higher the degree of injury, the shorter the latent period, after which the secondary phase commences with a restart of the inflammatory cascades and a continuation of the cytokine storm [5]. During this phase, trophic factors decrease, the amount of reactive oxygen species (ROS) and the mitochondrial membrane permeability increase, and intracellular calcium concentration rises, all creating a self-sustaining destructive landscape [21,23,24]. This landscape facilitates a pro-excitatory environment, which, combined with the increased glutamate resulting from neuronal loss, leads to massive depolarisation waves and seizure activity that further exacerbate neuronal loss [25–27].

Approximately three days after the hypoxic injury, the third phase begins [28]. Depending on the injury severity, it either tilts toward a reparatory phase, with the regeneration of damaged tissues, or, conversely, tilts toward continuing the inflammatory processes and aggravating the existing lesions [29–32].

PA produces different patterns of neurological injuries, depending on the severity and duration of oxygen deprivation, as well as the infant's gestational age [5,33]. In preterm neonates, PA leads to a form of white matter injury called periventricular leukomalacia (PVL), which involves oligodendrocyte death and hindered myelination [34–37]. PVL presents with two distinct forms: cystic and diffuse [34,35]. In full-term neonates, acute sentinel PA damages the grey matter in the thalamus, basal ganglia, and brainstem, while repeated moderate hypoxic events lead to a "watershed" pattern, primarily affecting cortical regions [38,39]. However, depending on the context, neonates can also present multiple patterns simultaneously [5,38,40].

These lesions accompanying PA may lead to outcomes ranging from mild symptoms to neonatal hypoxic-ischemic encephalopathy (HIE), with severe hypoxic events potentially resulting in death [41]. Unfortunately, the neuronal damage often comes with dramatic long-term consequences: cerebral palsy, intellectual disability, psychiatric disorders, epilepsy, neurodevelopmental disorders (e.g., ADHD), learning disabilities, motor disabilities, etc., thus posing a great burden on the individual, as well as on society [40–48].

Although extensive work has focused on the pathophysiological cascades triggered by PA and the efficacy of therapeutic hypothermia and pharmacotherapy, important gaps remain in understanding why some newborns are more susceptible to brain damage than others [5,33]. The severity and persistence of PA outcomes underscore the need to investigate additional prenatal factors. In this regard, the maternal microbiome, long overlooked, has surfaced as a key player capable of influencing foetal health through immune regulation and metabolic pathways [7,9,49].

3. The Maternal Gut Microbiome: Composition, Functions, and Impact on Foetal Development

The maternal microbiota represents a complex consortium of microorganisms colonizing various maternal body sites, including the gastrointestinal tract, conjunctiva, airways, vagina, breast tissue, skin, and oral cavity [50]. Among these niches, the gut microbiota is the most extensively studied and undergoes significant shifts during pregnancy, driven by hormonal, metabolic, and immunological changes [51,52]. These adaptations promote enhanced energy storage and immune modulation [14,53].

In the gut, *Firmicutes* (split into several groups in the Genome Taxonomy Database) and *Bacteroidetes* (*Bacteroidota*) phyla comprise approximately 90% of the total bacterial population, with *Actinobacteria* (*Actinomycetota*), *Proteobacteria* (*Pseudomonadota*), *Verrucomicrobia*, and archaeal members such as *Euryarchaeota* accounting for the remaining 10% [51,54]. A balanced, eubiotic microbiota is characterized by diverse species, a beneficial ratio of key taxa, and vital functional capabilities, such as short-chain fatty acids (SCFAs) production, immune system education, maintenance of epithelial barrier integrity, vitamin and nutrient synthesis, and competitive exclusion of pathogens [55,56].

Despite this complexity, deviations from a balanced microbiome or dysbiosis can be categorized into three main forms: type 1, with the loss of beneficial microbes (e.g., *Lactobacillus* in the vaginal microbiome, *Bifidobacterium* in the gut); type 2, with an overgrowth of potentially harmful species (e.g., increased *Proteobacteria* during pregnancy) and type 3, with an overall reduction in microbial diversity, leading to compromised functional capacity and resilience [56].

3.1. Factors Influencing Microbiome Composition

Multiple factors influence the maternal microbiome during pregnancy, leading to unique microbial signatures that can impact both maternal and foetal health (Table 1).

These influences create a dynamic, ever-shifting microbial landscape in the mother, which, in turn, shapes foetal development, systemic inflammation, the mother's health, and the maturation of the foetal gut–brain axis [10–14,57–63].

Table 1. Factors Influencing Maternal Microbiome Composition During Pregnancy.

Factor	Effect on Microbiome	Impact Type	References
High-fat diet	↓ Bacteroidetes, ↑ Firmicutes, ↑ pro-inflammatory taxa	Dysbiotic	[64,65]
High-fiber diet	↑ short-chain fatty acids (SCFA)-producers, ↑ Bifidobacterium, ↑ diversity	Beneficial	[66–68]
Alcohol consumption	↓ beneficial bacteria, altered metabolic pathways	Dysbiotic	[69]
Antibiotic use	↓ diversity, loss of key commensals, ↑ resistant strains	Dysbiotic	[63,70]
Probiotic supplementation	↑ specific beneficial strains, ↑ barrier function	Beneficial	[71]
Iron supplementation	Altered Bacteroidetes to Firmicutes ratio, ↑ pathobionts	Variable	[72]
Elevated body mass index (BMI)/Obesity	↓ diversity, ↑ Firmicutes, altered SCFA production	Dysbiotic	[73,74]
Insulin resistance	Altered glucose metabolism pathways, ↓ butyrate producers	Dysbiotic	[75]
Pre-existing gastrointestinalconditions: inflammatory bowel disease (IBD)	↓ diversity, ↑ Proteobacteria, ↑ inflammatory markers	Dysbiotic	[76]
Smoking	↓ beneficial anaerobes, ↑ opportunistic pathogens	Dysbiotic	[77,78]
Regular exercise	↑ diversity, ↑ SCFA production, ↑ Akkermansia	Beneficial	[79,80]
Psychological stress	↑ Proteobacteria, ↓ Lactobacillus, altered barrier function	Dysbiotic	[13,81]
Pregnancy progression	↑ Proteobacteria (with pregnancy progression), ↓ diversity (adaptive)	Adaptive	[51,52,82]
Previous pregnancies	Enhanced microbial stability, faster adaptation	Beneficial	[83]
Maternal age	Age-dependent diversity changes, altered metabolic capacity	Variable	[51]
Residential environment	Urban vs. rural differences in diversity and composition	Variable	[51]

↑ increased/higher levels; ↓ decreased/lower levels.

3.2. Maternal Health Implications

Maternal gut microbiome shapes maternal health throughout pregnancy (Figure 1). Dysregulated maternal gut communities are associated with elevated proinflammatory cytokines, altered oestrogen metabolism, and impaired metabolic health [84,85]. Gestational dysbiosis is associated with conditions such as preeclampsia and hypertension [13,86,87]. Probiotic administration decreases blood glucose levels, lowers gestational diabetes markers, reduces insulin levels, and improves glucose tolerance; however, some studies argue for minimal or no probiotic effect [86,88].

Beyond maternal metabolic health, gut microbes can modulate the maternal immune system by influencing populations of granulocytes and regulatory T cells (T_{reg}) [13]. Diets rich in fibre, known to improve the gut microbiota, are associated with lower proportions of proinflammatory immune cells. At the same time, dysbiosis increases M1 macrophages and proinflammatory cytokines (e.g., TNF-α, IL-1β, IL-6, IL-17), and reduces anti-inflammatory cytokines such as IL-10 [13]. Moreover, dysbiosis can lead to maternal immune activation, with important consequences for both maternal and foetal health [89]. Some studies conclude that dysbiosis increases the risk of preterm birth; however, the evidence remains conflicting [90–93].

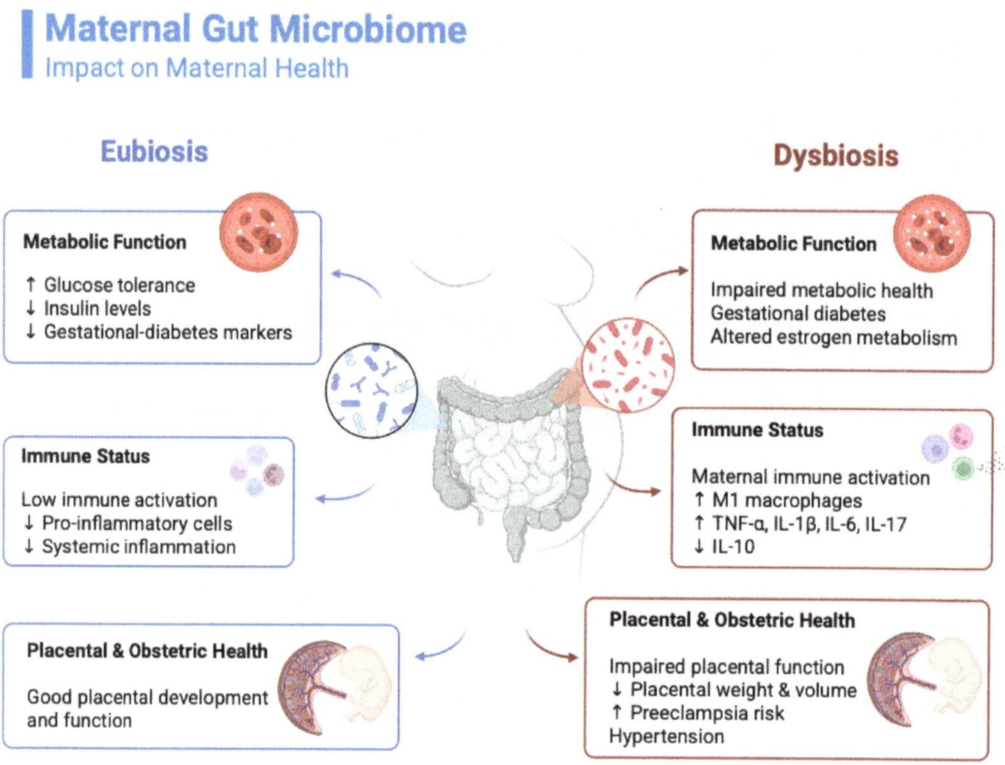

Figure 1. Influence of the maternal gut microbiome on the metabolic function, immune, and placental health. Left: A balanced maternal gut microbiome (eubiosis) is associated with improved glucose tolerance, reduced insulin levels, and lower markers of gestational diabetes. It is characterized by low systemic inflammation, with decreased pro-inflammatory cytokines and controlled populations of granulocytes and T-regulatory cells, as well as optimal placental development. Right: In contrast, dysbiosis is associated with impaired metabolic function (gestational diabetes, insulin resistance, altered oestrogen metabolism), heightened immune activation (elevated levels of M1 macrophages, Tumour Necrosis Factor alpha (TNF-α), Interleukin-1 beta (IL-1β), Interleukin-6 (IL-6), Interleukin-17 (IL-17); reduced levels of Interleukin-10 (IL-10); maternal immune activation), and compromised placental function (lower placental weight and volume; increased risk of preeclampsia and hypertension). Created in BioRender. Racovita, A. (2025) https://BioRender.com/75vubaj.

Maternal microbiota influences placental development, and dysbiosis is accompanied by lower placental weight and volume, as well as impaired maternal-foetal circulation interfaces [94]. These conditions are attenuated by SCFAs administration [94] and certain bacterial species (e.g., *Bifidobacterium breve*) [95]. Although a wave of scientific discoveries pinpointed that the placenta might present a microbiome of its own, scientific consensus tilts towards no microbiome [73,94–98].

3.3. Direct Effects on Foetal Development

Aside from maternal health, pregnancy well-being, and placental development, the maternal gut microbiota influences the development of foetal immune and nervous systems, and the infant's future metabolic health (Figure 2) [86,88].

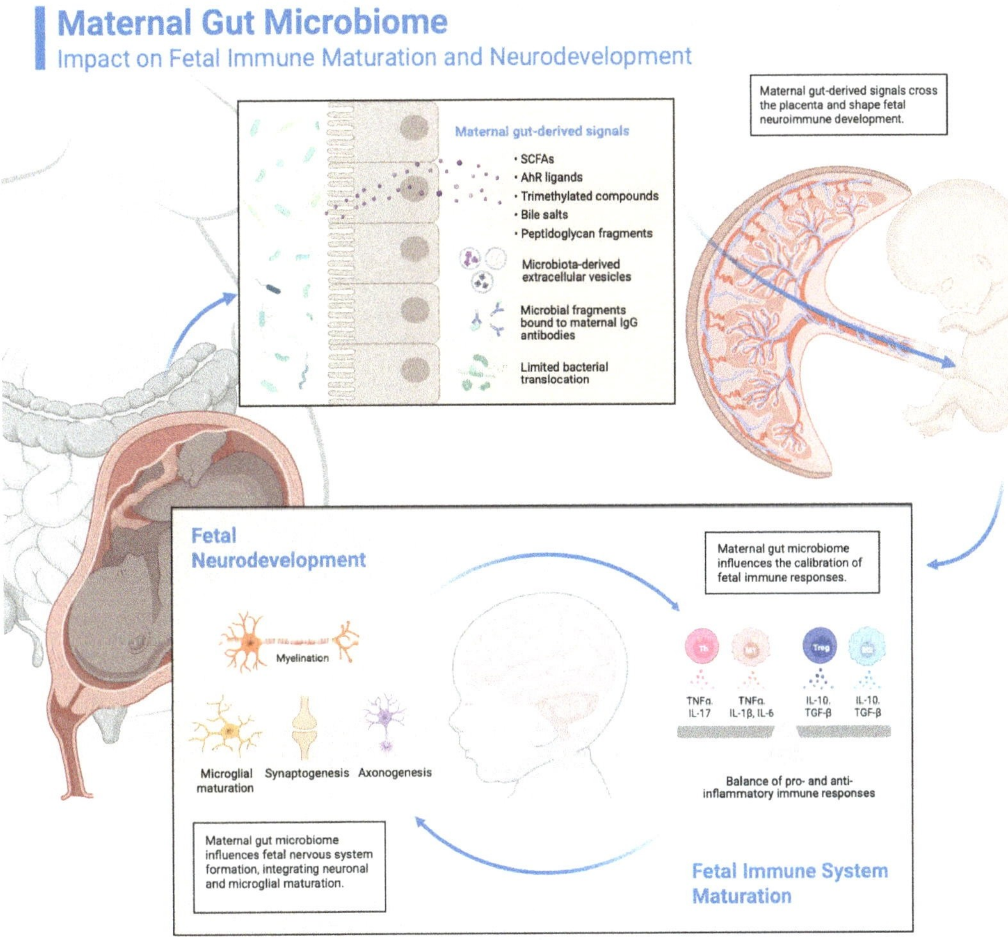

Figure 2. Impact of the maternal gut microbiome-derived signals on fetal immune and neurodevelopment. Bioactive microbial metabolites, namely short-chain fatty acids (SCFAs), aryl hydrocarbon receptor ligands, bile salts, trimethylated compounds, and peptidoglycan fragments, together with IgG-bound microbial antigens, microbiota-derived extracellular vesicles, and trace amounts of bacteria, cross the placental barrier and modulate fetal neuroimmune development. These signals regulate the balance between pro-inflammatory cytokines (TNF-α, IL-17, IL-1β, IL-6) and anti-inflammatory cytokines (IL-10, TGF-β), promoting microglial maturation, synaptogenesis, axonogenesis, and myelination. Created in BioRender. Racovita, A. (2025) https://BioRender.com/ye58bcx.

3.3.1. Immune System Maturation

From in utero to early childhood, the immune system undergoes a set of fine-tuned, sequential developmental processes [99] that evolve in parallel with the maternal gut microbiota, engaging in a bidirectional interplay [100] that shapes the immune cell populations. Dysbiosis can alter normal immune system development and can shift the balance either towards more immune reactivity with increased chances to develop autoimmune diseases or atopy, or towards a less reactive immune system, incapable of defending itself against threats [13,100–107]. For example, a maternal high-fat diet, known to induce dysbiosis, increases the expression of receptors that compromise the intestinal mucus barrier (e.g., LRRC19) [108], leading to increased bacterial translocation and heightened inflammatory state. Also, it upregulates proinflammatory RORγt-positive cells (precursors of IL-17-producing cells) [58,109] and elevates the levels of TNF-α and IL-6 [110]. In contrast, probiotics and prebiotic administration push the immune balance toward a more tolerogenic state, marked by an increase in regulatory T cells (T_{regs}) and in IL-10 and TGF-β levels [111,112].

The mechanisms underlying immune system maturation involve four primary pathways: (1) microbial metabolites—SCFAs [100,107,113–116], aryl hydrocarbon receptor (AhR) ligands [10], bile salts [117], peptidoglycan fragments [118], trimethylated microbial metabolites [10], (2) IgG-bound microbial antigens [119], (3) microbiota derived extracellular vesicles [120] and (4) bacterial translocation [58].

Immune system maturation, although an autonomous process, is closely intertwined with neurodevelopment [121,122], each influencing the other's course [123].

3.3.2. Neurodevelopmental Processes

Maternal microbial communities and their metabolites impact the fetal neural circuits' development, such as thalamocortical axons [57]. Germ-free animal models revealed altered gene expression related to myelination and axon development, highlighting the pivotal role of microbiota in neurodevelopment [10]. In addition, differentiation of GABAergic neurons appears to depend on AhR signaling [124], which is activated by microbial metabolites in the foetus.

Recently, the role of the maternal gut-fetal brain axis in fetal neurodevelopment has been increasingly considered [14]. Dysbiotic conditions lead to maternal immune activation, which has been linked to autism-like phenotypes in offspring, via IL-17-producing Th17 cells [63,89,125–127]. In addition, specific bacterial taxa (e.g., *Fusobacteria*) in the maternal gut have also been associated with better fine motor outcomes in children [128] and enhanced neurogenesis [129].

Maternal diet, which is known to impact microbiome, can either exacerbate [130] or mitigate [131] long-term neurodevelopmental deficits. Antibiotic use during gestation, known to induce dysbiosis, alters hippocampal development and emotional regulation in offspring [9,70], suggesting that the maternal microbiome profoundly impacts fetal brain health and development.

4. Neuroprotective Mechanisms Mediated by the Maternal Microbiome

The response of the immature brain to metabolic deprivation conditions, such as hypoxia/ischemia, involves mechanisms still to be revealed, which could explain why some newborns walk away from the same hypoxic insult while others live with lifelong disability [132–135]. The maternal microbiome could be an important factor influencing this response: dysbiosis magnifies cytokine-driven injury [136], while eubiosis, by promoting anti-inflammatory mechanisms, redox buffering, and reparatory signals (Figure 3), blunts the secondary wave of neuronal loss that could determine whether recovery or lifelong disability prevails [137].

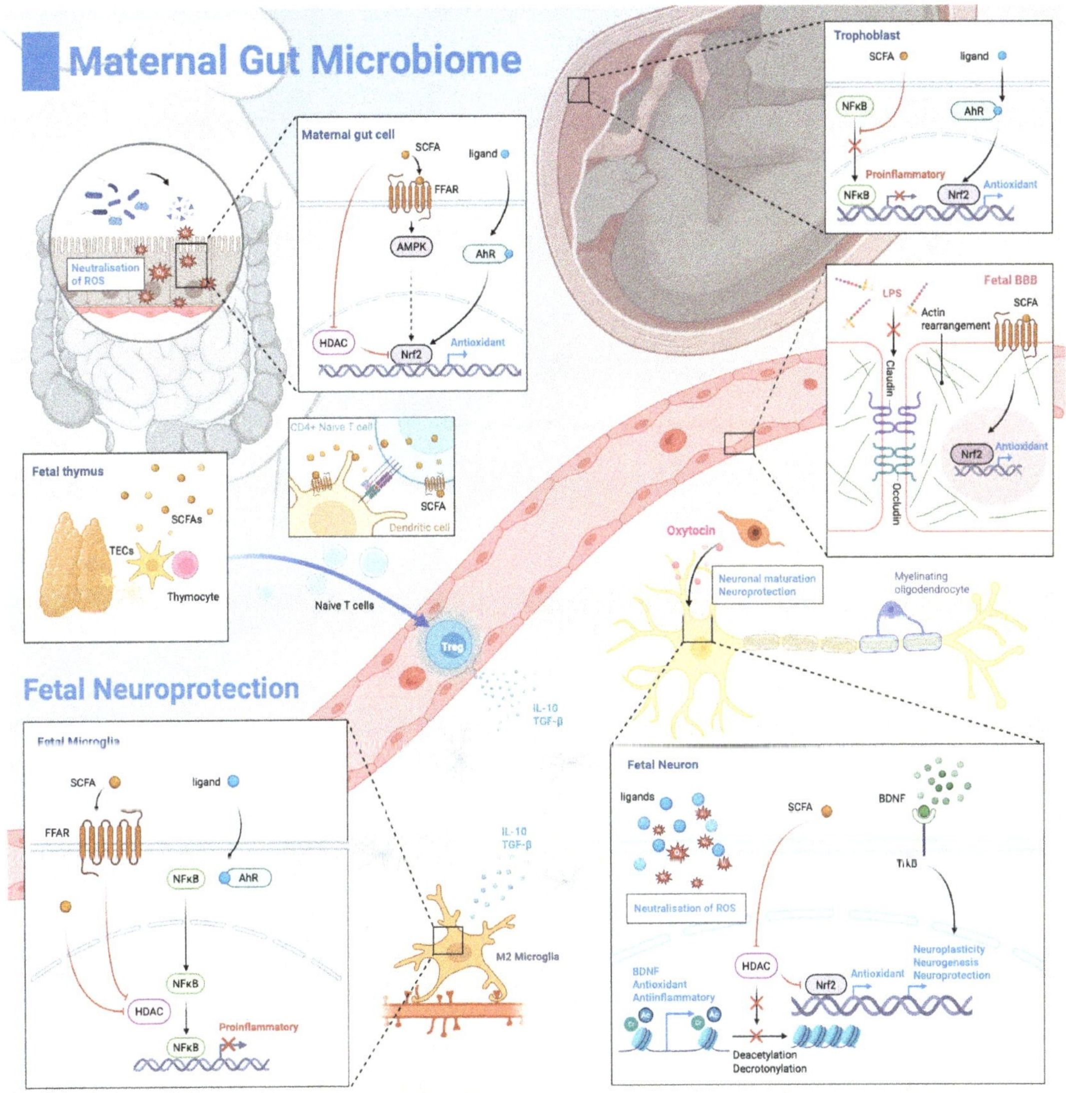

Figure 3. Neuroprotective mechanisms mediated by the maternal gut microbiome. Maternal gut microbiome-derived metabolites confer neuroprotection against perinatal hypoxic-ischemic injury through four integrated pathways spanning the maternal gut-placental-fetal brain axis. **(1)** Anti-inflammatory mechanisms: Short-chain fatty acids (SCFAs) inhibit histone deacetylase (HDAC) activity, promote M2 microglial polarization, reduce neuroinflammation, and strengthen gut-placental-blood-brain barriers. SCFAs expand regulatory T cells (T_{regs}) by priming thymic epithelial cells (TECs) to select T_{reg} precursor thymocytes, dendritic cells to promote differentiation into T_{regs}, and, consequently, suppress inflammatory cascades. Tryptophan-derived indole metabolites activate aryl hydrocarbon receptor (AhR), suppressing NF-κB p65 phosphorylation in microglia and decreasing

IL-1β and TNF-α expression; **(2)** Antioxidant defense: eubiosis prevents ROS leaks into the circulation and contributes to upregulation of junction proteins. At the placental level, microbiota-derived metabolites directly neutralise ROS and lower oxidative stress. Through FFAR activation and Nrf2 signaling, the antioxidant capacity of the fetal BBB is increased, leading to better integrity and protection against ROS activity. In neurons, by HDAC inhibition, SCFAs induce Nrf2 activation that enhances antioxidant enzyme activity and lowers ferroptosis. AhR ligands also directly buffer some of the ROS species. Consecutive to NF-κB inhibition, M2 microglia produce less ROS, mitigating the oxidative stress; **(3)** Neural repair and neurogenesis: Microbiome metabolites maintain brain-derived neurotrophic factor (BDNF) expression through H3 crotonylation via HDAC inhibition. This promotes neurogenesis, oligodendrocyte maturation, and protective glial responses that enhance tissue preservation, myelination, and adequate pruning during the recovery phase; **(4)** Oxytocin system modulation: Specific maternal gut bacteria (particularly *Lactobacillus reuteri*) enhance hypothalamic oxytocin expression. During hypoxic stress, oxytocin facilitates the critical gamma-aminobutyric acid (GABA) switch from excitatory to inhibitory signaling, reducing excitotoxicity and providing direct neuroprotection against neuronal loss. Created in BioRender. Racovita, A. (2025) https://BioRender.com/wc5q5uf.

4.1. Anti-Inflammatory Mechanisms

Within hours following a hypoxic-ischemic insult, a secondary cytokine storm unleashes, with increased concentrations of IL-1β, TNF-α, and IL-6. These cytokines amplify the initial lesion size, and their levels hold predictive power for outcome forecast [138–141].

Maternal gut metabolites, mainly SCFAs and indole derivatives, can dampen that cytokine surge at several, hierarchically layered checkpoints: (i) epigenetic silencing of NF-κB inside microglia [142], (ii) tightening of gut-placenta-blood/cerebrospinal fluid barriers that stop endotoxin leakage [143,144], and (iii) prenatal expansion of T_{regs} that pre-set a tolerogenic baseline [145].

SCFAs inhibit the histone deacetylase (HDAC), which reduces the expression of inflammatory cytokines [146] and prevents NF-κB translocation, both directly, as well as through HDAC inhibition [147], thus steering microglia towards an Arg-1/IL-10 rich M2 anti-inflammatory phenotype [142,148]. In a middle cerebral artery occlusion stroke model in adult mice, butyrate, a key SCFA, almost halved infarcted volume when administered alone [149]. Moreover, in a model of neuronal damage due to cardiac arrest, butyrate attenuated the microglial inflammation through the NF-κB pathway [150] and reduced brain inflammation through activation of the JAK-STAT pathway [151]. Studies both in the adult and neonate found that the microglia respond similarly to SCFAs, by reducing NF-κB translocation, thus providing an anti-inflammatory and protective response to hypoxia [152].

SCFAs bind G-protein-coupled receptors (GPCRs) in brain endothelial and intestinal cells, providing a second layer of protection. GPCR activation increases the expression of claudin 5 and occludin in both cell types, thus tightening the barriers and limiting endotoxin accumulation in the brain [143,153]. Furthermore, SCFAs induce actin cytoskeleton rearrangements in a blood-brain barrier (BBB) in vitro model, contributing to improved BBB integrity [144]. A structurally sound BBB promotes a lower microglial activation, thus limiting the initial lesions and inflammation [154,155].

In parallel, SCFAs boost T_{reg} cells' expansion [113,156] in the thymic and placental tissues, which lowers the fetal IL-6/IL-1β basal levels and increases the anti-inflammatory cytokine IL-10 production. SCFAs may exert these effects by directly inhibiting HDACs in naïve T cells, activating free fatty acid receptors (FFARs) on both dendritic cells and naïve T cells, and priming thymic epithelial cells to select T_{reg} cells [113,156,157].

Gut-derived indole derivatives cross the placenta and reach fetal brain tissues [158]. In animal models of ischemia, they affect microglia through AhR activation, which suppresses p65 phosphorylation and decreases IL-1β/TNF-α expression through the NF-κB pathway [159–161]. As shown in a stroke model, reduction of NF-κB translocation provides an anti-inflammatory and protective response to hypoxia in both adult and neonate mouse brains [152]. Experimental evidence shows that AhR-ligands interact with microglia in the fetal brain and reduce brain inflammation [162].

Epigenetic silencing of NF-κB, GPCR-driven BBB tightening, systemic T_{reg} expansion, and AhR activation form together a multilayered anti-inflammatory buffer that shields the newborn brain against PA damage.

4.2. Antioxidant Pathways: Maternal Gut Barrier, Placenta, Blood-Brain-Barrier, Glial Cells, and Neurons

The hypoxic injury begins with a primary burst of ROS and a larger oxidative aftershock, which result in mitochondrial failure and excitotoxicity, which together form a self-amplifying destructive loop [18]. Beyond classical apoptosis and necrosis, ferroptosis, an iron-dependent, lipid-peroxidation-driven form of regulated cell death, has emerged as an additional contributor in neonatal HIE [163,164]. Moreover, experimental data link Nrf2 activation to lower ferroptosis and improved outcomes in PA [163,165,166].

The maternal gut microbiome shapes antioxidant defences in both the mother and foetus, playing a key role in antioxidant capacity transfer to the foetus [167,168]. In mice, the maternal microbiome drives distinct metabolomics profiles in the placenta, foetal gut, and foetal brain, confirming transplacental transfer of microbial metabolites [158].

This antioxidant "defence line" spans several interconnected compartments along the maternal-foetal axis: maternal gut, placenta, epithelial barriers (BBB and placento-fetal), neurons, and glial cells.

Within the maternal gut, eubiosis sustains intestinal barrier function and neutralises ROS, preventing leakage of pro-oxidant and inflammatory molecules into the circulation [169]. A well-colonized maternal gut contributes to antioxidant protection in two complementary ways: first, gut bacteria neutralise reactive oxygen species with antioxidant enzymes—NADH peroxidase and SOD—as well as with metabolites with ROS-scavenging activity [170,171]; second, microbial metabolites increase antioxidant defence through Nrf2 signalling and upregulate tight-junction proteins [169–171]. SCFAs bind FFAR, inhibit HDAC, and activate AMPK, which increases Nrf2 nuclear translocation and induces antioxidant defences [170]. AhR ligands increase Nrf2 nuclear translocation and induce tight junction proteins, as well as increased enzymatic antioxidant defences [169].

In the placenta, maternal antioxidant reserves and microbiota-derived metabolites help neutralize reactive oxygen species before they reach the foetus, supporting glutathione balance and redox homeostasis [172,173], thus protecting the placenta and the foetus [174]. AhR ligands induce Nrf2 translocation, which enhances enzymatic antioxidant systems and provides antioxidant buffers themselves through interaction with IDO (indoleamine 2,3-dioxygenase) [172]. Direct placental Nrf2 activation by SCFAs in vivo remains to be demonstrated; although current evidence supports SCFA transfer and sensing, with anti-inflammatory effects through NF-κB that lower oxidative stress at the placenta [94,174].

In the foetal BBB, an increased permeability allows for more ROS increased oxidative damage in PA [175]. The microbiome regulates maturation of the BBB, controlling the expression of tight junction proteins: occludins and claudins [176]. Because SCFAs reach foetal circulation, FFAR activation and Nrf2 signalling could enhance antioxidant capacity and maintain BBB integrity [154,177].

In neurons, SCFAs protect against oxidative ischemic damage in animal models [178,179]. SCFAs inhibit neuronal HDAC [180], increase Nrf2 nuclear translocation, enhance antioxidant enzyme activity, and lower ferroptosis [181]. Additionally, they act through the BDNF-Akt3 pathway to reduce oxidative stress and apoptosis [178]. AhR ligands could exhibit direct antioxidant effects. Thus, tryptophan metabolites related to gut microbiota, which are endogenous AhR ligands, attenuated neuronal damage in an ischemic hippocampus model [182].

Glial cells, including microglia and astrocytes, further contribute by fine-tuning inflammatory responses and limiting oxidative injury, roles that have been demonstrated in ischemia and related injury models [159,165]. Mechanistically, this could involve NF-κB inhibition, which triggers M2 phenotype with lower ROS production, both through SCFA and AhR ligands [150,159], although direct evidence in PA is currently lacking.

Together, these layers form a coordinated antioxidant network that may critically influence vulnerability or resilience to oxidative injury in PA.

4.3. Neurogenesis, Neuroplasticity, and Neural Repair

The maternal microbiota modulates neurotrophic factor expression through pathways that lead to increased antioxidant enzymes. Short-chain fatty acids, particularly butyrate, enhance brain-derived neurotrophic factor (BDNF) expression by direct histone deacetylase inhibition and increased H3 crotonylation, which together amplify antioxidant gene transcription [183–185]. This SCFA-mediated BDNF upregulation activates the PI3K/AKT signaling cascade, which phosphorylates and inactivates pro-apoptotic proteins BDNF and increases antioxidant expression [186], as shown in a model of ischemia reperfusion injury [178]. In contrast, maternal high-fat diet-induced dysbiosis correlates with reduced offspring BDNF levels, along with functional impairments [187–190].

By promoting a strong anti-inflammatory phenotype in microglia, adequate synaptic pruning and development occur [191], leading to a greater neuronal reserve and improved wiring.

In dysbiotic animal models (induced through high-fat diet), butyrate improves mitochondrial function in the cerebral cortex and the synaptic fraction [192], enhances neuroplasticity [193], and shows an impressive neuroprotective effect in ischemia [142,194], all increasing the neural functional reserve. One possible mechanism underlying this protective effect could be the ability of sodium butyrate to stimulate neurogenesis, as demonstrated in rat models of perinatal asphyxia [142,195].

In addition, the maternal gut microbiome affects oligodendrocyte activation and myelination processes in the offspring's central nervous system [196]. Through HDAC inhibition in precursor oligodendrocytes, SCFAs guide their maturation and proper myelination, creating additional resistance against PA [196]. Moreover, SCFAs reduce brain inflammation, lowering the risk of precursor oligodendrocyte disruption.

Maternal eubiosis likely primes the foetus towards a reparative phenotype during phase 3, providing direct neuroprotection through enhanced neurogenesis and neural repair, thus improving neuronal reserve and mitigating the PA effects.

4.4. The Maternal Microbiome, Oxytocin Signaling, and Perinatal Asphyxia—Direct Neuroprotection?

The maternal microbiome's neuroprotective effects extend beyond direct metabolite production to include modulation of critical neuropeptide systems, particularly oxytocin (OXT) [197]. We believe this neuroendocrine pathway represents an underexplored mechanism through which the maternal gut bacteria influence fetal resilience to hypoxic injury.

Maternal gut dysbiosis, particularly from high-fat diets, depletes beneficial species like *Lactobacillus reuteri*, correlating with diminished hypothalamic OXT expression in offspring and impaired stress regulation [197]. Crucially, *L. reuteri* supplementation during pregnancy enhances maternal OXT signaling via the *vagus* nerve stimulation [198,199], with strain-specific effects not replicated by other *Lactobacillus* species [200].

The microbiome-OXT connection is particularly relevant for PA outcomes, as OXT facilitates the critical transition from excitatory to inhibitory GABA signaling in the fetal brain, a key protective mechanism against hypoxia-induced excitotoxicity [201]. In experimental PA models, OXT administration reduces neuronal loss, seizure severity, and hippocampal damage, with efficacy dependent on injury severity [202–205].

The neuroprotective capacity is enhanced by OXT's co-release with arginine vasopressin during perinatal stress, which suppresses excessive neuronal excitability during oxygen deprivation [206–208]. Further, germ-free and antibiotic-exposed [209] animal models display altered OXT signaling and social deficits reminiscent of neurodevelopmental disorders, whereas offspring treatment with L. reuteri reversed maternal high-fat diet-induced social and synaptic deficits [197], suggesting that early microbial colonization is essential for normal oxytocinergic development.

Therefore, maternal dysbiosis may compromise this endogenous neuroprotective system, increase HIE severity, and disrupt OXT signalling. Microbiome-based interventions that target OXT could enhance fetal neuroprotection against PA.

5. The Maternal Gut Microbiome Dysbiosis and Perinatal Asphyxia Outcomes

While maternal eubiosis provides multiple layers of neuroprotection against hypoxic-ischemic injury, dysbiosis creates a contrasting scenario of increased vulnerability. Disrupted maternal microbial communities compromise the very pathways that normally shield the developing brain from oxidative damage, inflammatory cascades, and metabolic dysfunction.

This section examines the experimental evidence demonstrating how the maternal microbiome imbalance amplifies the pathophysiological processes underlying PA, transforming what might be a manageable hypoxic insult into severe, lasting neurological injury.

5.1. Mechanisms of Dysbiosis-Induced Neuronal Vulnerability

PA creates a storm of oxidative vulnerability that, amplified by maternal (and subsequent fetal) dysbiosis, amplifies PA's neurological injury. This section examines how dysbiosis-derived endotoxins amplify the oxidative damage cascade specific to hypoxic-ischemic events.

As dysbiosis results in a higher lipopolysaccharide (LPS) concentration in the bloodstream [155], LPS can amplify the oxidative and proinflammatory stress [210,211], with increased ROS production, TNF-α level, and antioxidant enzymes, as shown in a model of PA, combined with LPS exposure. LPS action seems to also depend on timing [211], while even hypothermia could bring minimal benefit against a LPS sensitised neural tissue [212].

Mitochondrial dysfunction from PA is another factor of ROS and triggers multiple forms of cellular death, including ferroptosis [18]. Dysbiosis reduces bacterial metabolites that promote a tolerogenic and anti-inflammatory status, and increase oxidative stress and pro-oxidative and inflammatory cytokines such as IL-6, IL-1β, and TNF-α [181,213], thus amplifying the effect of PA.

By allowing more LPS into the bloodstream, dysbiosis affects BBB integrity [176,214–216], leading to enhanced neuroinflammation and, ultimately, more severe outcomes.

Lastly, by increasing pro-inflammatory cytokines and decreasing beneficial bacterial metabolites, dysbiosis transforms microglia into a pro-inflammatory phenotype with increased ROS production and neuronal damage [217,218].

5.2. Experimental Findings

Detrimental microbiome alterations consistently exacerbate PA-induced brain damage (Table 2). Empirical findings reveal that a high-fat gestational diet linked to gut dysbiosis heightens neuroinflammation and cellular injury in the hippocampus of rat pups subjected to PA, as shown by increased TNF-α, IL-1β, and S100B [219]. Similarly, antibiotic-induced maternal dysbiosis amplifies PA vulnerability and exacerbates brain injuries, with altered neurodevelopmental reflexes and increased hippocampal S100B [9], while neonatal antibiotic exposure reduces SCFA-producing bacteria and promotes astrocytic gliosis and microglial activation during perinatal ischemia [7].

Conversely, interventions that display neuroprotective effects have been shown to positively affect microbiome homeostasis (Table 2). Nutritional supplements, including trans-resveratrol [220], citicoline [221], and omega-3 PUFAs [222], not only provide direct neuroprotection through mechanisms such as SIRT1 activation and membrane stabilization but also promote beneficial bacteria that may protect against PA [223–225].

Table 2. Experimental Evidence Linking Microbiome Modulation to Perinatal Asphyxia (PA) Outcomes.

Intervention	**Model**	**Primary Action**	**Microbiome Effect**	**Perinatal Asphyxia (PA) Outcomes**	**Key Markers**	**References**
High-fat diet	Sprague-Dawley (SD) rats, PA P7	Metabolic dysregulation	↓ Bacteroidetes, ↑ Firmicutes/Bacteroidetes ratio	↑ Hippocampal injury	↑ TNF-α, IL-1β, S100B	[219]
Gestational antibiotics	SD rats, PA birth	Microbiome depletion #	↓ Diversity, ↓ *Lactobacillus/Bifidobacterium*	↑ Brain injury, altered reflexes	↑ S100B	[9]
Neonatal antibiotics	Mice, PA	Microbiome disruption #	↓ short-chain fatty acid (SCFA) producers	↑ Neuronal damage, gliosis	↑ GFAP, Iba1	[7]
Omega-3 PUFA	Mice, HI P9	Anti-inflammatory	↑ Butyrate producers 1*	↓ Injury volume (5 weeks)	↓ NF-κB, apoptosis	[222]
Resveratrol	SD rats, PA P7	SIRT1 activation	↑ *Lactobacillus* and *Bifidobacterium* 2*	↓ Hippocampal damage	↓ IL-1β, TNF-α	[220]
Citicoline	SD rats, PA P7	Membrane stabilization	Maintains homeostasis 3*	Preserved hippocampus	↓ Inflammation	[221]
Lactoferrin	Rats, PA	↑ Nrf2, ↓ ferroptosis	↑ *Bifidobacterium* and *Lactobacillus* 4*	↓ Neuronal death	↓ Ferroptosis	[226,227]
SCFAs	SD rats, PA P7	Metabolic dysregulation	Direct metabolite	↓ 30% infarct, ↑ neurogenesis	↓ IL-1β, COX-2	[194]
Probiotics †	SD rats, PA birth	Microbiome depletion #	↑ *Lactobacillus* and *Bifidobacterium*, ↑ SCFAs	Blood-brain barrier (BBB) protection	↓ Microglial activation	[8]

↑ increased/higher levels; ↓ decreased/lower levels; # Primary mechanism is microbiome-related; * Microbiome effects documented separately from PA studies: 1* [225]; 2* [223]; 3* [224]; 4* [227]; † *L. acidophilus* + *B. infantis*; IL-1β inflammatory model.

Lactoferrin supplementation reduces PA-induced neuronal death through Nrf2-mediated ferroptosis inhibition [226], while also promoting beneficial bacteria and inhibiting harmful ones [227].

Direct microbiome-targeted interventions provide compelling evidence for causality. SCFA administration after neonatal hypoxia reduces brain infarct volume by 30%, decreases IL-1β and COX-2 expression, and enhances neurogenesis through histone crotonylation and HDAC inhibition [142,185,194,195].

Maternal probiotic supplementation with *Lactobacillus acidophilus* and *Bifidobacterium infantis* protects blood-brain barrier integrity and reduces microglial activation even in inflammatory challenge models [8], suggesting that microbiome optimization creates resilience against multiple injury pathways.

These experimental findings demonstrate that the maternal microbiome represents a modifiable factor influencing offspring vulnerability to hypoxic-ischemic injury, with interventions showing efficacy through both direct neuroprotective mechanisms and indirect microbiome-mediated pathways.

5.3. Vertical Transmission and Neonatal Gut-Brain Axis Disruption

Neonates born to mothers with gut microbiome dysbiosis may vertically inherit an altered microbial community [50], potentially predisposing them to an impaired gut-brain axis function and heightened susceptibility to PA-induced neurological injury [136]. While infant gut microbiome colonization timing remains debated, most evidence indicates postnatal establishment [97], with the maternal gut microbiome serving as the primary inoculum source [50]. Multiple factors shape this vertical transmission, including maternal health status, delivery mode, breastfeeding practices, and maternal care habits [49,228,229].

This inherited dysbiosis creates a foundation of vulnerability that compounds PA pathophysiology. Because of the infant gut's rich vascularization and the underdeveloped fetal circulatory system, gastrointestinal tissues show increased vulnerability to PA [230,231]. Animal models of global hypoxemia or mesenteric artery occlusion report upregulation of IL-1 and IL-17, gut barrier disruption, and increased bacterial translocation, reflected by higher blood LPS levels [49,232,233].

Furthermore, hypoxemic conditions consistently alter intestinal microbiome composition, with murine models showing increased Firmicutes and decreased Bacteroides and Proteobacteria populations [234,235]. This demonstrates that hypoxemia directly modifies gut microbiome composition, establishing a self-reinforcing cycle where inherited dysbiosis increases PA susceptibility while PA-induced hypoxemia further destabilizes the gut microbiome [49,236].

6. Therapeutic Interventions Targeting the Maternal Microbiome

Given the profound impact of the maternal gut microbiome on fetal neurodevelopment and resilience to hypoxic injury, therapeutic strategies targeting maternal microbial communities represent a promising avenue for preventing or mitigating perinatal asphyxia outcomes. These interventions range from simple dietary modifications to sophisticated microbiome-engineering approaches, each offering distinct mechanisms for enhancing neuroprotection through modulation of oxidative stress and inflammatory pathways.

6.1. Dietary Modifications

Maternal diet shapes its microbiome, and it also shapes the infant gut microbiome [237,238]. Despite appearing relatively straightforward, dietary interventions demonstrate significant therapeutic potential.

Dietary fibre serves as the primary substrate for SCFA production, with demonstrated anti-inflammatory effects [239,240]. High-fibre diets during pregnancy increase SCFA-producing bacteria, enhancing SCFAs levels [67,241]. Among the most studied, the Mediterranean diet, rich in vegetables, fibres, and antioxidants, seems to provide a more anti-inflammatory microbiome [242,243], with effects reverberating into early childhood, hence protection against PA [237]. A randomized control trial analysing the impact of the Mediterranean diet showed better neurodevelopment outcomes at the 2-year mark (Bayley-III) [244].

Alongside fibres, another important aspect of dietary modifications could be related to the amount of polyphenols found in the Mediterranean diet. While having an impact on the gut microbiome and on the inflammatory reverberations of the gut microbiome [224,245], polyphenols serve as antioxidants and have been found to reduce the impact of brain ischemic damage in animal models [220,246–249]. Polyphenols enhance the development of SCFAs-producing bacteria, especially *Akkermansia* and *Bifidobacterium*, and inhibit potentially harmful bacteria growth [250–252]. Dietary polyphenols are metabolized mainly in the colon into bioactive components, such as equol, urolithin, and protocatechuic acid, by several bacterial species (*Bifidobacterium* sp., *Lactobacillus* sp., *Bacteroides* sp., etc.) [251,253]. Through the interaction with the maternal microbiome, resveratrol, transformed into 4-hydroxyphenylacetic acid, acts through the SIRT1 pathway and prevents obesity [254].

Omega-3 fatty acid supplementation deserves special emphasis, with maternal DHA-enriched diets preventing neuronal apoptosis and reducing brain injury volume in HIE models through anti-inflammatory effects and microglial NF-κB signalling suppression [222,255], an effect that was present at 5 weeks postnatal. Alongside direct effects in animal studies, in humans, omega-3 has been associated with fewer pre-term births and better neonatal health [256]. Omega-3 fatty acids also influence maternal microbiome and the inflammatory status of the placenta and the offspring, respectively [225]. These supplements maintain the intestinal wall integrity and modulate host immune response through modulating the gut composition, promoting butyrate-producing bacteria and Bacteroidetes sp. [257].

Fermented foods such as kombucha, yogurt, and kimchi show promise for establishing a balanced maternal microbiome [258]. Furthermore, these foods have been reported to offer benefits for pregnancy overall [259]. These effects likely stem from their complex composition; while serving as a vehicle for probiotics, these foods also supply essential macronutrients (carbohydrates, proteins, fats), alongside other bioactive compounds such as prebiotics and microbial metabolites [259,260]. A review on fermented foods shows the benefits on immune system maturation, brain development and maternal health, and infant microbiota [261].

In addition, maternal administration of lactoferrin improved outcomes in PA [226] through Nrf2 signalling, increasing antioxidant buffer and reducing ferroptosis. In addition, lactoferrin modulates gut diversity and promotes beneficial bacterial growth (e.g., *Bifidobacterium* and *Lactobacillus*) [227].

Conversely, a Western-style high-fat/low-fibre dietary pattern promotes dysbiosis, elevates maternal and placental cytokine levels, and predisposes offspring to neuroinflammation [187].

6.2. Emerging Microbiome-Based Therapeutic Strategies

Alongside dietary and lifestyle interventions, targeted manipulation of the maternal gut microbiome through prebiotics, probiotics, postbiotics, and synbiotics represents an expanding area of research with significant implications for maternal–infant health and neurodevelopment. These interventions are distinct yet complementary, each influencing microbial composition and function via different mechanisms.

6.2.1. Prebiotics: Enhancing Endogenous Microbial Metabolism

Prebiotics are defined as selectively fermentable, non-digestible food components, most notably galacto-oligosaccharides (GOS) and fructo-oligosaccharides (FOS), that stimulate the growth and activity of beneficial gut microorganisms. During pregnancy, prebiotic supplementation alters microbial fermentation patterns, leading to increased production of SCFAs such as acetate, propionate, and butyrate. These metabolites cross the placental barrier and exert systemic effects, enhancing antioxidant defense and modulating immune tolerance in the offspring [112,262]. Notably, prebiotics induce a tolerogenic immune shift by promoting regulatory B and T cells expansion and elevating acetate and butyrate concentrations, metabolites with potent anti-inflammatory functions [112,263].

Such effects may enhance fetal resilience to hypoxic-ischemic insults, although direct evidence in the context of PA is currently lacking.

6.2.2. Probiotics: Augmenting Beneficial Microbial Populations

Probiotics, live microorganisms that confer health benefits when administered in adequate amounts, provide a more direct strategy to enhance maternal gut composition. Supplementation during gestation with *Lactobacillus* and *Bifidobacterium* species lowers proinflammatory cytokines, preserves the BBB, and enhances oligodendrocyte maturation following neonatal challenges [8]. Mechanistically, probiotics increase antioxidant capacity, improve immune response, suppress inflammatory cascades, and produce antimicrobial peptides that inhibit pathogenic bacteria [264–267].

Strain-specific effects have also been observed. For example, *Clostridium butyricum* enhances SCFAs production and mitigates ischemia/reperfusion injury in rodent models [179], while *Lactobacillus reuteri* modulates OXT signalling, increasing hypothalamic oxytocin expression and improving GABA-mediated neuroprotection under hypoxic stress [199,266]. In addition, probiotic interventions have reduced neuronal damage in experimental models of perinatal inflammation [8], and are clinically employed in conditions such as necrotizing enterocolitis (NEC), a pathology that can follow PA [268].

In maternal populations, probiotics have been used to reduce complications, including gestational diabetes, preeclampsia, and infection-related outcomes [111], factors that may indirectly influence PA severity if unmanaged. Nevertheless, while preclinical findings are promising, human data remain limited and inconsistent, highlighting the need for large-scale trials [269].

6.2.3. Postbiotics: Delivering Microbial Metabolites Directly

Postbiotics are non-viable bacterial components or metabolites, such as SCFAs, that offer an alternative route to harness microbial benefits without live organisms. Supplementation with SCFAs has demonstrated safety and efficacy in attenuating hypoxic-ischemic brain lesions, improving BBB function, reducing oxidative stress, and suppressing inflammation in animal studies [137,154,194]. Additional evidence suggests that postbiotics may ameliorate placental dysfunction and immune dysregulation associated with dysbiosis [12,94].

6.2.4. Synbiotics and Other Advanced Strategies

Synbiotics, defined as combinations of probiotics and prebiotics, represent an evolving therapeutic concept. By supplying beneficial microorganisms alongside their preferred substrates, synbiotics may synergistically enhance colonization and functional outputs. While PA-specific studies are absent, synbiotics have improved clinical outcomes in NEC and reduced allergy incidence in infants [270]. These findings suggest potential neuroprotective benefits under hypoxic stress conditions. Maternal synbiotic supplementation during gestation and lactation significantly enhances antioxidant capacity, upregulates mitochondrial function–related genes, and increases the abundance of beneficial intestinal microbiota in offspring, suggesting a synergistic benefit over probiotics alone [271].

Faecal microbiota transplantation (FMT), although still experimental in pregnancy and neonates, is another emerging modality. By transferring processed stool from healthy donors, FMT restores microbial diversity and metabolic function. Evidence supports its efficacy in metabolic syndrome and inflammatory conditions across animal and human studies [272–274], but its safety and utility in PA-related contexts require rigorous evaluation.

6.2.5. Addressing Harmful Metabolites and Novel Pharmacological Targets

Not all microbial metabolites are beneficial. Compounds such as trimethylamine (TMA) and trimethylamine N-oxide (TMAO) are implicated in adverse cardiovascular programming and offspring hypertension [275,276]. Inhibitors like 3,3-dimethyl-1-butanol (DMB) suppress TMA/TMAO formation, mitigating these risks [276,277].

Furthermore, pharmacological targeting of the microbial-sensing pathways shows promise. FFARs agonists, such as 4-CMBT, or AhR ligands like indole-3-propionic acid (IPA), have demonstrated anti-inflammatory and neuroprotective effects in preclinical settings [159,278,279]. Such strategies, though untested in PA, may complement microbiome-directed therapies to enhance foetal resilience to hypoxia.

In summary, microbiome-based interventions, including prebiotics, probiotics, postbiotics, synbiotics, and emerging pharmacological approaches, offer exciting opportunities to improve maternal and neonatal outcomes. While animal models provide strong mechanistic evidence, clinical translation remains at an early stage. Addressing safety, timing, and strain-specific effects will be essential to realize the full potential of these therapies in preventing or mitigating the PA-related neurodevelopmental injury.

7. Current Challenges and Future Research Directions

Understanding the maternal microbiome's influence on PA remains a developing area of investigation. While animal models have provided initial insights into the maternal microbiome–foetal brain axis, human research remains limited. This is compounded by methodological variability, ethical concerns, and the absence of clear clinical guidelines for microbiome-based interventions during pregnancy or neonatal care. Bridging these gaps is crucial for translating preclinical findings into effective and safe perinatal neuroprotective strategies.

7.1. Methodological and Scientific Gaps

Most studies to date have focused on the maternal gut microbiome, excluding other potentially influential microbial niches such as the vaginal, oral, skin, and placental microbiota. These ecosystems may play important and interactive roles in foetal immune and neurodevelopment, particularly in modulating the neonatal response to hypoxic stress. The exclusion of these sites from current studies limits our understanding of the broader maternal microbial landscape and its role in perinatal outcomes.

In addition, this review employed a narrative approach, which, while suitable for conceptual exploration, lacks the methodological rigor of systematic reviews and may not capture the full spectrum of available evidence. The majority of research investigating maternal gut microbiome influences on offspring development originates from rodent models, which enable microbiome manipulation through antibiotic depletion, gnotobiotic designs, and maternal immune activation [9,57,125]. Large animal models have validated the impact of maternal gut microbiome on infant development, both in cattle and swine [280], as well as in non-human primates [281–283]. Since perinatal asphyxia models are well-established in rodents, large animals, and primates, these species offer valuable platforms for investigating maternal microbiome influences on hypoxic-ischemic outcomes [142,284–286]. However, few studies have directly examined the interplay between maternal gut microbiome manipulation and neonatal hypoxic-ischemic injury, representing a critical research gap that these established models are well-positioned to address.

There is also a lack of longitudinal human studies examining maternal microbiome dynamics across pregnancy and their association with neonatal outcomes following PA. Without such studies, the ability to draw causal links or develop predictive microbial biomarkers remains limited.

7.2. *Ethical and Safety Concerns*

The use of microbiome-modulating therapies during pregnancy and in neonates introduces several ethical and safety concerns. Although many probiotics contain strains that are naturally present in the human gut or used in fermented foods, such as *Lactobacillus*, *Bifidobacterium*, *Streptococcus*, *Saccharomyces*, and *Bacillus*, their safety in vulnerable populations remains incompletely understood [287].

Evidence suggests that maternal probiotic use is relatively safe. A systematic review by Jarde et al. found no increased risk of miscarriage, foetal malformation, premature birth, low birth weight, or Caesarean section associated with probiotic administration during pregnancy. Therefore, targeting the maternal microbiome may represent a safer alternative to direct postnatal probiotic supplementation, particularly in preterm or at-risk neonates [90].

In contrast, neonatal probiotic use is more controversial. Case reports have documented severe adverse outcomes, including bacteraemia, fungemia, and endocarditis, in immunocompromised or preterm infants treated with probiotics [288–291], while others consider the direct administration rather beneficial [292]. In 2023, the U.S. Food and Drug Administration (FDA) issued a formal warning following reports of infections and deaths in preterm infants who received probiotics to prevent necrotizing enterocolitis [293]. These incidents underscore the need for stringent safety protocols and regulatory oversight when considering microbiota-based interventions in neonatology.

Beyond individual-level risks, public health concerns also emerge. O'Doherty et al. argue that microbial interventions may inadvertently affect others through microbial shedding or environmental transmission. For example, altering the microbiome of a pregnant individual may influence the microbial exposure of her infant, household members, or even broader community contacts. These findings highlight the importance of ethical scrutiny and the development of responsible policies around microbiome modulation, especially during critical developmental windows [294].

Taken together, these findings reinforce the need for internationally accepted standards and clinical guidelines governing the design, testing, and application of microbiome-modulating therapies in perinatal medicine.

7.3. *Clinical Evidence and Limitations*

While there are many therapeutic microbiome-targeted options, few have been studied in the context of PA and even fewer possess a straight from the bench to bedside pathway that would allow human trials soon.

Several limitations hinder clinical translation, including mechanistic uncertainty regarding how microbiome modulation leads to improved neurodevelopmental outcomes in PA, population-specific variability in maternal microbiome composition, the absence of defined timing and dosing windows for intervention during pregnancy or postpartum, and substantial regulatory and safety hurdles in ensuring maternal–foetal safety for novel therapeutics.

Furthermore, animal models, while invaluable for mechanistic insights, do not fully replicate human perinatal physiology and immune–microbiome interactions, which complicates extrapolation to clinical practice. As a result, while the therapeutic potential of microbiome-based interventions in PA is promising, progress toward clinical application will require rigorous, multi-centre studies employing standardized protocols, validated biomarkers, and long-term neurodevelopmental follow-up.

7.4. Future Research Recommendations

Future research should expand beyond the maternal gut microbiome to explore the contribution of other maternal microbial niches. The vaginal and placental microbiota are likely to play central roles in foetal immune education and inflammatory priming. Their influence on susceptibility to perinatal hypoxia-related brain injury remains underexplored and deserves targeted investigation.

Mechanistic studies must also evolve. Integrating high-throughput multi-omics approaches, such as metagenomics, metabolomics, transcriptomics, and immune profiling, with histological and imaging assessments of the foetal brain will allow a better understanding of how maternal microbiome signals are transduced across the placental barrier. The placenta itself is emerging as a critical intermediary in this signalling network and should be prioritized in future experimental designs.

Human clinical research must shift toward longitudinal, standardized protocols. Prospective cohort studies tracking microbiome profiles across pregnancy and associating them with neurodevelopmental outcomes in infants exposed to PA are essential. These studies should incorporate both structural imaging and behavioural assessments to evaluate the functional significance of microbial influences on neurodevelopment.

Randomized controlled trials are urgently needed to evaluate the safety and efficacy of probiotics and related interventions during pregnancy. Given current safety concerns surrounding neonatal probiotic use, maternal-targeted strategies may offer a safer route to modulate the infant microbiome indirectly. However, rigorous adverse event monitoring and stratification by maternal and infant risk factors will be crucial in any interventional study.

Finally, incorporating microbiome-based screening into routine prenatal care may enhance risk stratification for perinatal brain injury. Identifying microbial signatures predictive of adverse outcomes could support early intervention and inform the development of personalized therapeutic strategies.

In summary, the maternal microbiome represents a promising yet underutilized target for influencing the outcome of perinatal hypoxia/asphyxia. At this moment, no intervention that addresses the microbiome is currently in practice for mitigating PA effects [295]. Addressing current methodological, ethical, and clinical limitations through multidisciplinary research is essential to realize the translational potential of microbiome-based interventions in perinatal neuroprotection.

8. Conclusions

This narrative review shows the critical role of the maternal gut microbiome in shaping fetal immune and neurodevelopmental trajectories, with growing evidence linking its composition to neonatal vulnerability or resilience in the context of perinatal hypoxia/asphyxia. Maternal gut eubiosis may enhance brain resilience to perinatal asphyxia through potent antioxidant and anti-inflammatory stimuli that reverberate across the gut-placenta-brain axis, and through promotion of neurogenesis, neuroplasticity, and oxytocinergic pathways. Conversely, dysbiosis may exacerbate hypoxic-ischemic brain injury by priming a hyperreactive immune system, increasing oxidative stress, and negatively influencing neurodevelopmental trajectories.

The therapeutic landscape offers promising opportunities through dietary modifications, prebiotic, probiotic, and postbiotic therapeutics. However, significant challenges remain, including methodological limitations in current research, ethical considerations surrounding maternal interventions, optimal timing strategies, and the critical need for thorough safety data in pregnancy populations.

While animal studies support these associations, human data remain limited. Further research is needed to establish causality, identify key microbial mediators, and evaluate the safety and efficacy of microbiome-targeted interventions during pregnancy and the perinatal period. Bridging mechanistic insights from animal models to longitudinal human studies will be essential to translate these findings into safe, effective interventions in obstetric and neonatal practice. Therefore, harnessing the maternal microbiome as a therapeutic target could offer a promising avenue to improve perinatal neuroprotection and long-term neurodevelopmental outcomes. Ultimately, integrating maternal microbiome modulation into perinatal care—alongside established approaches such as therapeutic hypothermia—may provide a transformative strategy to reduce the burden of hypoxic-ischemic brain injury.

Author Contributions: Conceptualization, V.-P.M. and M.I.I.; methodology, V.-P.M.; validation, V.-P.M., M.I.I., C.M.D.Z., A.M.C., A.-T.C., I.-A.D., C.C., G.G.P. and A.-M.Z.; formal analysis, V.-P.M.; investigation, V.-P.M.; resources, V.-P.M. and A.-M.Z.; data curation, V.-P.M. and M.I.I.; writing and original draft preparation, V.-P.M., M.I.I., C.M.D.Z., A.M.C., A.R., A.-T.C., I.-A.D., C.C., G.G.P. and A.-M.Z.; writing, review and editing, all authors; visualization, A.R.; supervision, G.G.P. and A.-M.Z.; project administration, V.-P.M. All authors have read and agreed to the published version of the manuscript.

Funding: The APC was funded by Carol Davila University of Medicine and Pharmacy.

Institutional Review Board Statement: Not applicable.

Informed Consent Statement: Not applicable.

Data Availability Statement: Not applicable.

Acknowledgments: The authors express gratitude to Mihai Moldovan (Department of Neuroscience, University of Copenhagen; Department of Clinical Neurophysiology, Rigshospitalet, Copenhagen, Denmark; Department of Physiology—Neuroscience "Carol Davila" University of Medicine and Pharmacy, Bucharest, Romania) for valuable discussions regarding the structural organization and conceptual framework of this manuscript. Figures in this manuscript were created with BioRender.com. The authors used ChatGPT (OpenAI GPT4o, o1-pro and o3-pro) and Claude (Anthropic Claude Sonnet 4.0, Opus 4.0 & Opus 4.1) to assist with text editing (to improve grammar, prose clarity, and stylistic consistency) and manuscript organization (brainstorming ideas for table of contents and better flow for complex sections). AI tools were not used for literature review, data interpretation, scientific reasoning, or generating conclusions. All outputs were reviewed and revised by the authors, who take full responsibility for the scientific accuracy of the final content.

Conflicts of Interest: The authors declare no conflicts of interest.

Abbreviations

The following abbreviations are used in this manuscript:

4-CMBT	4-Chloro-α-(1-methylethyl)-N-2-thiazolyl-benzeneacetamide
AhR	Aryl hydrocarbon receptor
AMPK	5′ adenosine monophosphate-activated protein kinase
BBB	Blood-brain barrier
BDNF	Brain-derived neurotrophic factor
BEVs	Bacterial extracellular vesicles
BMI	Body mass index
COX-2	Cyclooxygenase-2
CSF	Cerebrospinal fluid
DHA	Docosahexaenoic acid
DMB	3,3-dimethyl-1-butanol
F/B	Firmicutes/Bacteroidetes ratio
FDA	U.S. Food and Drug Administration
FFAR	Free fatty acid receptor
FFAR2	Free fatty acid receptor 2
FFAR3	Free fatty acid receptor 3
FMT	Faecal microbiota transplantation
FOS	Fructo-oligosaccharides
GABA	Gamma-aminobutyric acid
GFAP	Glial fibrillary acidic protein
GOS	Galacto-oligosaccharides
GPCRs	G protein-coupled receptors
H3	Histone 3
HDAC	Histone deacetylase
HI	Hypoxia-ischaemia
HIE	Hypoxic-ischaemic encephalopathy
Iba1	Ionised calcium-binding adapter molecule 1
IBD	Inflammatory bowel disease
IDO	Indoleamine 2,3-dioxygenase
IgG	Immunoglobulin G
IL-1β	Interleukin-1 beta
IL-6	Interleukin-6
IL-10	Interleukin-10
IL-17	Interleukin-17
IPA	Indole-3-propionic acid
JAK-STAT	Janus kinase/Signal Transducer and Activator of Transcription
LPS	Lipopolysaccharide
LRRC19	Leucine Rich Repeat Containing 19
MCAO	Middle cerebral artery occlusion
MGM	Maternal gut microbiota
MIA	Maternal immune activation
NADH	Nicotinamide adenine dinucleotide
NEC	Necrotising enterocolitis
NF-κB	Nuclear factor kappa B
Nrf2	Nuclear factor erythroid 2-related factor 2
OXT	Oxytocin
P7/P9	Postnatal day 7/9
PA	Perinatal asphyxia
PI3K/AKT	Phosphoinositide 3-kinase/Protein kinase B
PKC	Protein kinase C

PUFA	Polyunsaturated fatty acids
PVL	Periventricular leukomalacia
ROS	Reactive oxygen species
S100B	S100 calcium-binding protein B
SCFAs	Short-chain fatty acids
SD	Sprague-Dawley
SIRT1	Sirtuin 1
SOD	Superoxide dismutase
TECs	Thymic epithelial cells
TGF-β	Transforming growth factor-beta
TMA	Trimethylamine
TMAO	Trimethylamine N-oxide
TMG	Trimethylglycine
TNF-α	Tumour necrosis factor-alpha
T_{regs}	Regulatory T cells

References

1. Liu, L.; Oza, S.; Hogan, D.; Chu, Y.; Perin, J.; Zhu, J.; Lawn, J.E.; Cousens, S.; Mathers, C.; Black, R.E. Global, Regional, and National Causes of under-5 Mortality in 2000–15: An Updated Systematic Analysis with Implications for the Sustainable Development Goals. *Lancet* **2016**, *388*, 3027–3035, Erratum in *Lancet* **2017**, *389*, 1884. [CrossRef]
2. Kurinczuk, J.J.; White-Koning, M.; Badawi, N. Epidemiology of Neonatal Encephalopathy and Hypoxic-Ischaemic Encephalopathy. *Early Hum. Dev.* **2010**, *86*, 329–338. [CrossRef]
3. Oza, S.; Lawn, J.E.; Hogan, D.R.; Mathers, C.; Cousens, S.N. Neonatal Cause-of-Death Estimates for the Early and Late Neonatal Periods for 194 Countries: 2000–2013. *Bull. World Health Organ.* **2015**, *93*, 19–28. [CrossRef]
4. Cannavò, L.; Perrone, S.; Gitto, E. Brain-Oriented Strategies for Neuroprotection of Asphyxiated Newborns in the First Hours of Life. *Pediatr. Neurol.* **2023**, *143*, 44–49. [CrossRef]
5. Lear, B.A.; Zhou, K.Q.; Dhillon, S.K.; Lear, C.A.; Bennet, L.; Gunn, A.J. Preventive, Rescue and Reparative Neuroprotective Strategies for the Fetus and Neonate. *Semin. Fetal Neonatal Med.* **2024**, *29*, 101542. [CrossRef]
6. Wassink, G.; Davidson, J.O.; Dhillon, S.K.; Zhou, K.; Bennet, L.; Thoresen, M.; Gunn, A.J. Therapeutic Hypothermia in Neonatal Hypoxic-Ischemic Encephalopathy. *Curr. Neurol. Neurosci. Rep.* **2019**, *19*, 2. [CrossRef] [PubMed]
7. Drobyshevsky, A.; Synowiec, S.; Goussakov, I.; Fabres, R.; Lu, J.; Caplan, M. Intestinal Microbiota Modulates Neuroinflammatory Response and Brain Injury after Neonatal Hypoxia-Ischemia. *Gut Microbes* **2024**, *16*, 2333808. [CrossRef] [PubMed]
8. Lu, J.; Lu, L.; Yu, Y.; Baranowski, J.; Claud, E.C. Maternal Administration of Probiotics Promotes Brain Development and Protects Offspring's Brain from Postnatal Inflammatory Insults in C57/BL6J Mice. *Sci. Rep.* **2020**, *10*, 8178. [CrossRef] [PubMed]
9. Ionescu, M.I.; Maria Catrina, A.; Dogaru, I.A.; Catalina Barbalata, D.; Ciotei, C.; Haidoiu, C.; Suhaianu, V.; Gradisteanu Pircalabioru, G.; O'Mahony, S.M.; Zagrean, A.-M. MICROBIOME: The Trials and Errors of Developing an Experimental Model to Study the Impact of Maternal Gut Microbiome Disruption on Perinatal Asphyxia. *Reprod. Fertil.* **2024**, *5*, e240050. [CrossRef]
10. Husso, A.; Pessa-Morikawa, T.; Koistinen, V.M.; Kärkkäinen, O.; Kwon, H.N.; Lahti, L.; Iivanainen, A.; Hanhineva, K.; Niku, M. Impacts of Maternal Microbiota and Microbial Metabolites on Fetal Intestine, Brain, and Placenta. *BMC Biol.* **2023**, *21*, 207. [CrossRef]
11. Abdel-Haq, R.; Schlachetzki, J.C.M.; Glass, C.K.; Mazmanian, S.K. Microbiome-Microglia Connections via the Gut-Brain Axis. *J. Exp. Med.* **2019**, *216*, 41–59. [CrossRef] [PubMed]
12. Qin, X.; Zhang, M.; Chen, S.; Tang, Y.; Cui, J.; Ding, G. Short-Chain Fatty Acids in Fetal Development and Metabolism. *Trends Mol. Med.* **2024**, *31*, 625–639. [CrossRef]
13. Lu, X.; Shi, Z.; Jiang, L.; Zhang, S. Maternal Gut Microbiota in the Health of Mothers and Offspring: From the Perspective of Immunology. *Front. Immunol.* **2024**, *15*, 1362784. [CrossRef]
14. Sajdel-Sulkowska, E.M. The Impact of Maternal Gut Microbiota during Pregnancy on Fetal Gut-Brain Axis Development and Life-Long Health Outcomes. *Microorganisms* **2023**, *11*, 2199. [CrossRef]
15. Bennet, L. Sex, Drugs and Rock and Roll: Tales from Preterm Fetal Life. *J. Physiol.* **2017**, *595*, 1865–1881. [CrossRef]
16. Giussani, D.A. The Fetal Brain Sparing Response to Hypoxia: Physiological Mechanisms: Fetal Brain Sparing. *J. Physiol.* **2016**, *594*, 1215–1230. [CrossRef] [PubMed]
17. Jensen, A.; Berger, R. Fetal Circulatory Responses to Oxygen Lack. *J. Dev. Physiol.* **1991**, *16*, 181–207.

18. Okazaki, K.; Nakamura, S.; Koyano, K.; Konishi, Y.; Kondo, M.; Kusaka, T. Neonatal Asphyxia as an Inflammatory Disease: Reactive Oxygen Species and Cytokines. *Front. Pediatr.* **2023**, *11*, 1070743. [CrossRef]
19. Szydlowska, K.; Tymianski, M. Calcium, Ischemia and Excitotoxicity. *Cell Calcium* **2010**, *47*, 122–129. [CrossRef]
20. Hartings, J.A.; Shuttleworth, C.W.; Kirov, S.A.; Ayata, C.; Hinzman, J.M.; Foreman, B.; Andrew, R.D.; Boutelle, M.G.; Brennan, K.C.; Carlson, A.P.; et al. The Continuum of Spreading Depolarizations in Acute Cortical Lesion Development: Examining Leão's Legacy. *J. Cereb. Blood Flow Metab.* **2017**, *37*, 1571–1594. [CrossRef] [PubMed]
21. Wassink, G.; Gunn, E.R.; Drury, P.P.; Bennet, L.; Gunn, A.J. The Mechanisms and Treatment of Asphyxial Encephalopathy. *Front. Neurosci.* **2014**, *8*, 40. [CrossRef]
22. Iwata, O.; Iwata, S.; Thornton, J.S.; De Vita, E.; Bainbridge, A.; Herbert, L.; Scaravilli, F.; Peebles, D.; Wyatt, J.S.; Cady, E.B.; et al. "Therapeutic Time Window" Duration Decreases with Increasing Severity of Cerebral Hypoxia-Ischaemia under Normothermia and Delayed Hypothermia in Newborn Piglets. *Brain Res.* **2007**, *1154*, 173–180. [PubMed]
23. Hostetter, M.K. Society for Pediatric Research Presidential Address 1994: Yeast as Metaphor. *Pediatr. Res.* **1994**, *36*, 692–698. [CrossRef]
24. Thornton, C.; Leaw, B.; Mallard, C.; Nair, S.; Jinnai, M.; Hagberg, H. Cell Death in the Developing Brain after Hypoxia-Ischemia. *Front. Cell. Neurosci.* **2017**, *11*, 248. [CrossRef]
25. Rodriguez, J.; Xie, C.; Li, T. Inhibiting the Interaction between Apoptosisinducing Factor and Cyclophilin A Prevents Brain Injury in Neonatal Mice after Hypoxia-Ischemia. *Neuropharmacology* **2020**, *171*, 108088. [CrossRef]
26. Shaw, J.C.; Crombie, G.K.; Zakar, T.; Palliser, H.K.; Hirst, J.J. Perinatal Compromise Contributes to Programming of GABAergic and Glutamatergic Systems Leading to Long-Term Effects on Offspring Behaviour. *J. Neuroendocrinol.* **2020**, *32*, e12814. [CrossRef]
27. Shaw, J.C.; Berry, M.J.; Dyson, R.M.; Crombie, G.K.; Hirst, J.J.; Palliser, H.K. Reduced Neurosteroid Exposure Following Preterm Birth and Its' Contribution to Neurological Impairment: A Novel Avenue for Preventative Therapies. *Front. Physiol.* **2019**, *10*, 599. [CrossRef]
28. Fleiss, B.; Gressens, P. Tertiary Mechanisms of Brain Damage: A New Hope for Treatment of Cerebral Palsy? *Lancet Neurol.* **2012**, *11*, 556–566. [CrossRef]
29. Bennet, L.; Dhillon, S.; Lear, C.A.; van den Heuij, L.; King, V.; Dean, J.M.; Wassink, G.; Davidson, J.O.; Gunn, A.J. Chronic Inflammation and Impaired Development of the Preterm Brain. *J. Reprod. Immunol.* **2018**, *125*, 45–55. [CrossRef] [PubMed]
30. Dhillon, S.K.; Gunn, E.R.; Lear, B.A.; King, V.J.; Lear, C.A.; Wassink, G.; Davidson, J.O.; Bennet, L.; Gunn, A.J. Cerebral Oxygenation and Metabolism after Hypoxia-Ischemia. *Front. Pediatr.* **2022**, *10*, 925951. [CrossRef] [PubMed]
31. Buser, J.R.; Maire, J.; Riddle, A.; Gong, X.; Nguyen, T.; Nelson, K.; Luo, N.L.; Ren, J.; Struve, J.; Sherman, L.S.; et al. Arrested Preoligodendrocyte Maturation Contributes to Myelination Failure in Premature Infants. *Ann. Neurol.* **2012**, *71*, 93–109. [CrossRef]
32. Dean, J.M.; Mcclendon, E.; Hansen, K.; Azimi-Zonooz, A.; Chen, K.; Riddle, A.; Gong, X.; Sharifnia, E.; Hagen, M.; Ahmad, T.; et al. Prenatal Cerebral Ischemia Disrupts MRIdefined Cortical Microstructure through Disturbances in Neuronal Arborization. *Sci. Transl. Med.* **2013**, *5*, 168–175. [CrossRef] [PubMed]
33. Mota-Rojas, D.; Villanueva-García, D.; Solimano, A.; Muns, R.; Ibarra-Ríos, D.; Mota-Reyes, A. Pathophysiology of Perinatal Asphyxia in Humans and Animal Models. *Biomedicines* **2022**, *10*, 347. [CrossRef] [PubMed]
34. Hinojosa-Rodríguez, M.; Harmony, T.; Carrillo-Prado, C.; Van Horn, J.D.; Irimia, A.; Torgerson, C.; Jacokes, Z. Clinical Neuroimaging in the Preterm Infant: Diagnosis and Prognosis. *NeuroImage Clin.* **2017**, *16*, 355–368. [CrossRef]
35. Banker, B.Q.; Larroche, J.C. Periventricular Leukomalacia of Infancy. A Form of Neonatal Anoxic Encephalopathy. *Arch. Neurol.* **1962**, *7*, 386–410. [CrossRef]
36. Schneider, J.; Miller, S.P. Preterm Brain Injury: White Matter Injury. *Handb. Clin. Neurol.* **2019**, *162*, 155–172.
37. Volpe, J.J.; Kinney, H.C.; Jensen, F.E.; Rosenberg, P.A. The Developing Oligodendrocyte: Key Cellular Target in Brain Injury in the Premature Infant. *Int. J. Dev. Neurosci.* **2011**, *29*, 423–440. [CrossRef] [PubMed]
38. Nikas, I.; Dermentzoglou, V.; Theofanopoulou, M.; Theodoropoulos, V. Parasagittal Lesions and Ulegyria in Hypoxic-Ischemic Encephalopathy: Neuroimaging Findings and Review of the Pathogenesis. *J. Child Neurol.* **2008**, *23*, 51–58. [CrossRef]
39. Rivkin, M.J. Hypoxic-Ischemic Brain Injury in the Term Newborn. Neuropathology, Clinical Aspects, and Neuroimaging. *Clin. Perinatol.* **1997**, *24*, 607–625. [CrossRef]
40. Groenendaal, F.; de Vries, L.S. Fifty Years of Brain Imaging in Neonatal Encephalopathy Following Perinatal Asphyxia. *Pediatr. Res.* **2017**, *81*, 150–155. [CrossRef]
41. Shah, P.; Anvekar, A.; McMichael, J.; Rao, S. Outcomes of Infants with Apgar Score of Zero at 10 Min: The West Australian Experience. *Arch. Dis. Child. Fetal Neonatal Ed.* **2015**, *100*, F492–F494. [CrossRef]
42. Milner, K.M.; Neal, E.F.G.; Roberts, G.; Steer, A.C.; Duke, T. Long-Term Neurodevelopmental Outcome in High Risk Newborns in Resource-Limited Settings: A Systematic Review of the Literature. *Paediatr. Int. Child Health* **2015**, *35*, 227–242. [CrossRef]

43. Omizzolo, C.; Scratch, S.E.; Stargatt, R.; Kidokoro, H.; Thompson, D.K.; Lee, K.J.; Cheong, J.; Neil, J.; Inder, T.E.; Doyle, L.W.; et al. Neonatal Brain Abnormalities and Memory and Learning Outcomes at 7 Years in Children Born Very Preterm. *Memory* **2014**, *22*, 605–615. [CrossRef]
44. Stipdonk, L.W.; Franken, M.-C.J.P.; Dudink, J. Language Outcome Related to Brain Structures in School-Aged Preterm Children: A Systematic Review. *PLoS ONE* **2018**, *13*, e0196607, Erratum in *PLoS ONE* **2018**, *13*, e0203298.
45. Wheelock, M.D.; Austin, N.C.; Bora, S.; Eggebrecht, A.T.; Melzer, T.R.; Woodward, L.J.; Smyser, C.D. Altered Functional Network Connectivity Relates to Motor Development in Children Born Very Preterm. *NeuroImage* **2018**, *183*, 574–583. [CrossRef]
46. Lee, B.L.; Gano, D.; Rogers, E.E.; Xu, D.; Cox, S.; James Barkovich, A.; Li, Y.; Ferriero, D.M.; Glass, H.C. Long-Term Cognitive Outcomes in Term Newborns with Watershed Injury Caused by Neonatal Encephalopathy. *Pediatr. Res.* **2022**, *92*, 505–512. [CrossRef]
47. Ahearne, C.E.; Boylan, G.B.; Murray, D.M. Short and Long Term Prognosis in Perinatal Asphyxia: An Update. *World J. Clin. Pediatr.* **2016**, *5*, 67–74. [CrossRef] [PubMed]
48. de Haan, M.; Wyatt, J.S.; Roth, S.; Vargha-Khadem, F.; Gadian, D.; Mishkin, M. Brain and Cognitive-Behavioural Development after Asphyxia at Term Birth. *Dev. Sci.* **2006**, *9*, 350–358. [CrossRef]
49. Matara, D.-I.; Pouliakis, A.; Xanthos, T.; Sokou, R.; Kafalidis, G.; Iliodromiti, Z.; Boutsikou, T.; Iacovidou, N.; Salakos, C. Microbial Translocation and Perinatal Asphyxia/Hypoxia: A Systematic Review. *Diagnostics* **2022**, *12*, 214. [CrossRef] [PubMed]
50. Ferretti, P.; Pasolli, E.; Tett, A.; Asnicar, F.; Gorfer, V.; Fedi, S.; Armanini, F.; Truong, D.T.; Manara, S.; Zolfo, M.; et al. Mother-to-Infant Microbial Transmission from Different Body Sites Shapes the Developing Infant Gut Microbiome. *Cell Host Microbe* **2018**, *24*, 133–145.e5. [CrossRef] [PubMed]
51. Yang, H.; Guo, R.; Li, S.; Liang, F.; Tian, C.; Zhao, X.; Long, Y.; Liu, F.; Jiang, M.; Zhang, Y.; et al. Systematic Analysis of Gut Microbiota in Pregnant Women and Its Correlations with Individual Heterogeneity. *NPJ Biofilms Microbiomes* **2020**, *6*, 32. [CrossRef] [PubMed]
52. Smid, M.C.; Ricks, N.M.; Panzer, A.; Mccoy, A.N.; Azcarate-Peril, M.A.; Keku, T.O.; Boggess, K.A. Maternal Gut Microbiome Biodiversity in Pregnancy. *Am. J. Perinatol.* **2018**, *35*, 24–30.
53. Koren, O.; Goodrich, J.K.; Cullender, T.C.; Spor, A.; Laitinen, K.; Bäckhed, H.K.; Gonzalez, A.; Werner, J.J.; Angenent, L.T.; Knight, R.; et al. Host Remodeling of the Gut Microbiome and Metabolic Changes during Pregnancy. *Cell* **2012**, *150*, 470–480. [CrossRef]
54. Parks, D.H.; Chuvochina, M.; Chaumeil, P.-A.; Rinke, C.; Mussig, A.J.; Hugenholtz, P. A Complete Domain-to-Species Taxonomy for Bacteria and Archaea. *Nat. Biotechnol.* **2020**, *38*, 1079–1086, Erratum in *Nat. Biotechnol.* **2020**, *38*, 1098. [CrossRef]
55. Buttó, L.F.; Haller, D. Dysbiosis in Intestinal Inflammation: Cause or Consequence. *Int. J. Med. Microbiol.* **2016**, *306*, 302–309. [CrossRef]
56. DeGruttola, A.K.; Low, D.; Mizoguchi, A.; Mizoguchi, E. Current Understanding of Dysbiosis in Disease in Human and Animal Models. *Inflamm. Bowel Dis.* **2016**, *22*, 1137–1150. [CrossRef]
57. Vuong, H.E.; Pronovost, G.N.; Williams, D.W.; Coley, E.J.L.; Siegler, E.L.; Qiu, A.; Kazantsev, M.; Wilson, C.J.; Rendon, T.; Hsiao, E.Y. The Maternal Microbiome Modulates Fetal Neurodevelopment in Mice. *Nature* **2020**, *586*, 281–286. [CrossRef]
58. Ruiz-Triviño, J.; Álvarez, D.; Cadavid J, Á.P.; Alvarez, A.M. From Gut to Placenta: Understanding How the Maternal Microbiome Models Life-Long Conditions. *Front. Endocrinol.* **2023**, *14*, 1304727. [CrossRef] [PubMed]
59. Faienza, M.F.; Urbano, F.; Anaclerio, F.; Moscogiuri, L.A.; Konstantinidou, F.; Stuppia, L.; Gatta, V. Exploring Maternal Diet-Epigenetic-Gut Microbiome Crosstalk as an Intervention Strategy to Counter Early Obesity Programming. *Curr. Issues Mol. Biol.* **2024**, *46*, 4358–4378. [CrossRef] [PubMed]
60. Ziętek, M.; Celewicz, Z.; Szczuko, M. Short-Chain Fatty Acids, Maternal Microbiota and Metabolism in Pregnancy. *Nutrients* **2021**, *13*, 1244. [CrossRef]
61. Wu, J.; Wang, K.; Wang, X.; Pang, Y.; Jiang, C. The Role of the Gut Microbiome and Its Metabolites in Metabolic Diseases. *Protein Cell* **2021**, *12*, 360–373. [CrossRef]
62. Beckers, K.F.; Sones, J.L. Maternal Microbiome and the Hypertensive Disorder of Pregnancy, Preeclampsia. *Am. J. Physiol. Heart Circ. Physiol.* **2020**, *318*, H1–H10. [CrossRef] [PubMed]
63. Deady, C.; FitzGerard, J.; Kara, N.; Mazzocchi, M.; O'Mahony, A.; Ionescu, M.I.; Shanahan, R.; Kelly, S.; Crispie, F.; Cotter, P.D.; et al. Maternal Immune Activation and Antibiotics Affect Offspring Neurodevelopment, Behaviour, and Microbiome. *Brain Behav. Immun. Health* **2025**, *48*, 101065. [CrossRef]
64. Mann, P.E.; Huynh, K.; Widmer, G. Maternal High Fat Diet and Its Consequence on the Gut Microbiome: A Rat Model. *Gut Microbes* **2018**, *9*, 143–154. [CrossRef] [PubMed]
65. Gohir, W.; Whelan, F.J.; Surette, M.G.; Moore, C.; Schertzer, J.D.; Sloboda, D.M. Pregnancy-Related Changes in the Maternal Gut Microbiota Are Dependent upon the Mother's Periconceptional Diet. *Gut Microbes* **2015**, *6*, 310–320. [CrossRef]

66. Grant, E.T.; Boudaud, M.; Muller, A.; Macpherson, A.J.; Desai, M.S. Maternal Diet and Gut Microbiome Composition Modulate Early-Life Immune Development. *EMBO Mol. Med.* **2023**, *15*, e17241. [CrossRef]
67. Barrientos, G.; Ronchi, F.; Conrad, M.L. Nutrition during Pregnancy: Influence on the Gut Microbiome and Fetal Development. *Am. J. Reprod. Immunol.* **2024**, *91*, e13802. [CrossRef]
68. García-Mantrana, I.; Selma-Royo, M.; González, S.; Parra-Llorca, A.; Martínez-Costa, C.; Collado, M.C. Distinct Maternal Microbiota Clusters Are Associated with Diet during Pregnancy: Impact on Neonatal Microbiota and Infant Growth during the First 18 Months of Life. *Gut Microbes* **2020**, *11*, 962–978. [CrossRef]
69. Wang, Y.; Xie, T.; Wu, Y.; Liu, Y.; Zou, Z.; Bai, J. Impacts of Maternal Diet and Alcohol Consumption during Pregnancy on Maternal and Infant Gut Microbiota. *Biomolecules* **2021**, *11*, 369. [CrossRef]
70. Degroote, S.; Hunting, D.J.; Baccarelli, A.A.; Takser, L. Maternal Gut and Fetal Brain Connection: Increased Anxiety and Reduced Social Interactions in Wistar Rat Offspring Following Peri-Conceptional Antibiotic Exposure. *Prog. Neuro-Psychopharmacol. Biol. Psychiatry* **2016**, *71*, 76–82. [CrossRef] [PubMed]
71. Chen, Y.; Li, Z.; Tye, K.D.; Luo, H.; Tang, X.; Liao, Y.; Wang, D.; Zhou, J.; Yang, P.; Li, Y.; et al. Probiotic Supplementation during Human Pregnancy Affects the Gut Microbiota and Immune Status. *Front. Cell. Infect. Microbiol.* **2019**, *9*, 254. [CrossRef] [PubMed]
72. Dekker Nitert, M.; Gomez-Arango, L.F.; Barrett, H.L.; McIntyre, H.D.; Anderson, G.J.; Frazer, D.M.; Callaway, L.K. Iron Supplementation Has Minor Effects on Gut Microbiota Composition in Overweight and Obese Women in Early Pregnancy. *Br. J. Nutr.* **2018**, *120*, 283–289. [CrossRef] [PubMed]
73. Nuriel-Ohayon, M.; Neuman, H.; Koren, O. Microbial Changes during Pregnancy, Birth, and Infancy. *Front. Microbiol.* **2016**, *7*, 1031. [CrossRef] [PubMed]
74. Li, M.; Zhang, G.; Cui, L.; Zhang, L.; Zhou, Q.; Mu, C.; Chi, R.; Zhang, N.; Ma, G. Dynamic Changes in Gut Microbiota during Pregnancy among Chinese Women and Influencing Factors: A Prospective Cohort Study. *Front. Microbiol.* **2023**, *14*, 1114228. [CrossRef]
75. Gomez-Arango, L.F.; Barrett, H.L.; McIntyre, H.D.; Callaway, L.K.; Morrison, M.; Dekker Nitert, M.; SPRING Trial Group. Connections between the Gut Microbiome and Metabolic Hormones in Early Pregnancy in Overweight and Obese Women. *Diabetes* **2016**, *65*, 2214–2223. [CrossRef]
76. Torres, J.; Hu, J.; Seki, A.; Eisele, C.; Nair, N.; Huang, R.; Tarassishin, L.; Jharap, B.; Cote-Daigneault, J.; Mao, Q.; et al. Infants Born to Mothers with IBD Present with Altered Gut Microbiome That Transfers Abnormalities of the Adaptive Immune System to Germ-Free Mice. *Gut* **2020**, *69*, 42–51. [CrossRef]
77. Moossavi, S.; Sepehri, S.; Robertson, B.; Bode, L.; Goruk, S.; Field, C.J.; Lix, L.M.; de Souza, R.J.; Becker, A.B.; Mandhane, P.J.; et al. Composition and Variation of the Human Milk Microbiota Are Influenced by Maternal and Early-Life Factors. *Cell Host Microbe* **2019**, *25*, 324–335.e4. [CrossRef]
78. Peng, Y.; Tun, H.M.; Ng, S.C.; Wai, H.K.-F.; Zhang, X.; Parks, J.; Field, C.J.; Mandhane, P.; Moraes, T.J.; Simons, E.; et al. Maternal Smoking during Pregnancy Increases the Risk of Gut Microbiome-Associated Childhood Overweight and Obesity. *Gut Microbes* **2024**, *16*, 2323234. [CrossRef]
79. Santarossa, S.; Sitarik, A.R.; Cassidy-Bushrow, A.E.; Comstock, S.S. Prenatal Physical Activity and the Gut Microbiota of Pregnant Women: Results from a Preliminary Investigation. *Phys. Act. Nutr.* **2023**, *27*, 1–7. [CrossRef]
80. Bressa, C.; Bailén-Andrino, M.; Pérez-Santiago, J.; González-Soltero, R.; Pérez, M.; Montalvo-Lominchar, M.G.; Maté-Muñoz, J.L.; Domínguez, R.; Moreno, D.; Larrosa, M. Differences in Gut Microbiota Profile between Women with Active Lifestyle and Sedentary Women. *PLoS ONE* **2017**, *12*, e0171352. [CrossRef]
81. Ryan, N.; O'Mahony, S.; Leahy-Warren, P.; Philpott, L.; Mulcahy, H. The Impact of Perinatal Maternal Stress on the Maternal and Infant Gut and Human Milk Microbiomes: A Scoping Review. *PLoS ONE* **2025**, *20*, e0318237. [CrossRef]
82. Baker, M. Pregnancy Alters Resident Gut Microbes. *Nature* **2012**, *488*, 123. [CrossRef]
83. Kennedy, K.M.; Plagemann, A.; Sommer, J.; Hofmann, M.; Henrich, W.; Barrett, J.F.R.; Surette, M.G.; Atkinson, S.; Braun, T.; Sloboda, D.M. Parity Modulates Impact of BMI and Gestational Weight Gain on Gut Microbiota in Human Pregnancy. *Gut Microbes* **2023**, *15*, 2259316. [CrossRef]
84. Qi, X.; Yun, C.; Pang, Y.; Qiao, J. The Impact of the Gut Microbiota on the Reproductive and Metabolic Endocrine System. *Gut Microbes* **2021**, *13*, 1894070. [CrossRef]
85. Amato, K.R.; Pradhan, P.; Mallott, E.K.; Shirola, W.; Lu, A. Host-Gut Microbiota Interactions during Pregnancy. *Evol. Med. Public Health* **2024**, *12*, 7–23. [CrossRef]
86. DuPont, H.L.; Salge, M.M.H. The Importance of a Healthy Microbiome in Pregnancy and Infancy and Microbiota Treatment to Reverse Dysbiosis for Improved Health. *Antibiotics* **2023**, *12*, 1617. [CrossRef]
87. Chen, X.; Li, P.; Liu, M.; Zheng, H.; He, Y.; Chen, M.-X.; Tang, W.; Yue, X.; Huang, Y.; Zhuang, L.; et al. Gut Dysbiosis Induces the Development of Pre-Eclampsia through Bacterial Translocation. *Gut* **2020**, *69*, 513–522. [CrossRef]

88. Dunlop, A.L.; Mulle, J.G.; Ferranti, E.P.; Edwards, S.; Dunn, A.B.; Corwin, E.J. Maternal Microbiome and Pregnancy Outcomes That Impact Infant Health: A Review: A Review. *Adv. Neonatal Care* **2015**, *15*, 377–385. [CrossRef] [PubMed]
89. Bauman, M.D.; Iosif, A.-M.; Smith, S.E.P.; Bregere, C.; Amaral, D.G.; Patterson, P.H. Activation of the Maternal Immune System during Pregnancy Alters Behavioral Development of Rhesus Monkey Offspring. *Biol. Psychiatry* **2014**, *75*, 332–341. [CrossRef] [PubMed]
90. Jarde, A.; Lewis-Mikhael, A.-M.; Moayyedi, P.; Stearns, J.C.; Collins, S.M.; Beyene, J.; McDonald, S.D. Pregnancy Outcomes in Women Taking Probiotics or Prebiotics: A Systematic Review and Meta-Analysis. *BMC Pregnancy Childbirth* **2018**, *18*, 14. [CrossRef]
91. Rao, S.C.; Athalye-Jape, G.K.; Deshpande, G.C.; Simmer, K.N.; Patole, S.K. Probiotic Supplementation and Late-Onset Sepsis in Preterm Infants: A Meta-Analysis. *Pediatrics* **2016**, *137*, e20153684. [CrossRef] [PubMed]
92. Aceti, A.; Maggio, L.; Beghetti, I.; Gori, D.; Barone, G.; Callegari, M.L.; Fantini, M.P.; Indrio, F.; Meneghin, F.; Morelli, L.; et al. Probiotics Prevent Late-Onset Sepsis in Human Milk-Fed, Very Low Birth Weight Preterm Infants: Systematic Review and Meta-Analysis. *Nutrients* **2017**, *9*, 904. [CrossRef] [PubMed]
93. Buggio, L.; Somigliana, E.; Borghi, A.; Vercellini, P. Probiotics and Vaginal Microecology: Fact or Fancy? *BMC Womens Health* **2019**, *19*, 25. [CrossRef]
94. Pronovost, G.N.; Yu, K.B.; Coley-O'Rourke, E.J.L.; Telang, S.S.; Chen, A.S.; Vuong, H.E.; Williams, D.W.; Chandra, A.; Rendon, T.K.; Paramo, J.; et al. The Maternal Microbiome Promotes Placental Development in Mice. *Sci. Adv.* **2023**, *9*, eadk1887. [CrossRef]
95. Lopez-Tello, J.; Schofield, Z.; Kiu, R.; Dalby, M.J.; van Sinderen, D.; Le Gall, G.; Sferruzzi-Perri, A.N.; Hall, L.J. Maternal Gut Microbiota Bifidobacterium Promotes Placental Morphogenesis, Nutrient Transport and Fetal Growth in Mice. *Cell. Mol. Life Sci.* **2022**, *79*, 386. [CrossRef]
96. Perez-Muñoz, M.E.; Arrieta, M.-C.; Ramer-Tait, A.E.; Walter, J. A Critical Assessment of the "Sterile Womb" and "in Utero Colonization" Hypotheses: Implications for Research on the Pioneer Infant Microbiome. *Microbiome* **2017**, *5*, 48. [CrossRef]
97. Kennedy, K.M.; de Goffau, M.C.; Perez-Muñoz, M.E.; Arrieta, M.-C.; Bäckhed, F.; Bork, P.; Braun, T.; Bushman, F.D.; Dore, J.; de Vos, W.M.; et al. Questioning the Fetal Microbiome Illustrates Pitfalls of Low-Biomass Microbial Studies. *Nature* **2023**, *613*, 639–649. [CrossRef]
98. Xie, Z.; Chen, Z.; Chai, Y.; Yao, W.; Ma, G. Unveiling the Placental Bacterial Microbiota: Implications for Maternal and Infant Health. *Front. Physiol.* **2025**, *16*, 1544216. [CrossRef]
99. Ygberg, S.; Nilsson, A. The Developing Immune System—From Foetus to Toddler. *Acta Paediatr.* **2012**, *101*, 120–127. [CrossRef]
100. McDavid, A.; Laniewski, N.; Grier, A.; Gill, A.L.; Kessler, H.A.; Huyck, H.; Carbonell, E.; Holden-Wiltse, J.; Bandyopadhyay, S.; Carnahan, J.; et al. Aberrant Newborn T Cell and Microbiota Developmental Trajectories Predict Respiratory Compromise during Infancy. *iScience* **2022**, *25*, 104007. [CrossRef] [PubMed]
101. Gao, Y.; O'Hely, M.; Quinn, T.P.; Ponsonby, A.-L.; Harrison, L.C.; Frøkiær, H.; Tang, M.L.K.; Brix, S.; Kristiansen, K.; Burgner, D.; et al. Maternal Gut Microbiota during Pregnancy and the Composition of Immune Cells in Infancy. *Front. Immunol.* **2022**, *13*, 986340. [CrossRef] [PubMed]
102. Gonzalez-Perez, G.; Hicks, A.L.; Tekieli, T.M.; Radens, C.M.; Williams, B.L.; Lamousé-Smith, E.S.N. Maternal Antibiotic Treatment Impacts Development of the Neonatal Intestinal Microbiome and Antiviral Immunity. *J. Immunol.* **2016**, *196*, 3768–3779. [CrossRef]
103. Levan, S.R.; Stamnes, K.A.; Lin, D.L.; Panzer, A.R.; Fukui, E.; McCauley, K.; Fujimura, K.E.; McKean, M.; Ownby, D.R.; Zoratti, E.M.; et al. Elevated Faecal 12,13-DiHOME Concentration in Neonates at High Risk for Asthma Is Produced by Gut Bacteria and Impedes Immune Tolerance. *Nat. Microbiol.* **2019**, *4*, 1851–1861, Erratum in *Nat. Microbiol.* **2019**, *4*, 2020. [CrossRef]
104. Gao, Y.; Nanan, R.; Macia, L.; Tan, J.; Sominsky, L.; Quinn, T.P.; O'Hely, M.; Ponsonby, A.-L.; Tang, M.L.K.; Collier, F.; et al. The Maternal Gut Microbiome during Pregnancy and Offspring Allergy and Asthma. *J. Allergy Clin. Immunol.* **2021**, *148*, 669–678. [CrossRef]
105. de Moraes-Pinto, M.I.; Suano-Souza, F.; Aranda, C.S. Immune System: Development and Acquisition of Immunological Competence. *J. Pediatr. (Rio J.)* **2021**, *97* (Suppl. 1), S59–S66. [CrossRef]
106. Timm, S.; Schlünssen, V.; Olsen, J.; Ramlau-Hansen, C.H. Prenatal Antibiotics and Atopic Dermatitis among 18-Month-Old Children in the Danish National Birth Cohort. *Clin. Exp. Allergy* **2017**, *47*, 929–936. [CrossRef]
107. Faas, M.M.; Smink, A.M. Shaping Immunity: The Influence of the Maternal Gut Bacteria on Fetal Immune Development. *Semin. Immunopathol.* **2025**, *47*, 13. [CrossRef]
108. Sun, Y.; Huang, S.; Li, M.; Yang, Y.; Ma, J.; Xie, R.; Wang, J.; Zhao, Q.; Qin, S.; He, L.; et al. Maternal High-Fat Diet Disrupts Intestinal Mucus Barrier of Offspring by Regulating Gut Immune Receptor LRRC19. *Commun. Biol.* **2025**, *8*, 402. [CrossRef] [PubMed]
109. Babu, S.T.; Niu, X.; Raetz, M.; Savani, R.C.; Hooper, L.V.; Mirpuri, J. Maternal High-Fat Diet Results in Microbiota-Dependent Expansion of ILC3s in Mice Offspring. *JCI Insight* **2018**, *3*, e99223. [CrossRef]

110. Wallace, J.G.; Bellissimo, C.J.; Yeo, E.; Fei Xia, Y.; Petrik, J.J.; Surette, M.G.; Bowdish, D.M.E.; Sloboda, D.M. Obesity during Pregnancy Results in Maternal Intestinal Inflammation, Placental Hypoxia, and Alters Fetal Glucose Metabolism at Mid-Gestation. *Sci. Rep.* **2019**, *9*, 17621. [CrossRef]
111. Liang, W.; Feng, Y.; Yang, D.; Qin, J.; Zhi, X.; Wu, W.; Jie, Q. Oral Probiotics Increased the Proportion of Treg, Tfr, and Breg Cells to Inhibit the Inflammatory Response and Impede Gestational Diabetes Mellitus. *Mol. Med.* **2023**, *29*, 122. [CrossRef] [PubMed]
112. Brosseau, C.; Selle, A.; Duval, A.; Misme-Aucouturier, B.; Chesneau, M.; Brouard, S.; Cherbuy, C.; Cariou, V.; Bouchaud, G.; Mincham, K.T.; et al. Prebiotic Supplementation during Pregnancy Modifies the Gut Microbiota and Increases Metabolites in Amniotic Fluid, Driving a Tolerogenic Environment in Utero. *Front. Immunol.* **2021**, *12*, 712614. [CrossRef]
113. Hu, M.; Eviston, D.; Hsu, P.; Mariño, E.; Chidgey, A.; Santner-Nanan, B.; Wong, K.; Richards, J.L.; Yap, Y.A.; Collier, F.; et al. Decreased Maternal Serum Acetate and Impaired Fetal Thymic and Regulatory T Cell Development in Preeclampsia. *Nat. Commun.* **2019**, *10*, 3031. [CrossRef]
114. Tan, J.; McKenzie, C.; Potamitis, M.; Thorburn, A.N.; Mackay, C.R.; Macia, L. The Role of Short-Chain Fatty Acids in Health and Disease. *Adv. Immunol.* **2014**, *121*, 91–119.
115. Thorburn, A.N.; McKenzie, C.I.; Shen, S.; Stanley, D.; Macia, L.; Mason, L.J.; Roberts, L.K.; Wong, C.H.Y.; Shim, R.; Robert, R.; et al. Evidence That Asthma Is a Developmental Origin Disease Influenced by Maternal Diet and Bacterial Metabolites. *Nat. Commun.* **2015**, *6*, 7320. [CrossRef]
116. Furusawa, Y.; Obata, Y.; Fukuda, S.; Endo, T.A.; Nakato, G.; Takahashi, D.; Nakanishi, Y.; Uetake, C.; Kato, K.; Kato, T.; et al. Commensal Microbe-Derived Butyrate Induces the Differentiation of Colonic Regulatory T Cells. *Nature* **2013**, *504*, 446–450, Erratum in *Nature* **2014**, *506*, 254. [CrossRef]
117. Jain, N. The Early Life Education of the Immune System: Moms, Microbes and (Missed) Opportunities. *Gut Microbes* **2020**, *12*, 1824564. [CrossRef]
118. Humann, J.; Mann, B.; Gao, G.; Moresco, P.; Ramahi, J.; Loh, L.N.; Farr, A.; Hu, Y.; Durick-Eder, K.; Fillon, S.A.; et al. Bacterial Peptidoglycan Traverses the Placenta to Induce Fetal Neuroproliferation and Aberrant Postnatal Behavior. *Cell Host Microbe* **2016**, *19*, 388–399. [CrossRef]
119. Gomez de Agüero, M.; Ganal-Vonarburg, S.C.; Fuhrer, T.; Rupp, S.; Uchimura, Y.; Li, H.; Steinert, A.; Heikenwalder, M.; Hapfelmeier, S.; Sauer, U.; et al. The Maternal Microbiota Drives Early Postnatal Innate Immune Development. *Science* **2016**, *351*, 1296–1302. [CrossRef] [PubMed]
120. Kaisanlahti, A.; Turunen, J.; Byts, N.; Samoylenko, A.; Bart, G.; Virtanen, N.; Tejesvi, M.V.; Zhyvolozhnyi, A.; Sarfraz, S.; Kumpula, S.; et al. Maternal Microbiota Communicates with the Fetus through Microbiota-Derived Extracellular Vesicles. *Microbiome* **2023**, *11*, 249. [CrossRef]
121. Li, Q.; Barres, B.A. Microglia and Macrophages in Brain Homeostasis and Disease. *Nat. Rev. Immunol.* **2018**, *18*, 225–242. [CrossRef]
122. Zengeler, K.E.; Lukens, J.R. Innate Immunity at the Crossroads of Healthy Brain Maturation and Neurodevelopmental Disorders. *Nat. Rev. Immunol.* **2021**, *21*, 454–468. [CrossRef]
123. Yeo, X.Y.; Choi, Y.; Hong, Y.; Kwon, H.N.; Jung, S. Contemporary Insights into Neuroimmune Interactions across Development and Aging. *Front. Neurol.* **2025**, *16*, 1611124. [CrossRef] [PubMed]
124. Gohlke, J.M.; Stockton, P.S.; Sieber, S.; Foley, J.; Portier, C.J. AhR-Mediated Gene Expression in the Developing Mouse Telencephalon. *Reprod. Toxicol.* **2009**, *28*, 321–328. [CrossRef]
125. Kim, S.; Kim, H.; Yim, Y.S.; Ha, S.; Atarashi, K.; Tan, T.G.; Longman, R.S.; Honda, K.; Littman, D.R.; Choi, G.B.; et al. Maternal Gut Bacteria Promote Neurodevelopmental Abnormalities in Mouse Offspring. *Nature* **2017**, *549*, 528–532. [CrossRef]
126. Kim, E.; Paik, D.; Ramirez, R.N.; Biggs, D.G.; Park, Y.; Kwon, H.-K.; Choi, G.B.; Huh, J.R. Maternal Gut Bacteria Drive Intestinal Inflammation in Offspring with Neurodevelopmental Disorders by Altering the Chromatin Landscape of CD4+ T Cells. *Immunity* **2022**, *55*, 145–158.e7. [CrossRef]
127. Choi, G.B.; Yim, Y.S.; Wong, H.; Kim, S.; Kim, H.; Kim, S.V.; Hoeffer, C.A.; Littman, D.R.; Huh, J.R. The Maternal Interleukin-17a Pathway in Mice Promotes Autism-like Phenotypes in Offspring. *Science* **2016**, *351*, 933–939. [CrossRef]
128. Sun, Z.; Lee-Sarwar, K.; Kelly, R.S.; Lasky-Su, J.A.; Litonjua, A.A.; Weiss, S.T.; Liu, Y.-Y. Revealing the Importance of Prenatal Gut Microbiome in Offspring Neurodevelopment in Humans. *EBioMedicine* **2023**, *90*, 104491. [CrossRef]
129. Surzenko, N.; Pjetri, E.; Munson, C.A.; Friday, W.B.; Hauser, J.; Mitchell, E.S. Prenatal Exposure to the Probiotic Lactococcus Lactis Decreases Anxiety-like Behavior and Modulates Cortical Cytoarchitecture in a Sex Specific Manner. *PLoS ONE* **2020**, *15*, e0223395. [CrossRef]
130. Sarker, G.; Peleg-Raibstein, D. Maternal Overnutrition Induces Long-Term Cognitive Deficits across Several Generations. *Nutrients* **2018**, *11*, 7. [CrossRef] [PubMed]

131. Liu, X.; Li, X.; Xia, B.; Jin, X.; Zou, Q.; Zeng, Z.; Zhao, W.; Yan, S.; Li, L.; Yuan, S.; et al. High-Fiber Diet Mitigates Maternal Obesity-Induced Cognitive and Social Dysfunction in the Offspring via Gut-Brain Axis. *Cell Metab.* **2021**, *33*, 923–938.e6. [CrossRef]
132. Falsaperla, R.; Sciuto, S.; Gioè, D.; Sciuto, L.; Pisani, F.; Pavone, P.; Ruggieri, M. Mild Hypoxic-Ischemic Encephalopathy: Can Neurophysiological Monitoring Predict Unfavorable Neurological Outcome? A Systematic Review and Meta-Analysis. *Am. J. Perinatol.* **2023**, *40*, 833–838. [CrossRef]
133. Hayes, B.C.; Ryan, S.; McGarvey, C.; Mulvany, S.; Doherty, E.; Grehan, A.; Madigan, C.; Matthews, T.; King, M.D. Brain Magnetic Resonance Imaging and Outcome after Hypoxic Ischaemic Encephalopathy. *J. Matern. Fetal. Neonatal Med.* **2016**, *29*, 777–782. [CrossRef] [PubMed]
134. Törn, A.E.; Hesselman, S.; Johansen, K.; Ågren, J.; Wikström, A.-K.; Jonsson, M. Outcomes in Children after Mild Neonatal Hypoxic Ischaemic Encephalopathy: A Population-Based Cohort Study. *BJOG* **2023**, *130*, 1602–1609. [CrossRef]
135. DuPont, T.L.; Chalak, L.F.; Morriss, M.C.; Burchfield, P.J.; Christie, L.; Sánchez, P.J. Short-Term Outcomes of Newborns with Perinatal Acidemia Who Are Not Eligible for Systemic Hypothermia Therapy. *J. Pediatr.* **2013**, *162*, 35–41. [CrossRef]
136. Chen, A.; Teng, C.; Wei, J.; Wu, X.; Zhang, H.; Chen, P.; Cai, D.; Qian, H.; Zhu, H.; Zheng, X.; et al. Gut Microbial Dysbiosis Exacerbates Long-Term Cognitive Impairments by Promoting Intestinal Dysfunction and Neuroinflammation Following Neonatal Hypoxia-Ischemia. *Gut Microbes* **2025**, *17*, 2471015. [CrossRef]
137. Hsu, C.-Y.; Khachatryan, L.G.; Younis, N.K.; Mustafa, M.A.; Ahmad, N.; Athab, Z.H.; Polyanskaya, A.V.; Kasanave, E.V.; Mirzaei, R.; Karampoor, S. Microbiota-Derived Short Chain Fatty Acids in Pediatric Health and Diseases: From Gut Development to Neuroprotection. *Front. Microbiol.* **2024**, *15*, 1456793. [CrossRef]
138. Chakkarapani, A.A.; Aly, H.; Benders, M.; Cotten, C.M.; El-Dib, M.; Gressens, P.; Hagberg, H.; Sabir, H.; Wintermark, P.; Robertson, N.J.; et al. Therapies for Neonatal Encephalopathy: Targeting the Latent, Secondary and Tertiary Phases of Evolving Brain Injury. *Semin. Fetal Neonatal Med.* **2021**, *26*, 101256. [CrossRef] [PubMed]
139. Perlman, J.M. Intervention Strategies for Neonatal Hypoxic-Ischemic Cerebral Injury. *Clin. Ther.* **2006**, *28*, 1353–1365. [CrossRef]
140. Aly, H.; Khashaba, M.T.; El-Ayouty, M.; El-Sayed, O.; Hasanein, B.M. IL-1beta, IL-6 and TNF-Alpha and Outcomes of Neonatal Hypoxic Ischemic Encephalopathy. *Brain Dev.* **2006**, *28*, 178–182. [CrossRef]
141. O'Dea, M.I.; Kelly, L.A.; McKenna, E.; Strickland, T.; Hurley, T.P.; Butler, J.; Vavasseur, C.; El-Khuffash, A.F.; Miletin, J.; Fallah, L.; et al. Altered Cytokine Endotoxin Responses in Neonatal Encephalopathy Predict MRI Outcomes. *Front. Pediatr.* **2021**, *9*, 734540. [CrossRef]
142. Ziemka-Nalecz, M.; Jaworska, J.; Sypecka, J.; Polowy, R.; Filipkowski, R.K.; Zalewska, T. Sodium Butyrate, a Histone Deacetylase Inhibitor, Exhibits Neuroprotective/Neurogenic Effects in a Rat Model of Neonatal Hypoxia-Ischemia. *Mol. Neurobiol.* **2017**, *54*, 5300–5318. [CrossRef]
143. Lewis, K.; Lutgendorff, F.; Phan, V.; Söderholm, J.D.; Sherman, P.M.; McKay, D.M. Enhanced Translocation of Bacteria across Metabolically Stressed Epithelia Is Reduced by Butyrate. *Inflamm. Bowel Dis.* **2010**, *16*, 1138–1148. [CrossRef] [PubMed]
144. Knox, E.G.; Aburto, M.R.; Tessier, C.; Nagpal, J.; Clarke, G.; O'Driscoll, C.M.; Cryan, J.F. Microbial-Derived Metabolites Induce Actin Cytoskeletal Rearrangement and Protect Blood-Brain Barrier Function. *iScience* **2022**, *25*, 105648. [CrossRef]
145. Wojciech, L.; Tan, K.S.W.; Gascoigne, N.R.J. Taming the Sentinels: Microbiome-Derived Metabolites and Polarization of T Cells. *Int. J. Mol. Sci.* **2020**, *21*, 7740. [CrossRef]
146. Mancino, S.; Boraso, M.; Galmozzi, A.; Serafini, M.M.; De Fabiani, E.; Crestani, M.; Viviani, B. Dose-Dependent Dual Effects of HDAC Inhibitors on Glial Inflammatory Response. *Sci. Rep.* **2025**, *15*, 12262. [CrossRef] [PubMed]
147. Leus, N.G.; Zwinderman, M.R.; Dekker, F.J. Histone Deacetylase 3 (HDAC 3) as Emerging Drug Target in NF-KB-Mediated Inflammation. *Curr. Opin. Chem. Biol.* **2016**, *33*, 160–168. [CrossRef]
148. Caetano-Silva, M.E.; Rund, L.; Hutchinson, N.T.; Woods, J.A.; Steelman, A.J.; Johnson, R.W. Inhibition of Inflammatory Microglia by Dietary Fiber and Short-Chain Fatty Acids. *Sci. Rep.* **2023**, *13*, 2819. [CrossRef]
149. Zhou, Z.; Xu, N.; Matei, N.; McBride, D.W.; Ding, Y.; Liang, H.; Tang, J.; Zhang, J.H. Sodium Butyrate Attenuated Neuronal Apoptosis via GPR41/Gβγ/PI3K/Akt Pathway after MCAO in Rats. *J. Cereb. Blood Flow Metab.* **2021**, *41*, 267–281. [CrossRef]
150. Sun, J.; Lu, L.; Lian, Y.; Xu, S.; Zhu, Y.; Wu, Y.; Lin, Q.; Hou, J.; Li, Y.; Yu, Z. Sodium Butyrate Attenuates Microglia-Mediated Neuroinflammation by Modulating the TLR4/MyD88/NF-KB Pathway and Microbiome-Gut-Brain Axis in Cardiac Arrest Mice. *Mol. Brain* **2025**, *18*, 13. [CrossRef] [PubMed]
151. Zhang, R.; Chen, X.; Shen, Q.; Liu, L.; Hou, X.; Liu, N. Sodium Butyrate Improves the Effects of Brain Injury in a Small-for-Gestational-Age Rat Model by Activating the JAK1/STAT3 Pathway. *J. Neuropathol. Exp. Neurol.* **2025**, nlaf085. [CrossRef]
152. Dietz, R.M.; Wright, C.J. Oxidative Stress Diseases Unique to the Perinatal Period: A Window into the Developing Innate Immune Response. *Am. J. Reprod. Immunol.* **2018**, *79*, e12787. [CrossRef] [PubMed]

153. Silva, Y.P.; Bernardi, A.; Frozza, R.L. The Role of Short-Chain Fatty Acids from Gut Microbiota in Gut-Brain Communication. *Front. Endocrinol.* **2020**, *11*, 25. [CrossRef]
154. Hoyles, L.; Snelling, T.; Umlai, U.-K.; Nicholson, J.K.; Carding, S.R.; Glen, R.C.; McArthur, S. Microbiome-Host Systems Interactions: Protective Effects of Propionate upon the Blood-Brain Barrier. *Microbiome* **2018**, *6*, 55. [CrossRef] [PubMed]
155. Wei, J.; Chen, A.; Huang, D.; Teng, C.; Cai, D.; Wu, X.; Wang, T.; Hu, W.; Huang, Z.; Wang, P.; et al. Gut Microbiome-Derived Lipopolysaccharides Aggravate Cognitive Impairment via TLR4-Mediated Inflammatory Signaling in Neonatal Rats Following Hypoxic-Ischemic Brain Damage. *Brain Behav. Immun.* **2025**, *127*, 4–24. [CrossRef] [PubMed]
156. Kaisar, M.M.M.; Pelgrom, L.R.; van der Ham, A.J.; Yazdanbakhsh, M.; Everts, B. Butyrate Conditions Human Dendritic Cells to Prime Type 1 Regulatory T Cells via Both Histone Deacetylase Inhibition and G Protein-Coupled Receptor 109A Signaling. *Front. Immunol.* **2017**, *8*, 1429. [CrossRef]
157. Tao, R.; de Zoeten, E.F.; Ozkaynak, E.; Chen, C.; Wang, L.; Porrett, P.M.; Li, B.; Turka, L.A.; Olson, E.N.; Greene, M.I.; et al. Deacetylase Inhibition Promotes the Generation and Function of Regulatory T Cells. *Nat. Med.* **2007**, *13*, 1299–1307. [CrossRef]
158. Pessa-Morikawa, T.; Husso, A.; Kärkkäinen, O.; Koistinen, V.; Hanhineva, K.; Iivanainen, A.; Niku, M. Maternal Microbiota-Derived Metabolic Profile in Fetal Murine Intestine, Brain and Placenta. *BMC Microbiol.* **2022**, *22*, 46. [CrossRef]
159. Peesh, P.; Blasco-Conesa, M.P.; El Hamamy, A.; Khan, R.; Guzman, G.U.; Honarpisheh, P.; Mohan, E.C.; Goodman, G.W.; Nguyen, J.N.; Banerjee, A.; et al. Benefits of Equilibrium between Microbiota- and Host-Derived Ligands of the Aryl Hydrocarbon Receptor after Stroke in Aged Male Mice. *Nat. Commun.* **2025**, *16*, 1767. [CrossRef]
160. Miyamoto, K.; Sujino, T.; Kanai, T. The Tryptophan Metabolic Pathway of the Microbiome and Host Cells in Health and Disease. *Int. Immunol.* **2024**, *36*, 601–616. [CrossRef]
161. Honarpisheh, P.; Bryan, R.M.; McCullough, L.D. Aging Microbiota-Gut-Brain Axis in Stroke Risk and Outcome. *Circ. Res.* **2022**, *130*, 1112–1144. [CrossRef] [PubMed]
162. Wang, T.; Chen, B.; Luo, M.; Xie, L.; Lu, M.; Lu, X.; Zhang, S.; Wei, L.; Zhou, X.; Yao, B.; et al. Microbiota-Indole 3-Propionic Acid-Brain Axis Mediates Abnormal Synaptic Pruning of Hippocampal Microglia and Susceptibility to ASD in IUGR Offspring. *Microbiome* **2023**, *11*, 245. [CrossRef]
163. Li, Y.; Wang, T.; Sun, P.; Zhu, W.; Chen, Y.; Chen, M.; Yang, X.; Du, X.; Zhao, Y. Farrerol Alleviates Hypoxic-Ischemic Encephalopathy by Inhibiting Ferroptosis in Neonatal Rats via the Nrf2 Pathway. *Physiol. Res.* **2023**, *72*, 511–520. [CrossRef] [PubMed]
164. Peeples, E.S.; Genaro-Mattos, T.C. Ferroptosis: A Promising Therapeutic Target for Neonatal Hypoxic-Ischemic Brain Injury. *Int. J. Mol. Sci.* **2022**, *23*, 7420. [CrossRef] [PubMed]
165. Fang, Y.; Zheng, Y.; Gao, Q.; Pang, M.; Wu, Y.; Feng, X.; Tao, X.; Hu, Y.; Lin, Z.; Lin, W. Activation of the Nrf2/Keap1 Signaling Pathway Mediates the Neuroprotective Effect of Perillyl Alcohol against Cerebral Hypoxic-Ischemic Damage in Neonatal Rats. *Redox Rep.* **2024**, *29*, 2394714. [CrossRef]
166. Liu, J.-X.; Zheng, D.; Chen, L.; Chen, S.; Min, J.-W. Nuclear Factor Erythroid 2-Related Factor 2 as a Potential Therapeutic Target in Neonatal Hypoxic-Ischemic Encephalopathy. *J. Integr. Neurosci.* **2024**, *23*, 103. [CrossRef]
167. Li, Y.; Liu, H.; Zhang, L.; Yang, Y.; Lin, Y.; Zhuo, Y.; Fang, Z.; Che, L.; Feng, B.; Xu, S.; et al. Maternal Dietary Fiber Composition during Gestation Induces Changes in Offspring Antioxidative Capacity, Inflammatory Response, and Gut Microbiota in a Sow Model. *Int. J. Mol. Sci.* **2019**, *21*, 31. [CrossRef]
168. Zhang, J.; Wang, J.; Ma, Z.; Fu, Z.; Zhao, Y.; Zeng, X.; Lin, G.; Zhang, S.; Guan, W.; Chen, F. Enhanced Antioxidative Capacity Transfer between Sow and Fetus via the Gut-Placenta Axis with Dietary Selenium Yeast and Glycerol Monolaurate Supplementation during Pregnancy. *Antioxidants* **2024**, *13*, 141. [CrossRef]
169. Singh, R.; Chandrashekharappa, S.; Bodduluri, S.R.; Baby, B.V.; Hegde, B.; Kotla, N.G.; Hiwale, A.A.; Saiyed, T.; Patel, P.; Vijay-Kumar, M.; et al. Enhancement of the Gut Barrier Integrity by a Microbial Metabolite through the Nrf2 Pathway. *Nat. Commun.* **2019**, *10*, 89. [CrossRef]
170. Kunst, C.; Schmid, S.; Michalski, M.; Tümen, D.; Buttenschön, J.; Müller, M.; Gülow, K. The Influence of Gut Microbiota on Oxidative Stress and the Immune System. *Biomedicines* **2023**, *11*, 1388. [CrossRef]
171. Uchiyama, J.; Akiyama, M.; Hase, K.; Kumagai, Y.; Kim, Y.-G. Gut Microbiota Reinforce Host Antioxidant Capacity via the Generation of Reactive Sulfur Species. *Cell Rep.* **2022**, *38*, 110479. [CrossRef]
172. Xu, K.; Liu, G.; Fu, C. The Tryptophan Pathway Targeting Antioxidant Capacity in the Placenta. *Oxidative Medicine and Cellular Longevity* **2018**, *2018*, 1054797. [CrossRef]
173. Cömert, T.K.; Akpinar, F.; Erkaya, S.; Durmaz, B.; Durmaz, R. The Effect of Gestational Weight Gain on Serum Total Oxidative Stress, Total Antioxidant Capacity and Gut Microbiota. *Biosci. Microbiota Food Health* **2022**, *41*, 160–167. [CrossRef]
174. Du, X.; Elsabagh, M.; He, F.; Wu, H.; Zhang, B.; Fan, K.; Wang, M.; Zhang, H. Gut Microbiota and Its Metabolites Modulate Pregnancy Outcomes by Regulating Placental Autophagy and Ferroptosis. *Antioxidants* **2025**, *14*, 970. [CrossRef]

175. Jithoo, A.; Penny, T.R.; Pham, Y.; Sutherland, A.E.; Smith, M.J.; Petraki, M.; Fahey, M.C.; Jenkin, G.; Malhotra, A.; Miller, S.L.; et al. The Temporal Relationship between Blood-Brain Barrier Integrity and Microglial Response Following Neonatal Hypoxia Ischemia. *Cells* **2024**, *13*, 660. [CrossRef]
176. Braniste, V.; Al-Asmakh, M.; Kowal, C.; Anuar, F.; Abbaspour, A.; Tóth, M.; Korecka, A.; Bakocevic, N.; Ng, L.G.; Kundu, P.; et al. The Gut Microbiota Influences Blood-Brain Barrier Permeability in Mice. *Sci. Transl. Med.* **2014**, *6*, 263ra158, Erratum in *Sci. Transl. Med.* **2014**, *6*, 266er7. [CrossRef]
177. González-Bosch, C.; Boorman, E.; Zunszain, P.A.; Mann, G.E. Short-Chain Fatty Acids as Modulators of Redox Signaling in Health and Disease. *Redox Biol.* **2021**, *47*, 102165. [CrossRef]
178. Sun, J.; Wang, F.; Li, H.; Zhang, H.; Jin, J.; Chen, W.; Pang, M.; Yu, J.; He, Y.; Liu, J.; et al. Neuroprotective Effect of Sodium Butyrate against Cerebral Ischemia/Reperfusion Injury in Mice. *BioMed Res. Int.* **2015**, *2015*, 395895. [CrossRef]
179. Sun, J.; Ling, Z.; Wang, F.; Chen, W.; Li, H.; Jin, J.; Zhang, H.; Pang, M.; Yu, J.; Liu, J. Clostridium Butyricum Pretreatment Attenuates Cerebral Ischemia/Reperfusion Injury in Mice via Anti-Oxidation and Anti-Apoptosis. *Neurosci. Lett.* **2016**, *613*, 30–35. [CrossRef] [PubMed]
180. Park, M.J.; Sohrabji, F. The Histone Deacetylase Inhibitor, Sodium Butyrate, Exhibits Neuroprotective Effects for Ischemic Stroke in Middle-Aged Female Rats. *J. Neuroinflamm.* **2016**, *13*, 300. [CrossRef] [PubMed]
181. Liu, X.; Tao, X.; Xiong, Z.; Wang, H.; Zeng, L. Mechanism of HDAC2 Regulating Nrf2 Acetylation Level in Neuronal Ferroptosis of Neonatal Rats with Hypoxic-Ischemic Brain Injury. *Brain Inj.* **2025**, *39*, 635–645. [CrossRef]
182. Hwang, I.K.; Yoo, K.-Y.; Li, H.; Park, O.K.; Lee, C.H.; Choi, J.H.; Jeong, Y.-G.; Lee, Y.L.; Kim, Y.-M.; Kwon, Y.-G.; et al. Indole-3-Propionic Acid Attenuates Neuronal Damage and Oxidative Stress in the Ischemic Hippocampus. *J. Neurosci. Res.* **2009**, *87*, 2126–2137. [CrossRef] [PubMed]
183. Basak, S.; Das, R.K.; Banerjee, A.; Paul, S.; Pathak, S.; Duttaroy, A.K. Maternal Obesity and Gut Microbiota Are Associated with Fetal Brain Development. *Nutrients* **2022**, *14*, 4515. [CrossRef]
184. Radford-Smith, D.E.; Probert, F.; Burnet, P.W.J.; Anthony, D.C. Modifying the Maternal Microbiota Alters the Gut-Brain Metabolome and Prevents Emotional Dysfunction in the Adult Offspring of Obese Dams. *Proc. Natl. Acad. Sci. USA* **2022**, *119*, e2108581119. [CrossRef] [PubMed]
185. He, X.; Zhang, T.; Zeng, Y.; Pei, P.; Liu, Y.; Jia, W.; Zhao, H.; Bi, M.; Wang, S. Sodium Butyrate Mediates Histone Crotonylation and Alleviated Neonatal Rats Hypoxic-Ischemic Brain Injury through Gut-Brain Axis. *Front. Microbiol.* **2022**, *13*, 993146. [CrossRef]
186. Chen, A.; Xiong, L.-J.; Tong, Y.; Mao, M. The Neuroprotective Roles of BDNF in Hypoxic Ischemic Brain Injury. *Biomed. Rep.* **2013**, *1*, 167–176. [CrossRef]
187. Urbonaite, G.; Knyzeliene, A.; Bunn, F.S.; Smalskys, A.; Neniskyte, U. The Impact of Maternal High-Fat Diet on Offspring Neurodevelopment. *Front. Neurosci.* **2022**, *16*, 909762. [CrossRef]
188. Fusco, S.; Spinelli, M.; Cocco, S.; Ripoli, C.; Mastrodonato, A.; Natale, F.; Rinaudo, M.; Livrizzi, G.; Grassi, C. Maternal Insulin Resistance Multigenerationally Impairs Synaptic Plasticity and Memory via Gametic Mechanisms. *Nat. Commun.* **2019**, *10*, 4799. [CrossRef] [PubMed]
189. Page, K.C.; Jones, E.K.; Anday, E.K. Maternal and Postweaning High-Fat Diets Disturb Hippocampal Gene Expression, Learning, and Memory Function. *Am. J. Physiol. Regul. Integr. Comp. Physiol.* **2014**, *306*, R527–R537. [CrossRef]
190. Niculescu, M.D.; Lupu, D.S. High Fat Diet-Induced Maternal Obesity Alters Fetal Hippocampal Development. *Int. J. Dev. Neurosci.* **2009**, *27*, 627–633. [CrossRef]
191. Mastenbroek, L.J.M.; Kooistra, S.M.; Eggen, B.J.L.; Prins, J.R. The Role of Microglia in Early Neurodevelopment and the Effects of Maternal Immune Activation. *Semin. Immunopathol.* **2024**, *46*, 1. [CrossRef]
192. Cavaliere, G.; Catapano, A.; Trinchese, G.; Cimmino, F.; Penna, E.; Pizzella, A.; Cristiano, C.; Lama, A.; Crispino, M.; Mollica, M.P. Butyrate Improves Neuroinflammation and Mitochondrial Impairment in Cerebral Cortex and Synaptic Fraction in an Animal Model of Diet-Induced Obesity. *Antioxidants* **2022**, *12*, 4. [CrossRef]
193. Jiang, Y.; Li, K.; Li, X.; Xu, L.; Yang, Z. Sodium Butyrate Ameliorates the Impairment of Synaptic Plasticity by Inhibiting the Neuroinflammation in 5XFAD Mice. *Chem. Biol. Interact.* **2021**, *341*, 109452. [CrossRef]
194. Jaworska, J.; Ziemka-Nalecz, M.; Sypecka, J.; Zalewska, T. The Potential Neuroprotective Role of a Histone Deacetylase Inhibitor, Sodium Butyrate, after Neonatal Hypoxia-Ischemia. *J. Neuroinflamm.* **2017**, *14*, 34. [CrossRef]
195. Jaworska, J.; Zalewska, T.; Sypecka, J.; Ziemka-Nalecz, M. Effect of the HDAC Inhibitor, Sodium Butyrate, on Neurogenesis in a Rat Model of Neonatal Hypoxia-Ischemia: Potential Mechanism of Action. *Mol. Neurobiol.* **2019**, *56*, 6341–6370. [CrossRef]
196. Wang, Y.; Zhu, J.; Zou, N.; Zhang, L.; Wang, Y.; Zhang, M.; Wang, C.; Yang, L. Pathogenesis from the Microbial-Gut-Brain Axis in White Matter Injury in Preterm Infants: A Review. *Front. Integr. Neurosci.* **2023**, *17*, 1051689. [CrossRef]
197. Buffington, S.A.; Di Prisco, G.V.; Auchtung, T.A.; Ajami, N.J.; Petrosino, J.F.; Costa-Mattioli, M. Microbial Reconstitution Reverses Maternal Diet-Induced Social and Synaptic Deficits in Offspring. *Cell* **2016**, *165*, 1762–1775. [CrossRef]

198. Poutahidis, T.; Kearney, S.M.; Levkovich, T.; Qi, P.; Varian, B.J.; Lakritz, J.R.; Ibrahim, Y.M.; Chatzigiagkos, A.; Alm, E.J.; Erdman, S.E. Microbial Symbionts Accelerate Wound Healing via the Neuropeptide Hormone Oxytocin. *PLoS ONE* **2013**, *8*, e78898. [CrossRef] [PubMed]
199. Sgritta, M.; Dooling, S.W.; Buffington, S.A.; Momin, E.N.; Francis, M.B.; Britton, R.A.; Costa-Mattioli, M. Mechanisms Underlying Microbial-Mediated Changes in Social Behavior in Mouse Models of Autism Spectrum Disorder. *Neuron* **2019**, *101*, 246–259.e6. [CrossRef]
200. Ibrahim, Y.M.; Kearney, S.M.; Levkovich, T.; Springer, A.; Mirabal, S.; Poutahidis, T.; Varian, B.J.; Lakritz, J.R.; Alm, E.J.; Erdman, S.E. Maternal Gut Microbes Control Offspring Sex and Survival. *J. Probiotics Health* **2014**, *2*, 1.
201. Tyzio, R.; Cossart, R.; Khalilov, I.; Minlebaev, M.; Hübner, C.A.; Represa, A.; Ben-Ari, Y.; Khazipov, R. Maternal Oxytocin Triggers a Transient Inhibitory Switch in GABA Signaling in the Fetal Brain during Delivery. *Science* **2006**, *314*, 1788–1792. [CrossRef]
202. Ceanga, M.; Spataru, A.; Zagrean, A.-M. Oxytocin Is Neuroprotective against Oxygen-Glucose Deprivation and Reoxygenation in Immature Hippocampal Cultures. *Neurosci. Lett.* **2010**, *477*, 15–18. [CrossRef]
203. Panaitescu, A.M.; Isac, S.; Pavel, B.; Ilie, A.S.; Ceanga, M.; Totan, A.; Zagrean, L.; Peltecu, G.; Zagrean, A.M. Oxytocin Reduces Seizure Burden and Hippocampal Injury in a Rat Model of Perinatal Asphyxia. *Acta Endocrinol.* **2018**, *14*, 315–319. [CrossRef]
204. Pâslaru, A.-C.; Călin, A.; Morozan, V.-P.; Stancu, M.; Tofan, L.; Panaitescu, A.M.; Zăgrean, A.-M.; Zăgrean, L.; Moldovan, M. Burst-Suppression EEG Reactivity to Photic Stimulation—A Translational Biomarker in Hypoxic–Ischemic Brain Injury. *Biomolecules* **2024**, *14*, 953. [CrossRef]
205. Ionescu, M.I.; Grigoras, I.-F.; Ionescu, R.-B.; Chitimus, D.M.; Haret, R.M.; Ianosi, B.; Ceanga, M.; Zagrean, A.-M. Oxytocin Exhibits Neuroprotective Effects on Hippocampal Cultures under Severe Oxygen-Glucose Deprivation Conditions. *Curr. Issues Mol. Biol.* **2024**, *46*, 6223–6236. [CrossRef]
206. Spoljaric, A.; Seja, P.; Spoljaric, I.; Virtanen, M.A.; Lindfors, J.; Uvarov, P.; Summanen, M.; Crow, A.K.; Hsueh, B.; Puskarjov, M.; et al. Vasopressin Excites Interneurons to Suppress Hippocampal Network Activity across a Broad Span of Brain Maturity at Birth. *Proc. Natl. Acad. Sci. USA* **2017**, *114*, E10819–E10828. [CrossRef] [PubMed]
207. Chen, J.; Aguilera, G. Vasopressin Protects Hippocampal Neurones in Culture against Nutrient Deprivation or Glutamate-Induced Apoptosis: Vasopressin Is Anti-Apoptotic in Hippocampal Neurone Cultures. *J. Neuroendocrinol.* **2010**, *22*, 1072–1081. [CrossRef]
208. Wellmann, S.; Benzing, J.; Cippà, G.; Admaty, D.; Creutzfeldt, R.; Mieth, R.A.; Beinder, E.; Lapaire, O.; Morgenthaler, N.G.; Haagen, U.; et al. High Copeptin Concentrations in Umbilical Cord Blood after Vaginal Delivery and Birth Acidosis. *J. Clin. Endocrinol. Metab.* **2010**, *95*, 5091–5096. [CrossRef] [PubMed]
209. Desbonnet, L.; Clarke, G.; Traplin, A.; O'Sullivan, O.; Crispie, F.; Moloney, R.D.; Cotter, P.D.; Dinan, T.G.; Cryan, J.F. Gut Microbiota Depletion from Early Adolescence in Mice: Implications for Brain and Behaviour. *Brain Behav. Immun.* **2015**, *48*, 165–173.
210. Stigger, F.; Lovatel, G.; Marques, M.; Bertoldi, K.; Moysés, F.; Elsner, V.; Siqueira, I.R.; Achaval, M.; Marcuzzo, S. Inflammatory Response and Oxidative Stress in Developing Rat Brain and Its Consequences on Motor Behavior Following Maternal Administration of LPS and Perinatal Anoxia. *Int. J. Dev. Neurosci.* **2013**, *31*, 820–827. [CrossRef] [PubMed]
211. Eklind, S.; Mallard, C.; Arvidsson, P.; Hagberg, H. Lipopolysaccharide Induces Both a Primary and a Secondary Phase of Sensitization in the Developing Rat Brain. *Pediatr. Res.* **2005**, *58*, 112–116. [CrossRef]
212. Osredkar, D.; Thoresen, M.; Maes, E.; Flatebø, T.; Elstad, M.; Sabir, H. Hypothermia Is Not Neuroprotective after Infection-Sensitized Neonatal Hypoxic-Ischemic Brain Injury. *Resuscitation* **2014**, *85*, 567–572. [CrossRef]
213. Beckers, K.F.; Flanagan, J.P.; Sones, J.L. Microbiome and Pregnancy: Focus on Microbial Dysbiosis Coupled with Maternal Obesity. *Int. J. Obes.* **2024**, *48*, 439–448, Erratum in *Int. J. Obes.* **2025**, *49*, 1421. [CrossRef] [PubMed]
214. Kalyan, M.; Tousif, A.H.; Sonali, S.; Vichitra, C.; Sunanda, T.; Praveenraj, S.S.; Ray, B.; Gorantla, V.R.; Rungratanawanich, W.; Mahalakshmi, A.M.; et al. Role of Endogenous Lipopolysaccharides in Neurological Disorders. *Cells* **2022**, *11*, 4038. [CrossRef]
215. Banks, W.A.; Gray, A.M.; Erickson, M.A.; Salameh, T.S.; Damodarasamy, M.; Sheibani, N.; Meabon, J.S.; Wing, E.E.; Morofuji, Y.; Cook, D.G.; et al. Lipopolysaccharide-Induced Blood-Brain Barrier Disruption: Roles of Cyclooxygenase, Oxidative Stress, Neuroinflammation, and Elements of the Neurovascular Unit. *J. Neuroinflamm.* **2015**, *12*, 223. [CrossRef] [PubMed]
216. Zhao, J.; Bi, W.; Xiao, S.; Lan, X.; Cheng, X.; Zhang, J.; Lu, D.; Wei, W.; Wang, Y.; Li, H.; et al. Neuroinflammation Induced by Lipopolysaccharide Causes Cognitive Impairment in Mice. *Sci. Rep.* **2019**, *9*, 5790. [CrossRef] [PubMed]
217. Ostrem, B.E.L.; Domínguez-Iturza, N.; Stogsdill, J.A.; Faits, T.; Kim, K.; Levin, J.Z.; Arlotta, P. Fetal Brain Response to Maternal Inflammation Requires Microglia. *Development* **2024**, *151*, dev.202252. [CrossRef]
218. Yeh, H.; Ikezu, T. Transcriptional and Epigenetic Regulation of Microglia in Health and Disease. *Trends Mol. Med.* **2019**, *25*, 96–111. [CrossRef]

219. Isac, S.; Panaitescu, A.M.; Iesanu, M.; Grigoras, I.F.; Totan, A.; Udriste, A.; Cucu, N.; Peltecu, G.; Zagrean, L.; Zagrean, A.-M. Maternal High-Fat Diet Modifies the Immature Hippocampus Vulnerability to Perinatal Asphyxia in Rats. *Neonatology* **2018**, *114*, 355–361. [CrossRef]
220. Isac, S.; Panaitescu, A.M.; Spataru, A.; Iesanu, M.; Totan, A.; Udriste, A.; Cucu, N.; Peltecu, G.; Zagrean, L.; Zagrean, A.-M. Trans-Resveratrol Enriched Maternal Diet Protects the Immature Hippocampus from Perinatal Asphyxia in Rats. *Neurosci. Lett.* **2017**, *653*, 308–313. [CrossRef]
221. Isac, S.; Panaitescu, A.M.; Iesanu, M.I.; Zeca, V.; Cucu, N.; Zagrean, L.; Peltecu, G.; Zagrean, A.-M. Maternal Citicoline-Supplemented Diet Improves the Response of the Immature Hippocampus to Perinatal Asphyxia in Rats. *Neonatology* **2020**, *117*, 729–735. [CrossRef]
222. Zhang, W.; Hu, X.; Yang, W.; Gao, Y.; Chen, J. Omega-3 Polyunsaturated Fatty Acid Supplementation Confers Long-Term Neuroprotection against Neonatal Hypoxic-Ischemic Brain Injury through Anti-Inflammatory Actions. *Stroke* **2010**, *41*, 2341–2347. [CrossRef]
223. Guo, L.; Chen, Q.; Gao, Y.; Jiang, H.; Zhou, F.; Zhang, F.; Xu, M. CDP-Choline Modulates Cholinergic Signaling and Gut Microbiota to Alleviate DSS-Induced Inflammatory Bowel Disease. *Biochem. Pharmacol.* **2023**, *217*, 115845. [CrossRef] [PubMed]
224. Prakash, V.; Bose, C.; Sunilkumar, D.; Cherian, R.M.; Thomas, S.S.; Nair, B.G. Resveratrol as a Promising Nutraceutical: Implications in Gut Microbiota Modulation, Inflammatory Disorders, and Colorectal Cancer. *Int. J. Mol. Sci.* **2024**, *25*, 3370. [CrossRef] [PubMed]
225. Robertson, R.C.; Kaliannan, K.; Strain, C.R.; Ross, R.P.; Stanton, C.; Kang, J.X. Maternal Omega-3 Fatty Acids Regulate Offspring Obesity through Persistent Modulation of Gut Microbiota. *Microbiome* **2018**, *6*, 95. [CrossRef]
226. Carvalho, A.V.S.; Sanches, E.F.; Ribeiro, R.T.; Durán-Carabali, L.E.; Júnior, O.R.; Muniz, B.D.; Wajner, M.; Wyse, A.T.; Netto, C.A.; Sizonenko, S.V. Maternal Lactoferrin Supplementation Prevents Mitochondrial and Redox Homeostasis Dysfunction, and Improves Antioxidant Defenses through Nrf2 and UCP2 Signaling after Neonatal Hypoxia-Ischemia. *Free Radic. Biol. Med.* **2025**, *231*, 68–79. [CrossRef]
227. Zhao, C.; Chen, N.; Ashaolu, T.J. Prebiotic and Modulatory Evidence of Lactoferrin on Gut Health and Function. *J. Funct. Foods* **2023**, *108*, 105741. [CrossRef]
228. Coscia, A.; Bardanzellu, F.; Caboni, E.; Fanos, V.; Peroni, D.G. When a Neonate Is Born, so Is a Microbiota. *Life* **2021**, *11*, 148. [CrossRef] [PubMed]
229. Suárez-Martínez, C.; Santaella-Pascual, M.; Yagüe-Guirao, G.; Martínez-Graciá, C. Infant Gut Microbiota Colonization: Influence of Prenatal and Postnatal Factors, Focusing on Diet. *Front. Microbiol.* **2023**, *14*, 1236254. [CrossRef]
230. Iribarren, I.; Hilario, E.; Álvarez, A.; Alonso-Alconada, D. Neonatal Multiple Organ Failure after Perinatal Asphyxia. *An. Pediatr. (Engl. Ed.)* **2022**, *97*, 280.e1–280.e8. [CrossRef]
231. Gamsu, H.R.; Kempley, S.T. Enteral Hypoxia/Ischaemia and Necrotizing Enterocolitis. *Semin. Neonatol.* **1997**, *2*, 245–254. [CrossRef]
232. Nikiforou, M.; Willburger, C.; de Jong, A.E.; Kloosterboer, N.; Jellema, R.K.; Ophelders, D.R.M.G.; Steinbusch, H.W.M.; Kramer, B.W.; Wolfs, T. Global Hypoxia-Ischemia Induced Inflammation and Structural Changes in the Preterm Ovine Gut Which Were Not Ameliorated by Mesenchymal Stem Cell Treatment. *Mol. Med.* **2016**, *22*, 244–257. [CrossRef]
233. Guzmán-De La Garza, F.J.; Cámara-Lemarroy, C.R.; Ballesteros-Elizondo, R.G.; Alarcón-Galván, G.; Cordero-Pérez, P.; Fernández-Garza, N.E. Ketamine Reduces Intestinal Injury and Inflammatory Cell Infiltration after Ischemia/Reperfusion in Rats. *Surg. Today* **2010**, *40*, 1055–1062. [CrossRef] [PubMed]
234. Moreno-Indias, I.; Torres, M.; Sanchez-Alcoholado, L.; Cardona, F.; Almendros, I.; Gozal, D.; Montserrat, J.M.; Queipo-Ortuño, M.I.; Farré, R. Normoxic Recovery Mimicking Treatment of Sleep Apnea Does Not Reverse Intermittent Hypoxia-Induced Bacterial Dysbiosis and Low-Grade Endotoxemia in Mice. *Sleep* **2016**, *39*, 1891–1897. [CrossRef]
235. Moreno-Indias, I.; Torres, M.; Montserrat, J.M.; Sanchez-Alcoholado, L.; Cardona, F.; Tinahones, F.J.; Gozal, D.; Poroyko, V.A.; Navajas, D.; Queipo-Ortuño, M.I.; et al. Intermittent Hypoxia Alters Gut Microbiota Diversity in a Mouse Model of Sleep Apnoea. *Eur. Respir. J.* **2015**, *45*, 1055–1065. [CrossRef]
236. Chen, G.; Li, F.; Du, J. Change of Gut Microbiome Structure in Preterm Infants with Hypoxic Ischemic Encephalopathy Induced by Apnea. *Pediatr. Neonatol.* **2023**, *64*, 455–464. [CrossRef] [PubMed]
237. Lundgren, S.N.; Madan, J.C.; Emond, J.A.; Morrison, H.G.; Christensen, B.C.; Karagas, M.R.; Hoen, A.G. Maternal Diet during Pregnancy Is Related with the Infant Stool Microbiome in a Delivery Mode-Dependent Manner. *Microbiome* **2018**, *6*, 109. [CrossRef]
238. Ionescu, M.I.; Zahiu, C.D.M.; Vlad, A.; Galos, F.; Gradisteanu Pircalabioru, G.; Zagrean, A.-M.; O'Mahony, S.M. Nurturing Development: How a Mother's Nutrition Shapes Offspring's Brain through the Gut. *Nutr. Neurosci.* **2025**, *28*, 50–72. [CrossRef]

239. Nakajima, A.; Kaga, N.; Nakanishi, Y.; Ohno, H.; Miyamoto, J.; Kimura, I.; Hori, S.; Sasaki, T.; Hiramatsu, K.; Okumura, K.; et al. Maternal High Fiber Diet during Pregnancy and Lactation Influences Regulatory T Cell Differentiation in Offspring in Mice. *J. Immunol.* **2017**, *199*, 3516–3524. [CrossRef]
240. Nakajima, A.; Habu, S.; Kasai, M.; Okumura, K.; Ishikawa, D.; Shibuya, T.; Kobayashi, O.; Osada, T.; Ohkusa, T.; Watanabe, S.; et al. Impact of Maternal Dietary Gut Microbial Metabolites on an Offspring's Systemic Immune Response in Mouse Models. *Biosci. Microbiota Food Health* **2020**, *39*, 33–38. [CrossRef] [PubMed]
241. Fu, J.; Zheng, Y.; Gao, Y.; Xu, W. Dietary Fiber Intake and Gut Microbiota in Human Health. *Microorganisms* **2022**, *10*, 2507. [CrossRef]
242. Strobel, K.M.; Juul, S.E.; Hendrixson, D.T. Maternal Nutritional Status and the Microbiome across the Pregnancy and the Post-Partum Period. *Microorganisms* **2023**, *11*, 1569. [CrossRef]
243. Miller, C.B.; Benny, P.; Riel, J.; Boushey, C.; Perez, R.; Khadka, V.; Qin, Y.; Maunakea, A.K.; Lee, M.-J. Adherence to Mediterranean Diet Impacts Gastrointestinal Microbial Diversity throughout Pregnancy. *BMC Pregnancy Childbirth* **2021**, *21*, 558. [CrossRef]
244. Crovetto, F.; Nakaki, A.; Arranz, A.; Borras, R.; Vellvé, K.; Paules, C.; Boutet, M.L.; Castro-Barquero, S.; Freitas, T.; Casas, R.; et al. Effect of a Mediterranean Diet or Mindfulness-Based Stress Reduction during Pregnancy on Child Neurodevelopment: A Prespecified Analysis of the IMPACT BCN Randomized Clinical Trial. *JAMA Netw. Open* **2023**, *6*, e2330255. [CrossRef] [PubMed]
245. Duda-Chodak, A.; Tarko, T.; Satora, P.; Sroka, P. Interaction of Dietary Compounds, Especially Polyphenols, with the Intestinal Microbiota: A Review. *Eur. J. Nutr.* **2015**, *54*, 325–341. [CrossRef]
246. Roumes, H.; Sanchez, S.; Benkhaled, I.; Fernandez, V.; Goudeneche, P.; Perrin, F.; Pellerin, L.; Guillard, J.; Bouzier-Sore, A.-K. Neuroprotective Effect of Eco-Sustainably Extracted Grape Polyphenols in Neonatal Hypoxia-Ischemia. *Nutrients* **2022**, *14*, 773. [CrossRef]
247. Pontes, P.B.; Toscano, A.E.; Lacerda, D.C.; da Silva Araújo, E.R.; da Costa, P.C.T.; Alves, S.M.; de Brito Alves, J.L.; Manhães-de-Castro, R. Effectiveness of Polyphenols on Perinatal Brain Damage: A Systematic Review of Preclinical Studies. *Foods* **2023**, *12*, 2278. [CrossRef] [PubMed]
248. Loren, D.J.; Seeram, N.P.; Schulman, R.N.; Holtzman, D.M. Maternal Dietary Supplementation with Pomegranate Juice Is Neuroprotective in an Animal Model of Neonatal Hypoxic-Ischemic Brain Injury. *Pediatr. Res.* **2005**, *57*, 858–864. [CrossRef] [PubMed]
249. Dumont, U.; Sanchez, S.; Olivier, B.; Chateil, J.-F.; Deffieux, D.; Quideau, S.; Pellerin, L.; Beauvieux, M.-C.; Bouzier-Sore, A.-K.; Roumes, H. Maternal Alcoholism and Neonatal Hypoxia-Ischemia: Neuroprotection by Stilbenoid Polyphenols. *Brain Res.* **2020**, *1738*, 146798. [CrossRef]
250. Ray, S.K.; Mukherjee, S. Evolving Interplay between Dietary Polyphenols and Gut Microbiota-an Emerging Importance in Healthcare. *Front. Nutr.* **2021**, *8*, 634944. [CrossRef]
251. Gowd, V.; Karim, N.; Shishir, M.R.I.; Xie, L.; Chen, W. Dietary Polyphenols to Combat the Metabolic Diseases via Altering Gut Microbiota. *Trends Food Sci. Technol.* **2019**, *93*, 81–93. [CrossRef]
252. Wang, X.; Qi, Y.; Zheng, H. Dietary Polyphenol, Gut Microbiota, and Health Benefits. *Antioxidants* **2022**, *11*, 1212. [CrossRef]
253. Leng, P.; Wang, Y.; Xie, M. Ellagic Acid and Gut Microbiota: Interactions, and Implications for Health. *Food Sci. Nutr.* **2025**, *13*, e70133. [CrossRef]
254. Wang, P.; Wang, R.; Zhao, W.; Zhao, Y.; Wang, D.; Zhao, S.; Ge, Z.; Ma, Y.; Zhao, X. Gut Microbiota-Derived 4-Hydroxyphenylacetic Acid from Resveratrol Supplementation Prevents Obesity through SIRT1 Signaling Activation. *Gut Microbes* **2025**, *17*, 2446391. [CrossRef] [PubMed]
255. Suganuma, H.; Arai, Y.; Kitamura, Y.; Hayashi, M.; Okumura, A.; Shimizu, T. Maternal Docosahexaenoic Acid-Enriched Diet Prevents Neonatal Brain Injury: Effect of Docosahexaenoic Acid Diet. *Neuropathology* **2010**, *30*, 597–605. [CrossRef] [PubMed]
256. Middleton, P.; Gomersall, J.C.; Gould, J.F.; Shepherd, E.; Olsen, S.F.; Makrides, M. Omega-3 Fatty Acid Addition during Pregnancy. *Cochrane Libr.* **2018**, *11*, CD003402. [CrossRef]
257. Costantini, L.; Molinari, R.; Farinon, B.; Merendino, N. Impact of Omega-3 Fatty Acids on the Gut Microbiota. *Int. J. Mol. Sci.* **2017**, *18*, 2645. [CrossRef]
258. Wang, C.; Wei, S.; Liu, B.; Wang, F.; Lu, Z.; Jin, M.; Wang, Y. Maternal Consumption of a Fermented Diet Protects Offspring against Intestinal Inflammation by Regulating the Gut Microbiota. *Gut Microbes* **2022**, *14*, 2057779. [CrossRef]
259. Erçelik, H.C.; Kaya, V. The Effects of Fermented Food Consumption in Pregnancy on Neonatal and Infant Health: An Integrative Review. *J. Pediatr. Nurs.* **2024**, *75*, 173–179. [CrossRef]
260. Rezac, S.; Kok, C.R.; Heermann, M.; Hutkins, R. Fermented Foods as a Dietary Source of Live Organisms. *Front. Microbiol.* **2018**, *9*, 1785. [CrossRef]
261. Pandiyan, A.; Gurung, M.; Mulakala, B.K.; Ponniah, S.K.; Yeruva, L. The Role of Fermented Foods in Maternal Health during Pregnancy and Infant Health during the First 1,000 Days of Life. *Front. Nutr.* **2025**, *12*, 1581723. [CrossRef]

262. Jones, J.M.; Reinke, S.N.; Mousavi-Derazmahalleh, M.; Garssen, J.; Jenmalm, M.C.; Srinivasjois, R.; Silva, D.; Keelan, J.; Prescott, S.L.; Palmer, D.J.; et al. Maternal Prebiotic Supplementation during Pregnancy and Lactation Modifies the Microbiome and Short Chain Fatty Acid Profile of Both Mother and Infant. *Clin. Nutr.* **2024**, *43*, 969–980. [CrossRef]
263. Hogenkamp, A.; Thijssen, S.; van Vlies, N.; Garssen, J. Supplementing Pregnant Mice with a Specific Mixture of Nondigestible Oligosaccharides Reduces Symptoms of Allergic Asthma in Male Offspring. *J. Nutr.* **2015**, *145*, 640–646. [CrossRef]
264. Cuinat, C.; Stinson, S.E.; Ward, W.E.; Comelli, E.M. Maternal Intake of Probiotics to Program Offspring Health. *Curr. Nutr. Rep.* **2022**, *11*, 537–562. [CrossRef]
265. Musazadeh, V.; Faghfouri, A.H.; Zarezadeh, M.; Pakmehr, A.; Moghaddam, P.T.; Hamedi-Kalajahi, F.; Jahandideh, A.; Ghoreishi, Z. Remarkable Impacts of Probiotics Supplementation in Enhancing of the Antioxidant Status: Results of an Umbrella Meta-Analysis. *Front. Nutr.* **2023**, *10*, 1117387, Erratum in *Front. Nutr.* **2024**, *11*, 1371746. [CrossRef] [PubMed]
266. Jones, S.E.; Versalovic, J. Probiotic Lactobacillus Reuteri Biofilms Produce Antimicrobial and Anti-Inflammatory Factors. *BMC Microbiol.* **2009**, *9*, 35. [CrossRef]
267. Latif, A.; Shehzad, A.; Niazi, S.; Zahid, A.; Ashraf, W.; Iqbal, M.W.; Rehman, A.; Riaz, T.; Aadil, R.M.; Khan, I.M.; et al. Probiotics: Mechanism of Action, Health Benefits and Their Application in Food Industries. *Front. Microbiol.* **2023**, *14*, 1216674, Erratum in *Front. Microbiol.* **2024**, [15], 1378225. [CrossRef]
268. Grev, J.; Berg, M.; Soll, R. Maternal Probiotic Supplementation for Prevention of Morbidity and Mortality in Preterm Infants. *Cochrane Database Syst. Rev.* **2018**, *12*, CD012519. [CrossRef] [PubMed]
269. Keunen, K.; van Elburg, R.M.; van Bel, F.; Benders, M.J.N.L. Impact of Nutrition on Brain Development and Its Neuroprotective Implications Following Preterm Birth. *Pediatr. Res.* **2015**, *77*, 148–155. [CrossRef]
270. Sha, C.; Jin, Z.; Ku, S.Y.; Kogosov, A.S.; Yu, S.; Bergese, S.D.; Hsieh, H. Necrotizing Enterocolitis and Neurodevelopmental Impairments: Microbiome, Gut, and Brain Entanglements. *Biomolecules* **2024**, *14*, 1254. [CrossRef] [PubMed]
271. Wang, K.; Kong, X.; Azad, M.A.K.; Zhu, Q.; Xiong, L.; Zheng, Y.; Hu, Z.; Yin, Y.; He, Q. Maternal Probiotic or Synbiotic Supplementation Modulates Jejunal and Colonic Antioxidant Capacity, Mitochondrial Function, and Microbial Abundance in Bama Mini-Piglets. *Oxid. Med. Cell. Longev.* **2021**, *2021*, 6618874. [CrossRef] [PubMed]
272. Zhang, L.; Zhou, W.; Zhan, L.; Hou, S.; Zhao, C.; Bi, T.; Lu, X. Fecal Microbiota Transplantation Alters the Susceptibility of Obese Rats to Type 2 Diabetes Mellitus. *Aging* **2020**, *12*, 17480–17502. [CrossRef] [PubMed]
273. Vrieze, A.; Van Nood, E.; Holleman, F.; Salojärvi, J.; Kootte, R.S.; Bartelsman, J.F.W.M.; Dallinga-Thie, G.M.; Ackermans, M.T.; Serlie, M.J.; Oozeer, R.; et al. Transfer of Intestinal Microbiota from Lean Donors Increases Insulin Sensitivity in Individuals with Metabolic Syndrome. *Gastroenterology* **2012**, *143*, 913–916.e7, Erratum in *Gastroenterology* **2013**, *144*, 250. [CrossRef]
274. Proença, I.M.; Allegretti, J.R.; Bernardo, W.M.; de Moura, D.T.H.; Ponte Neto, A.M.; Matsubayashi, C.O.; Flor, M.M.; Kotinda, A.P.S.T.; de Moura, E.G.H. Fecal Microbiota Transplantation Improves Metabolic Syndrome Parameters: Systematic Review with Meta-Analysis Based on Randomized Clinical Trials. *Nutr. Res.* **2020**, *83*, 1–14. [CrossRef]
275. Huang, Y.-H.; Tain, Y.-L.; Hsu, C.-N. Maternal Supplementation of Probiotics, Prebiotics or Postbiotics to Prevent Offspring Metabolic Syndrome: The Gap between Preclinical Results and Clinical Translation. *Int. J. Mol. Sci.* **2022**, *23*, 10173. [CrossRef]
276. Wang, Z.; Roberts, A.B.; Buffa, J.A.; Levison, B.S.; Zhu, W.; Org, E.; Gu, X.; Huang, Y.; Zamanian-Daryoush, M.; Culley, M.K.; et al. Non-Lethal Inhibition of Gut Microbial Trimethylamine Production for the Treatment of Atherosclerosis. *Cell* **2015**, *163*, 1585–1595. [CrossRef]
277. Hsu, C.-N.; Chang-Chien, G.-P.; Lin, S.; Hou, C.-Y.; Tain, Y.-L. Targeting on Gut Microbial Metabolite Trimethylamine-N-Oxide and Short-Chain Fatty Acid to Prevent Maternal High-Fructose-Diet-Induced Developmental Programming of Hypertension in Adult Male Offspring. *Mol. Nutr. Food Res.* **2019**, *63*, e1900073. [CrossRef]
278. Milligan, G.; Barki, N.; Tobin, A.B. Chemogenetic Approaches to Explore the Functions of Free Fatty Acid Receptor 2. *Trends Pharmacol. Sci.* **2021**, *42*, 191–202. [CrossRef] [PubMed]
279. Zhao, Q.; Chen, T.; Ni, C.; Hu, Y.; Nan, Y.; Lin, W.; Liu, Y.; Zheng, F.; Shi, X.; Lin, Z.; et al. Indole-3-Propionic Acid Attenuates HI-Related Blood-Brain Barrier Injury in Neonatal Rats by Modulating the PXR Signaling Pathway. *ACS Chem. Neurosci.* **2022**, *13*, 2897–2912. [CrossRef]
280. Amat, S.; Dahlen, C.R.; Swanson, K.C.; Ward, A.K.; Reynolds, L.P.; Caton, J.S. Bovine Animal Model for Studying the Maternal Microbiome, in Utero Microbial Colonization and Their Role in Offspring Development and Fetal Programming. *Front. Microbiol.* **2022**, *13*, 854453. [CrossRef] [PubMed]
281. Daiy, K.; Wiley, K.; Allen, J.; Bailey, M.T.; Dettmer, A.M. Associations among Rearing Environment and the Infant Gut Microbiome with Early-Life Neurodevelopment and Cognitive Development in a Nonhuman Primate Model (Macaca Mulatta). *J. Dev. Orig. Health Dis.* **2025**, *16*, e1. [CrossRef]

282. Ma, J.; Prince, A.L.; Bader, D.; Hu, M.; Ganu, R.; Baquero, K.; Blundell, P.; Alan Harris, R.; Frias, A.E.; Grove, K.L.; et al. High-Fat Maternal Diet during Pregnancy Persistently Alters the Offspring Microbiome in a Primate Model. *Nat. Commun.* **2014**, *5*, 3889. [CrossRef]
283. Li, M.; Brokaw, A.; Furuta, A.M.; Coler, B.; Obregon-Perko, V.; Chahroudi, A.; Wang, H.-Y.; Permar, S.R.; Hotchkiss, C.E.; Golos, T.G.; et al. Non-Human Primate Models to Investigate Mechanisms of Infection-Associated Fetal and Pediatric Injury, Teratogenesis and Stillbirth. *Front. Genet.* **2021**, *12*, 680342. [CrossRef] [PubMed]
284. Collins, B.; Lemanski, E.A.; Wright-Jin, E. The Importance of Including Maternal Immune Activation in Animal Models of Hypoxic-Ischemic Encephalopathy. *Biomedicines* **2024**, *12*, 2559. [CrossRef]
285. Koehler, R.C.; Yang, Z.-J.; Lee, J.K.; Martin, L.J. Perinatal Hypoxic-Ischemic Brain Injury in Large Animal Models: Relevance to Human Neonatal Encephalopathy. *J. Cereb. Blood Flow Metab.* **2018**, *38*, 2092–2111. [CrossRef]
286. Jacobson Misbe, E.N.; Richards, T.L.; McPherson, R.J.; Burbacher, T.M.; Juul, S.E. Perinatal Asphyxia in a Nonhuman Primate Model. *Dev. Neurosci.* **2011**, *33*, 210–221. [CrossRef]
287. Doron, S.; Snydman, D.R. Risk and Safety of Probiotics. *Clin. Infect. Dis.* **2015**, *60* (Suppl. 2), S129–S134. [CrossRef] [PubMed]
288. Bertelli, C.; Pillonel, T.; Torregrossa, A.; Prod'hom, G.; Fischer, C.J.; Greub, G.; Giannoni, E. Bifidobacterium Longum Bacteremia in Preterm Infants Receiving Probiotics. *Clin. Infect. Dis.* **2015**, *60*, 924–927. [CrossRef] [PubMed]
289. Pillai, A.; Tan, J.; Paquette, V.; Panczuk, J. Does Probiotic Bacteremia in Premature Infants Impact Clinically Relevant Outcomes? A Case Report and Updated Review of Literature. *Clin. Nutr. ESPEN* **2020**, *39*, 255–259. [CrossRef]
290. Roy, U.; Jessani, L.G.; Rudramurthy, S.M.; Gopalakrishnan, R.; Dutta, S.; Chakravarty, C.; Jillwin, J.; Chakrabarti, A. Seven Cases of Saccharomyces Fungaemia Related to Use of Probiotics. *Mycoses* **2017**, *60*, 375–380. [CrossRef]
291. Farella, I.; Fortunato, M.; Martinelli, D.; De Carlo, C.; Sparapano, E.; Stolfa, S.; Romanelli, F.; De Laurentiis, V.; Martinotti, S.; Capozzi, L.; et al. Lactobacillus Rhamnosus Sepsis in a Preterm Infant Following Probiotic Administration: Challenges in Diagnosis. *Microorganisms* **2025**, *13*, 265. [CrossRef] [PubMed]
292. Underwood, M.A.; Umberger, E.; Patel, R.M. Safety and Efficacy of Probiotic Administration to Preterm Infants: Ten Common Questions. *Pediatr. Res.* **2020**, *88*, 48–55. [CrossRef] [PubMed]
293. Available online: https://www.fda.gov/media/172606/download?attachment (accessed on 3 May 2025).
294. O'Doherty, K.C.; Virani, A.; Wilcox, E.S. The Human Microbiome and Public Health: Social and Ethical Considerations. *Am. J. Public Health* **2016**, *106*, 414–420. [CrossRef] [PubMed]
295. Chan, N.H.; Hawkins, C.C.; Rodrigues, B.V.; Cornet, M.-C.; Gonzalez, F.F.; Wu, Y.W. Neuroprotection for Neonatal Hypoxic-Ischemic Encephalopathy: A Review of Novel Therapies Evaluated in Clinical Studies. *Dev. Med. Child Neurol.* **2025**, *67*, 591–599. [CrossRef]

antioxidants

MDPI

Review

Inhaled and Systemic Steroids for Bronchopulmonary Dysplasia: Targeting Inflammation and Oxidative Stress

Francesca Galletta [1], Alessandra Li Pomi [2], Sara Manti [1,*,†] and Eloisa Gitto [2,†]

1 Pediatric Unit, Department of Human Pathology in Adult and Developmental Age 'Gaetano Barresi', University of Messina, 98124 Messina, Italy; francygall.92@gmail.com

2 Neonatal and Pediatric Intensive Care Unit, Department of Clinical and Experimental Medicine, University of Messina, 98122 Messina, Italy; alessandra.lipomi92@gmail.com (A.L.P.); egitto@unime.it (E.G.)

* Correspondence: smanti@unime.it; Tel.: +39-090-2213-153

† These authors contributed equally to this work.

Abstract: Bronchopulmonary dysplasia (BPD) remains a significant complication of preterm birth, characterized by impaired alveolar and vascular development, chronic lung inflammation, and long-term respiratory morbidity. Corticosteroids, both systemic and inhaled, have been widely investigated as potential therapeutic agents due to their anti-inflammatory properties and their emerging role in modulating oxidative stress. This narrative review explores the current evidence regarding the use of inhaled and systemic corticosteroids in the prevention and management of BPD, analyzing their efficacy, safety profiles, and long-term outcomes. While systemic corticosteroids, particularly dexamethasone, have demonstrated benefits in reducing ventilator dependence and lung inflammation, concerns regarding adverse neurodevelopmental effects have limited their routine use. Inhaled steroids have been proposed as a safer alternative, but their role in altering the disease trajectory remains controversial. A better understanding of the optimal timing, dosage, and patient selection is essential to maximize benefits while minimizing risks. Future research should focus on optimizing dosing strategies and identifying subgroups of preterm infants who may derive the greatest benefit from corticosteroid therapy.

Keywords: bronchopulmonary dysplasia; systemic steroids; inhaled steroids; lung injury; inflammation; oxidative stress; newborn

1. Introduction

Bronchopulmonary dysplasia (BPD) is a chronic lung disease primarily affecting preterm infants, particularly those born before 27 weeks of gestation [1]. Despite advances in neonatal care, its prevalence remains high, with up to 75% of extremely preterm infants diagnosed with the condition [2]. BPD is currently defined by the need for supplemental oxygen or respiratory support at 36 weeks' postmenstrual age (PMA) and is categorized into three severity grades based on oxygen dependency and ventilation requirements [3]. Severe cases of BPD are associated with prolonged hospital stays, difficulties in weaning from respiratory support, feeding intolerance, and intermittent hypoxic episodes [4,5]. Many affected infants require home oxygen therapy upon discharge [5]. The long-term consequences of BPD extend into adolescence and adulthood, often leading to impaired lung function, recurrent respiratory infections, and neurodevelopmental challenges [6,7]. Furthermore, severe cases carry a significant mortality risk, with rates reaching 3% during the initial hospital stay and up to 20% within two years in infants requiring tracheostomy [8,9]. While initially described as a consequence of high oxygen exposure and prolonged mechanical ventilation, it is now understood as a multifactorial disease where immune dysregulation and abnormal lung development play critical roles [3].

1.1. The Role of Inflammation in the Pathogenesis of BPD

Inflammation plays a central role in the pathogenesis of BPD [Figure 1]. Both prenatal and postnatal factors contribute to the development of an exaggerated and persistent inflammatory response in the immature lung [10,11]. Particularly, one of the main prenatal factor contributors to BPD is intrauterine inflammation, often associated with conditions such as chorioamnionitis, which can elicit a fetal inflammatory response. Prenatal exposure to inflammation may cause early immune activation, predisposing the neonatal lung to exaggerated inflammatory responses after birth [10]. This immune hyperactivation contributes to the infiltration of inflammatory cells and the release of pro-inflammatory cytokines, including IL-1β, IL-6, IL-8, and TNF-α, which recruit and activate neutrophils and macrophages. These cells, in turn, release reactive oxygen species (ROS) and proteases, exacerbating alveolar and vascular injury [12]. Postnatal exposures to oxygen therapy, mechanical ventilation, and infections further exacerbate the inflammatory state, promoting endothelial dysfunction, increased vascular permeability, and interstitial edema [11,13]. The persistence of inflammation disrupts the normal processes of lung development, including alveolarization and angiogenesis, leading to impaired gas exchange and chronic lung dysfunction [13]. Studies indicate that inflammatory signaling is mediated by pattern recognition receptors (PRRs), such as toll-like receptors (TLRs) and the receptor for advanced glycation end-products (RAGE), which respond to danger-associated molecular patterns (DAMPs) and pathogen-associated molecular patterns (PAMPs) [11]. Moreover, recent microbiome studies suggest that an altered lung microbiota composition in preterm infants may contribute to heightened inflammation, with an increased prevalence of potentially pathogenic species such as *Ureaplasma* spp. and *Corynebacterium* spp., and a reduction in beneficial Lactobacilli. This dysbiosis may further exacerbate immune dysregulation and predispose preterm infants to BPD [14].

1.2. The Role of Oxidative Stress in the Pathogenesis of BPD

Oxidative stress (OS) is a central contributor to the pathogenesis of BPD, intricately linked to both inflammation and impaired lung development [Figure 1]. Preterm infants are particularly susceptible due to their abrupt shift from a hypoxic intrauterine environment to a hyperoxic extrauterine setting, compounded by mechanical ventilation and oxygen supplementation [12]. This transition challenges the immature antioxidant defense systems of the preterm lung, resulting in the excessive generation of ROS such as superoxide anions, hydrogen peroxide, and hydroxyl radicals. These ROS induce lipid peroxidation, protein oxidation, and DNA damage, leading to epithelial and endothelial cell injury, increased alveolar–capillary permeability, and interstitial edema [11]. Emerging evidence highlights a dynamic interplay between oxidative stress and inflammation, with ROS capable of activating transcription factors such as NF-κB and AP-1, further amplifying inflammatory response and perpetuating lung injury [11]. Furthermore, OS has been shown to modulate the expression of genes crucial for lung development, including peroxisome proliferator-activated receptor gamma (PPARγ) and β-catenin. The hyperoxia-induced downregulation of PPARγ impairs alveolar epithelial cell growth and differentiation, while persistent β-catenin activation is associated with abnormal extracellular matrix deposition and fibrosis [11]. Experimental models have demonstrated that the pharmacological modulation of these pathways can mitigate hyperoxia-induced lung injury [14]. Metabolomic and proteomic studies have identified specific oxidative stress biomarkers, such as lipid peroxidation products and oxidized proteins, in the early postnatal period of infants who subsequently developed BPD. These findings suggest that OS may begin antenatally, with oxidative damage markers detectable shortly after birth, and that early antioxidant strategies might be necessary for effective prevention [14]. Despite the role of these biomarkers in BPD onset and progression, their regulation in response to corticosteroid treatment has yet to be systematically characterized in clinical populations. While preclinical studies indicate that corticosteroids may influence oxidative stress mechanisms and related proteomic pathways, clinical investigations demonstrating the direct associations between steroid therapy and the shifts in metabolomic or proteomic profiles in neonates remain limited [14]. Further research is necessary to determine whether these biomarkers could serve as reliable tools to predict or monitor therapeutic response, ultimately supporting the development of more personalized corticosteroid-based interventions in the management of BPD [14].

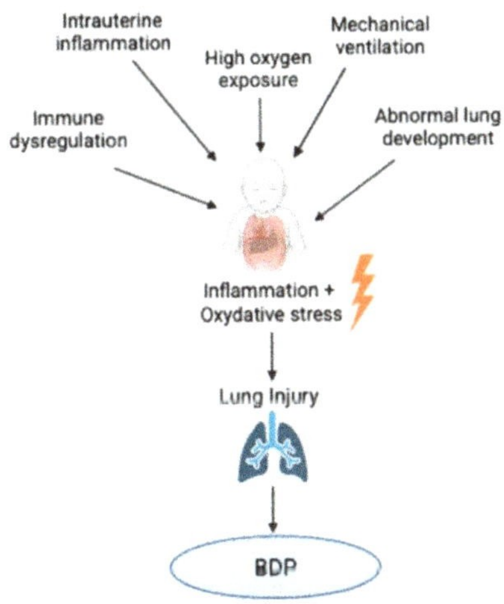

Figure 1. Conceptual model illustrating the proposed role of oxidative stress and inflammation in the pathogenesis of bronchopulmonary dysplasia (BPD) and the potential modulatory effects of corticosteroids. The diagram integrates current knowledge from experimental and clinical studies to depict how antenatal and postnatal factors trigger reactive oxygen species (ROS) production and pro-inflammatory cascades, contributing to lung injury and impaired alveolar development. Corticosteroids are shown to act at multiple levels, modulating inflammation and oxidative stress pathways.

1.3. Rationale for the Use of Steroids in the Management of BPD

Therapeutic strategies have focused on modulating inflammatory and oxidative stress responses to mitigate lung injury and promote normal pulmonary development. Among these, glucocorticoids have been considered for their ability to suppress inflammation and stabilize lung function [4]. Additionally, corticosteroids may provide a protective effect by not only dampening inflammation but also indirectly modulating oxidative stress. Their anti-inflammatory action includes the suppression of pro-oxidant cytokine release and the promotion of antioxidant pathways, potentially reducing ROS production and its downstream effects [15,16]. Some studies have shown that steroid administration is associated with decreased levels of oxidative stress biomarkers, suggesting a synergistic role between anti-inflammatory and antioxidant effects [15,16]. However, their use remains controversial, necessitating a careful evaluation of their benefits and potential risks. This narrative review aims to explore the role of corticosteroids as potential therapeutic approaches targeting lung inflammation and OS. We performed a literature search using the main scientific databases, including PubMed, Scopus,

Embase, and Google Scholar, selecting only relevant articles published in high-quality journals and written in English to ensure the reliability and scientific rigor of the review.

2. Systemic Steroids in BPD

The primary corticosteroids used for their anti-inflammatory effects in infants with BPD include dexamethasone, hydrocortisone, prednisolone, and methylprednisolone [17]. Although other corticosteroids like betamethasone and fludrocortisone have been considered, there is insufficient and inconsistent data supporting their use in evolving BPD, and they will not be further reviewed here [17].

2.1. Dexamethasone in BPD

Dexamethasone is among the most extensively studied corticosteroids in preterm neonates, particularly for its role in reducing BPD. Several randomized controlled trials (RCTs) have shown that early systemic dexamethasone administration, particularly within the first 7 days of life, reduced the incidence of BPD at 36 weeks PMA (RR 0.72, 95% CI 0.63–0.82; 17 studies, 2791 infants; high-certainty evidence), as well as the combined outcome of death or BPD (RR 0.88, 95% CI 0.81–0.95; 17 studies, 2791 infants; high-certainty evidence) [18–20]. However, the effect on mortality alone was not significant (RR 0.95, 95% CI 0.85–1.06; 31 studies, 4373 infants; high-certainty evidence), and the certainty of the evidence for BPD across all studies was downgraded to moderate due to suspected publication bias [18]. Moreover, dexamethasone facilitates extubation, with significant improvements noted as early as day 3 and sustained through day 10 [18]. Late administration (after seven days of age) has also shown consistent benefits, including a reduced need for rescue corticosteroids, lower extubation failure rates, and decreased home oxygen dependency [17,18]. Regarding the dosage used, the remote Dexamethasone: A Randomized Trial (DART) protocol used a cumulative dose of 0.89 mg/kg over 10 days in extremely low birth-weight infants, achieving a significant facilitation of extubation without increasing mortality or neurodevelopmental impairment [19]. However, more recent systematic reviews and meta-analyses indicated that moderately early initiation (between 8 and 14 days) with medium cumulative doses (2–4 mg/kg) appeared to provide the most favorable outcomes in preventing BPD, with up to a 57% relative risk reduction in some studies [20,21]. The Canadian Paediatric Society endorses the initiation of dexamethasone at 0.15–0.2 mg/kg/day for 7–10 days in neonates requiring mechanical ventilation beyond the first postnatal week [22]. This regimen aligns with evidence suggesting improved pulmonary outcomes when therapy begins between 8 and 14 days of age and maintains a cumulative dose between 2 and 4 mg/kg. In summary, longer treatment durations and moderate dosing regimens appear to be more effective than shorter or low-dose approaches [19,21]. However, despite these pulmonary benefits, dexamethasone therapy carries notable risks, particularly when initiated early or at higher doses. Its early use has been consistently linked with an increased risk of cerebral palsy (CP) and neuromotor impairment, even with courses as brief as three days [20]. Doyle et al. [23] found that early administration of dexamethasone raised the incidence of CP, while late administration did not. Zeng et al. reported that high-dose early regimens (>3 mg/kg) heightened this risk further [24]. Gastrointestinal complications are also a concern. The early administration of dexamethasone, especially in conjunction with indomethacin, has been associated with gastrointestinal perforation [25]. Other adverse events include hyperglycemia, hypertension, and, as recently reported, hypoglycemia, a less commonly recognized effect of dexamethasone in neonates, warranting closer glucose monitoring during treatment [23]. Infections remain a common complication, with some studies reporting incidences exceeding 50% [23,26]. Nevertheless, neither dexamethasone nor other systemic steroids, such as hydrocortisone, have been shown to increase the risk of necrotizing enterocolitis (NEC), retinopathy of prematurity (ROP), or sepsis in a consistent manner across studies [26]. Ultimately, the risk–benefit profile of dexamethasone is highly dependent on timing, dosage, and duration. While early administration may reduce BPD incidence, it increases the likelihood of neurodevelopmental harm. Conversely, delayed and moderately dosed regimens appear to preserve pulmonary efficacy while minimizing long-term adverse outcomes. In current clinical practices, dexamethasone is generally discouraged during the first week of life [27]. Its administration beyond this period should be limited to preterm infants who are at a high risk of developing BPD. One approach to identifying such high-risk neonates is using predictive tools, such as the neonatal BPD Outcome Estimator [28].

2.2. Hydrocortisone in BPD

Hydrocortisone, a corticosteroid with predominant mineralocorticoid activity, has also been widely investigated for its potential role in preventing BPD in preterm infants [17]. Emerging evidence suggests that early administration may offer meaningful benefits, particularly in infants at a high risk for pulmonary morbidity [17]. One of the most notable studies is the PREMILOC trial (N = 523), which demonstrated that low-dose hydrocortisone, administered within the first 24 h of life and continued for 10 days, significantly improved survival without BPD at 36 weeks PMA (60% vs. 51%, OR 1.48; 95% CI 1.02–2.16; $p = 0.04$) [29]. Importantly, the trial avoided early gastrointestinal perforation by prohibiting the use of prophylactic nonsteroidal anti-inflammatory drugs (NSAIDs) in the first 24 h. In contrast to dexamethasone, hydrocortisone was not associated with increased risks of CP or long-term neurodevelopmental impairment (NDI). In fact, a post hoc analysis of a subgroup of infants born at 24–25 weeks' gestation showed a significant neurological benefit (moderate-to-severe NDI: 2% in the hydrocortisone group vs. 18% in the placebo group; $p = 0.02$) [29]. The role of hydrocortisone in the later neonatal period has also been evaluated. The STOP-BPD trial (N = 372) assessed high-dose hydrocortisone in ventilator-dependent preterm infants between 7 and 14 days of age. While no significant difference was found in the composite outcome of death or BPD at 36 weeks' PMA (71% vs. 74%, $p = 0.54$), mortality alone was significantly lower in the hydrocortisone group (15.5% vs. 23.7%, $p = 0.048$) [30]. Similarly to what was observed in DART, hydrocortisone also facilitated extubation at multiple time points (Day 3: 16% vs. 7%, $p = 0.01$; Day 7: 46% vs. 22%, $p < 0.001$; and Day 14: 66% vs. 49%, $p = 0.001$) [30]. At two years corrected age, follow-up data were available for 96% of participants in the study. Among the infants assessed, the combined rate of death or NDI was 56.7% in the hydrocortisone group and 62.7% in the placebo group, with no statistically significant difference (adjusted risk difference −5.2%, 95% CI −15.1% to 4.6%; $p = 0.28$) [31]. Mortality rates were lower in the hydrocortisone group (21.5%) compared to the placebo (29.5%), though this difference did not reach significance ($p = 0.08$). Similarly, the incidence of NDI among survivors was comparable between groups, 43.9% in the hydrocortisone group versus 46.5% in the placebo group ($p = 0.68$), with no notable differences observed in the specific domains of the neurodevelopmental assessment [31]. In a multicenter RCT, Wattenberg et al. [32] enrolled 800 infants of <30 weeks' gestation who had been intubated for at least 7 days. The infants were randomized to receive hydrocortisone (4 mg/kg/day, tapered over 10 days) or the placebo. The primary efficacy outcome, survival without moderate/severe BPD at 36 weeks PMA, occurred in 16.6% of the hydrocortisone group and 13.2% of the placebo group (adjusted RR 1.27; 95% CI 0.93–1.74). At 22–26 months corrected age, survival without moderate/severe NDI was similar between groups (36.9% vs. 37.3%) [32]. Regarding its safety profile, hydrocortisone was generally well tolerated. In the Wattenberg study, hypertension requiring medication was more common in the hydrocortisone group (4.3% vs. 1.0%), while the incidence of other adverse events was comparable [32]. In contrast to dexamethasone, hydrocortisone has not been associated with an increased risk of cognitive impairment, as seen in data from the EPICE cohort [33]. A recent systematic review and meta-analysis by Doyle et al. [24] reported that hydrocortisone did not significantly reduce the risk of BPD compared to the placebo (RR = 0.98; 95% CI 0.87–1.10), speculating that the predominantly low cumulative doses (<2 mg/kg in most studies) may have been insufficient to elicit a strong anti-inflammatory effect. Similar findings were reported by van de Loo et al. [34], although Doyle did note a mortality benefit with early hydrocortisone use. A very recent meta-regression analysis compared systemic postnatal corticosteroids across 26 RCTs including 3700 infants [35]. The analysis highlighted a differential effect of corticosteroid type: dexamethasone was associated with improved survival, free of CP, in infants with a BPD risk of >70% but was potentially harmful in those with a BPD risk of <30% [35]. In contrast, no substantial benefit or harm was observed with hydrocortisone, suggesting a less pronounced or more context-dependent role. The timing of the treatment did not significantly influence dexamethasone outcomes. Additionally, a meta-analysis by Zhou et al. [36] found that early low-dose hydrocortisone may reduce the combined outcome of BPD or death in infants weighing <1000 g with a prior exposure to chorioamnionitis (OR 0.52; 95% CI 0.32–0.79; NNT = 6), but not in those without such exposure. No significant differences were observed in the individual outcomes of BPD or death regardless of their exposure status [36].

2.3. *Prednisolone and Methylprednisolone in BPD*

Prednisolone and methylprednisolone are synthetic glucocorticoids with similar anti-inflammatory potency and are often used interchangeably in clinical settings [17]. However, their application in the context of BPD remains limited to specific scenarios, and the available evidence for it is still sparse [17]. Unlike dexamethasone or hydrocortisone, these agents have primarily been investigated for late use, once BPD is already established, particularly in infants beyond 36 weeks PMA [17]. Retrospective data have suggested some clinical benefits. In a study by Bhandari et al. [37], prednisolone was administered to infants with established BPD (mean PMA ~38 weeks) who were receiving supplemental oxygen but were not mechanically ventilated. The findings showed that 63% of the treated infants were successfully weaned off supplemental oxygen, indicating a potential role for prednisolone in improving respiratory status. Similarly, Linafelter et al. [38] evaluated a longer course of prednisolone (≥30 days; median duration: 67 days) in infants with severe BPD, including both those on mechanical ventilation and those on noninvasive support. The study noted a significant reduction in respiratory support needs after the first week of treatment. Despite these initial improvements, the benefits appeared to stabilize beyond the first week, and extended therapy did not yield additional clinical gains. Furthermore, prolonged use was associated with notable adverse effects, most significantly impaired linear growth observed by week 4 of the treatment [17]. In Linafelter's cohort, although a high survival rate to discharge was reported (86%), more than half of the infants (57%) ultimately required a tracheostomy, underscoring the complexity and severity of their condition [38]. Importantly, no prospective RCTs have been conducted to assess the efficacy or safety of prednisolone or methylprednisolone in the prevention or early treatment of BPD [17].

2.4. *Current Recommendations*

A recent position statement from the Canadian Paediatric Society emphasized that the quality of evidence supporting the use of these corticosteroids in established BPD was very low, and therefore, their use could not be routinely recommended [22]. Notably, once

BPD has been established, the 2020 European Respiratory Society (ERS) guidelines on the long-term management of children with BPD do not recommend systemic corticosteroids (including dexamethasone or hydrocortisone) [39]. The task force found no strong or direct evidence supporting the benefit of systemic corticosteroids in improving clinically meaningful outcomes in this population. Moreover, concerns regarding the potential adverse effects, including growth suppression and negative neurodevelopmental consequences, contributed to the recommendation against their routine use [39]. As a result, systemic corticosteroids are discouraged in the long-term management of established BPD, particularly as standard therapy. However, in cases of severe disease, persistent symptoms, or frequent hospital readmissions unresponsive to other treatments (e.g., bronchodilators), a carefully monitored trial of corticosteroids may be considered on an individual basis, guided by clinical judgment [39] [Tables 1 and 2].

Table 1. Comparison between systemic and inhaled corticosteroids in the prevention and management of bronchopulmonary dysplasia in preterm infants.

	Systemic Corticosteroids	Inhaled Corticosteroids
Route of administration	Intravenous or oral	Inhaled (with or without surfactant)
Main agents	Dexamethasone, Hydrocortisone, Prednisolone	Budesonide, Beclomethasone, Fluticasone
Timing of use	Early (first week), Moderately early (8–14 days), Late (>14 days)	Early initiation (<7 days), some used later
Primary benefit	Effective reduction in BPD * incidence, improves extubation	Localized effect with reduced systemic exposure
Main risks	Neurodevelopmental impairment, gastrointestinal perforation, infections	Possible increased mortality (early use), technical delivery limitations
Neurodevelopmental concerns	Significant with early/high-dose use	Less pronounced or unclear
Evidence level	Moderate to High (Cochrane, RCTs)	Low to Moderate (NEUROSIS, Cochrane)
Use in BPD	Discouraged (ERS guidelines)	Discouraged (ERS guidelines)

* BPD: bronchopulmonary dysplasia.

Table 2. Summary of the timing, duration, main benefits, adverse effects, and current recommendations for systemic and inhaled corticosteroids used in the prevention and management of bronchopulmonary dysplasia (BPD) in preterm infants. This table synthesizes clinical trial evidence and current guidelines to support clinical decision-making.

Steroid	Timing of Initiation	Duration	Main Benefits	Risks	Recommendation	Unresolved Issues
Dexamethasone	Early (≤7 days) or Late (>7 days)	3–10 days	Reduced BPD, facilitates extubation	Neurodevelopmental impairment, GI perforation, infections	Use after 1st week in high-risk infants	Ideal timing in moderate-risk infants; long-term safety; biomarker-guided use
Hydrocortisone	Early (<24 h) or 7–14 days	10 days	Improved survival without BPD	Hypertension, uncertain NDI effect	Consider in early high-risk or 7–14 d intubated infants	Optimal dose; timing relative to chorioamnionitis; metabolic outcomes
Prednisolone	Late (>36 weeks PMA)	≥30 days	Improved O2 weaning in severe BPD	Growth suppression, need for tracheostomy	Limited use in severe, established BPD only	Comparative efficacy; initiation thresholds; endocrine effects
Budesonide (inhaled)	Early (<24 h)	Until 32 weeks PMA or off O2	Reduced BPD incidence	Possible increased mortality, unclear long-term safety	Not routine; promising with surfactant, more data needed	Long-term safety; neurodevelopmental impact; optimal delivery method

BPD: bronchopulmonary dysplasia; GI: gastrointestinal disease; NDI: neurodevelopmental impairment.

3. Inhaled Corticosteroids in BPD

Inhaled corticosteroids (ICSs) have been proposed as a strategy to reduce pulmonary inflammation in preterm infants while minimizing systemic exposure. The theoretical advantage lies in delivering the anti-inflammatory effect directly to the lungs, lowering the risk of adverse systemic effects commonly associated with systemic corticosteroids [17]. The current literature includes two distinct approaches: (1) prolonged ICS administration and (2) intratracheal instillation of corticosteroids in combination with a surfactant, typically as a single dose early in life [17].

3.1. Budesonide in BPD

Among studies on prolonged ICS use, the NEUROSIS trial is particularly noteworthy as a large, multicentric RCT involving 863 infants born before 28 weeks of gestation [40]. High-dose inhaled budesonide, administered within the first 24 h of life and continued until 32 weeks PMA or weaning off supplemental oxygen, significantly reduced the incidence of BPD compared to the placebo (27.8% vs. 38.0%, $p = 0.004$). However, the composite outcome of death or BPD was only borderline significant (40.0% vs. 46.3%, $p = 0.05$), and mortality alone was numerically higher in the budesonide group (16.9% vs. 13.6%, RR 1.24), although not statistically significant [40]. A long-term two-year follow-up revealed an increased mortality rate in the budesonide group (19.9% vs. 14.5%, RR 1.37, $p = 0.04$), raising concerns about the trade-off between reduced BPD and potential excess mortality. These findings underscore the importance of balancing the pulmonary benefits with overall survival outcomes [40]. A recent Cochrane review supports the notion that the early initiation of ICSs may reduce the risk of BPD and death at 36 weeks' PMA (RR 0.86, 95% CI 0.75–0.99), with no significant adverse effects reported [35]. The Bassler et al. [40,41] trial, which also evaluated early budesonide use, confirmed a reduction in BPD (27.8% vs. 38.0%; $p = 0.004$) and lower reintubation rates and patent ductus arteriosus (PDA) surgery. However, mortality was again slightly higher in the treatment group, though not statistically significant (16.9% vs. 13.6%). Despite these encouraging findings, ICS delivery in extremely preterm infants remains technically challenging due to low tidal volumes and weak inspiratory efforts, which limit pulmonary deposition [42]. Studies on intratracheal corticosteroid–surfactant administration involve a single or limited number of doses delivered early in life. However, commercial formulations are not yet widely available [42]. Particularly, in one randomized trial involving 265 very low birth-weight neonates, the combined surfactant–budesonide therapy significantly reduced the incidence of death or BPD (42.0% vs. 66.0%; RR 0.58; $p < 0.001$; NNT = 4.1), with no significant differences in their long-term growth or neurodevelopmental outcomes at 2–3 years [43]. A meta-analysis of two RCTs showed that this approach resulted in a 43% reduction in the incidence of BPD (RR 0.57; 95% CI 0.43–0.76; NNT = 5) and a 40% reduction in the composite outcome of death or BPD (RR 0.60; 95% CI 0.49–0.74; NNT = 3), without a statistically significant difference in mortality alone [44]. Furthermore, a recent randomized trial by Liu et al. [45] reinforced these findings, demonstrating that an early intratracheal instillation of budesonide combined with pulmonary surfactant significantly reduced the incidence and severity of BPD, decreased oxygen requirements and the duration of respiratory support, and did not increase glucocorticoid-related complications, supporting the potential safety and efficacy of this combined approach. However, the recently published PLUSS trial investigated the effect of early prophylactic intratracheal budesonide mixed with a surfactant in extremely preterm infants and found no statistically significant improvement in survival that was free of BPD at 36 weeks postmenstrual age, although a slight numerical benefit was observed [46]. Additionally, a meta-analysis including five randomized controlled trials

($n = 1515$) found no significant differences in neurodevelopmental impairment, mortality, or growth parameters between infants who received intratracheal corticosteroids and those who did not (RR for NDI: 0.90; 95% CI: 0.79–1.03; $p = 0.14$), suggesting a neutral safety profile [47].

3.2. Other ICSs

Data regarding the use of fluticasone and beclomethasone for the prevention of BPD in preterm infants are currently limited. Both are more commonly used in managing infants with BPD or those who have developed long-term respiratory complications rather than as preventive therapies [24]. A 2024 systematic review provided further clarity by comparing different inhaled corticosteroid regimens. While promising results were highlighted for budesonide, beclomethasone and fluticasone propionate did not demonstrate superior or inferior effects on the clinical outcomes when compared with the controls [48]. While ICSs are thought to present a lower systemic risk profile than systemic corticosteroids, several key limitations prevent their widespread recommendation. First, the quality of evidence is variable, with many studies being small, underpowered, or inconsistent in outcome reporting [49]. Secondly, long-term safety data, especially neurodevelopment-related, remain scarce. However, the existing Cochrane reviews suggest that ICSs initiated after seven days of life do not significantly increase the risk of major adverse events or neurodevelopmental impairments [49,50]. Although evidence ranges from low to very low certainty, no consistent safety concerns such as hypertension, sepsis, or cerebral palsy have emerged, and ICSs may reduce the need for systemic steroid exposure [49,50]. Importantly, the 2022 Cochrane review concluded that the routine early use of ICSs outside of clinical trials is not currently supported. Conversely, the late administration of ICSs appears to offer neither significant benefits nor issues [49]. However, comparative studies between inhaled and systemic corticosteroids are lacking, and no clear evidence exists to support the superiority of one route over the other [49,50].

3.3. Current Recommendations

Regarding children with established BPD, the 2020 ERS guidelines do not recommend the use of ICSs in infants beyond 36 weeks PMA or following hospital discharge [39]. This recommendation is based on low-certainty evidence, as the available data do not demonstrate a clear clinical benefit that outweighs the potential risks, particularly the concerns related to growth restriction and adverse neurodevelopmental outcomes with prolonged or repeated exposure. As a result, the routine use of ICSs is discouraged for the long-term management of BPD. However, in select clinical scenarios, such as in cases of severe BPD, persistent respiratory symptoms, or recurrent hospitalizations not responsive to bronchodilators, a carefully monitored, individualized trial of ICSs may be considered [39] [Tables 1 and 2].

4. Conclusions

BPD remains a major challenge in the care of preterm infants, driven by persistent inflammation and impaired lung development. Corticosteroids, both systemic and inhaled, have shown potential in modulating this inflammatory process and OS, but their use must be carefully balanced against potential adverse effects, particularly in terms of neurodevelopment outcomes. Systemic and inhaled corticosteroids continue to play a selective role in BPD management. While dexamethasone and hydrocortisone show potential in specific settings, and budesonide has yielded promising results, particularly when combined with a surfactant, the limitations in long-term safety data and guideline support restrict their routine use. The evidence for other steroid agents remains insufficient. The choice to use corticosteroids for BPD prevention or treatment must be individualized, weighing gestational age, clinical trajectory, and BPD risk. Current evidence underscores the need for larger, high-quality RCTs, particularly those evaluating long-term neurodevelopmental outcomes, biomarker-guided strategies, and optimal timing and dosage regimens to better define the safest and most effective steroid therapies for this vulnerable population.

Author Contributions: Conceptualization, E.G. and S.M.; investigation, F.G. and A.L.P.; resources, F.G. and A.L.P.; data curation, S.M.; writing—original draft preparation, F.G.; writing—review and editing, S.M.; supervision, E.G. All authors have read and agreed to the published version of the manuscript.

Funding: This research received no external funding.

Conflicts of Interest: The authors declare no conflicts of interest.

Abbreviations

The following abbreviations are used in this manuscript:

BPD	Bronchopulmonary dysplasia
CP	Cerebral palsy
ERS	European Respiratory Society
ICS	Inhaled corticosteroids
IL	Interleukin
NDI	Neurodevelopmental impairment
NEC	Necrotizing enterocolitis
NSAIDs	Nonsteroidal anti-inflammatory drugs
OS	Oxidative stress
PDA	Patent Ductus Arteriosus
PMA	Postmenstrual age
ROS	Reactive oxygen species
TNF	Tumor necrosis factor
VILI	Ventilator-induced lung injury

References

1. Jensen, E.A.; Dysart, K.; Gantz, M.G.; McDonald, S.; Bamat, N.A.; Keszler, M.; Kirpalani, H.; Laughon, M.M.; Poindexter, B.B.; Duncan, A.F.; et al. The Diagnosis of Bronchopulmonary Dysplasia in Very Preterm Infants. An Evidence-based Approach. *Am. J. Respir. Crit. Care Med.* **2019**, *200*, 751–759. [CrossRef] [PubMed]
2. Siffel, C.; Kistler, K.D.; Lewis, J.F.M.; Sarda, S.P. Global incidence of bronchopulmonary dysplasia among extremely preterm infants: A systematic literature review. *J. Matern. Fetal Neonatal Med.* **2021**, *34*, 1721–1731. [CrossRef] [PubMed]
3. Higgins, R.D.; Jobe, A.H.; Koso-Thomas, M.; Bancalari, E.; Viscardi, R.M.; Hartert, T.V.; Ryan, R.M.; Kallapur, S.G.; Steinhorn, R.H.; Konduri, G.G.; et al. Bronchopulmonary dysplasia: Executive summary of a workshop. *J. Pediatr.* **2018**, *197*, 300–308. [CrossRef] [PubMed]
4. Strashun, S.; Seliga-Siwecka, J.; Chioma, R.; Zielińska, K.; Włodarczyk, K.; Villamor, E.; Philip, R.K.; Assaf, N.A.; Pierro, M. Steroid use for established bronchopulmonary dysplasia: Study protocol for a systematic review and meta-analysis. *BMJ Open* **2022**, *12*, e059553. [CrossRef] [PubMed]
5. Martin, R.J.; Di Fiore, J.M.; Walsh, M.C. Hypoxic Episodes in Bronchopulmonary Dysplasia. *Clin. Perinatol.* **2015**, *42*, 825–838. [CrossRef] [PubMed]
6. Pérez-Tarazona, S.; Rueda Esteban, S.; García-García, M.L.; Arroyas Sanchez, M.; de Mir Messa, I.; Acevedo Valarezo, T.; Mesa Medina, O.; Callejón Callejón, A.; Canino Calderín, E.M.; Albi Rodriguez, S.; et al. Respiratory outcomes of "new" bronchopulmonary dysplasia in adolescents: A multicenter study. *Pediatr. Pulmonol.* **2021**, *56*, 1205–1214. [CrossRef] [PubMed]
7. Moschino, L.; Bonadies, L.; Baraldi, E. Lung growth and pulmonary function after prematurity and bronchopulmonary dysplasia. *Pediatr. Pulmonol.* **2021**, *56*, 3499–3508. [CrossRef] [PubMed]

8. House, M.; Nathan, A.; Bhuiyan, M.A.N.; Ahlfeld, S.K. Morbidity and respiratory outcomes in infants requiring tracheostomy for severe bronchopulmonary dysplasia. *Pediatr. Pulmonol.* **2021**, *56*, 2589–2596. [CrossRef] [PubMed]
9. Ruan, Q.; Wang, J.; Shi, Y. Clinical characteristics and outcomes until 2 years of age in preterm infants with typical chest imaging findings of bronchopulmonary dysplasia: A propensity score analysis. *Front. Pediatr.* **2021**, *9*, 712516. [CrossRef] [PubMed]
10. Liu, P.C.; Hung, Y.L.; Shen, C.M.; Chen, P.C.; Hsieh, W.S. Histological chorioamnionitis and its impact on respiratory outcome in very-low-birth-weight preterm infants. *Pediatr. Neonatol.* **2021**, *62*, 258–264. [CrossRef] [PubMed]
11. Teng, M.; Wu, T.J.; Jing, X.; Day, B.W.; Pritchard, K.A., Jr.; Naylor, S.; Teng, R.J. Temporal Dynamics of Oxidative Stress and Inflammation in Bronchopulmonary Dysplasia. *Int. J. Mol. Sci.* **2024**, *25*, 10145. [CrossRef] [PubMed]
12. Perrone, S.; Manti, S.; Petrolini, C.; Dell'Orto, V.G.; Boscarino, G.; Ceccotti, C.; Bertini, M.; Buonocore, G.; Esposito, S.M.R.; Gitto, E. Oxygen for the Newborn: Friend or Foe? *Children* **2023**, *10*, 579. [CrossRef] [PubMed]
13. Perrone, S.; Manti, S.; Buttarelli, L.; Petrolini, C.; Boscarino, G.; Filonzi, L.; Gitto, E.; Esposito, S.M.R.; Nonnis Marzano, F. Vascular Endothelial Growth Factor as Molecular Target for Bronchopulmonary Dysplasia Prevention in Very Low Birth Weight Infants. *Int. J. Mol. Sci.* **2023**, *24*, 2729. [CrossRef] [PubMed]
14. Capasso, L.; Vento, G.; Loddo, C.; Tirone, C.; Iavarone, F.; Raimondi, F.; Dani, C.; Fanos, V. Oxidative Stress and Bronchopulmonary Dysplasia: Evidences From Microbiomics, Metabolomics, and Proteomics. *Front. Pediatr.* **2019**, *7*, 30. [CrossRef] [PubMed]
15. Camm, E.J.; Tijsseling, D.; Richter, H.G.; Adler, A.; Hansell, J.A.; Derks, J.B.; Cross, C.M.; Giussani, D.A. Oxidative stress in the developing brain: Effects of postnatal glucocorticoid therapy and antioxidants in the rat. *PLoS ONE* **2011**, *6*, e21142. [CrossRef] [PubMed]
16. Garrud, T.A.C.; Giussani, D.A. Combined Antioxidant and Glucocorticoid Therapy for Safer Treatment of Preterm Birth. *Trends Endocrinol. Metab.* **2019**, *30*, 258–269. [CrossRef] [PubMed]
17. Htun, Z.T.; Schulz, E.V.; Desai, R.K.; Marasch, J.L.; McPherson, C.C.; Mastrandrea, L.D.; Jobe, A.H.; Ryan, R.M. Postnatal steroid management in preterm infants with evolving bronchopulmonary dysplasia. *J. Perinatol.* **2021**, *4*, 1783–1796. [CrossRef] [PubMed]
18. Onland, W.; van de Loo, M.; Offringa, M.; van Kaam, A. Systemic corticosteroid regimens for prevention of bronchopulmonary dysplasia in preterm infants. *Cochrane Database Syst. Rev.* **2023**, *3*, CD010941. [CrossRef] [PubMed]
19. Rademaker, K.J.; Groenendaal, F.; van Bel, F.; de Vries, L.S.; Uiterwaal, C.S.P.M. The DART study of low-dose dexamethasone therapy. *Pediatrics* **2007**, *120*, 689–690. [CrossRef] [PubMed]
20. Raghuveer, T.S.; Zackula, R.E.; Lakhotia, R.; Binder, S.A. Systemic steroids and bronchopulmonary dysplasia: A systematic review and meta-analysis. *J. Perinatol.* **2024**. [CrossRef] [PubMed]
21. Ramaswamy, V.V.; Bandyopadhyay, T.; Nanda, D.; Bandiya, P.; Ahmed, J.; Garg, A.; Roehr, C.C.; Nangia, S. Assessment of Postnatal Corticosteroids for the Prevention of Bronchopulmonary Dysplasia in Preterm Neonates: A Systematic Review and Network Meta-analysis. *JAMA Pediatr.* **2021**, *175*, e206826. [CrossRef] [PubMed]
22. Lemyre, B.; Dunn, M.; Thebaud, B. Postnatal corticosteroids to prevent or treat bronchopulmonary dysplasia in preterm infants. *Paediatr. Child Health* **2020**, *25*, 322–331. [CrossRef] [PubMed]
23. Doyle, L.W.; Cheong, J.L.; Hay, S.; Manley, B.J.; Halliday, H.L. Late (≥7 days) systemic postnatal corticosteroids for prevention of bronchopulmonary dysplasia in pre- term infants. *Cochrane Database Syst. Rev.* **2021**, *11*, Cd001145. [PubMed]
24. Zeng, L.N.; Tian, J.H.; Song, F.J.; Li, W.R.; Jiang, L.C.; Gui, G.; Zhang, Y.; Ge, L.; Shi, J.; Sun, X.; et al. Corticosteroids for the prevention of bronchopulmonary dysplasia in preterm infants: A network meta- analysis. *Arch. Dis. Child. Fetal Neonatal Ed.* **2018**, *103*, F506–F511. [CrossRef] [PubMed]
25. Szabó, H.; Baraldi, E.; Colin, A.A. Corticosteroids in the prevention and treatment of infants with bronchopulmonary dysplasia: Part, I. systemic corticosteroids. *Pediatr. Pulmonol.* **2022**, *57*, 600–608. [CrossRef] [PubMed]
26. Padam, S.; Newby, B.; Wang, L.L. Dexamethasone Use to Reduce Mechanical Ventilation and Bronchopulmonary Dysplasia in the Neonatal Intensive Care Unit. *Can. J. Hosp. Pharm.* **2024**, *77*, e3523. [CrossRef] [PubMed]
27. Doyle, L.W. Postnatal corticosteroids to prevent or treat broncho- pulmonary dysplasia. *Neonatology* **2021**, *118*, 244–251. [CrossRef] [PubMed]
28. NICHD Neonatal Research Network, Neonatal BPD Outcome Estimator. Available online: https://neonatal.rti.org/index.cfm?fuseaction=BPDCalculator.start (accessed on 5 April 2025).
29. Baud, O.; Maury, L.; Lebail, F.; Ramful, D.; El Moussawi, F.; Nicaise, C.; Zupan-Simunek, V.; Coursol, A.; Beuchée, A.; Bolot, P.; et al. Effect of early low-dose hydro- cortisone on survival without bronchopulmonary dysplasia in ex- tremely preterm infants (PREMILOC): A double-blind, placebo-controlled, multicentre, randomised trial. *Lancet* **2016**, *387*, 1827–1836. [CrossRef] [PubMed]
30. Onland, W.; Cools, F.; Kroon, A.; Rademaker, K.; Merkus, M.P.; Dijk, P.H.; van Straaten, H.L.; Te Pas, A.B.; Mohns, T.; Bruneel, E.; et al. Effect of hydrocortisone therapy initiated 7 to 14 days after birth on mortality or bronchopulmonary dysplasia among very preterm infants receiving mechanical ventilation: A randomized clinical trial. *JAMA* **2019**, *321*, 354–363. [CrossRef] [PubMed]

31. Halbmeijer, N.M.; Onland, W.; Cools, F.; Swarte, R.; van der Heide-Jalving, M.; Merkus, M.P.; van Kaam, A.H.; STOP-BPDTrial Collaborators. Effect of Systemic Hydrocortisone Initiated 7 to 14 Days After Birth in Ventilated Preterm Infants on Mortality Neurodevelopment at 2 Years' Corrected Age: Follow-up of a Randomized Clinical Trial. *JAMA* **2021**, *326*, 355–357. [CrossRef] [PubMed]
32. Watterberg, K.L.; Walsh, M.C.; Li, L.; Chawla, S.; D'Angio, C.T.; Goldberg, R.N.; Hintz, S.R.; Laughon, M.M.; Yoder, B.A.; Kennedy, K.A.; et al. NICHD Neonatal Research Network. Hydrocortisone to Improve Survival without Bronchopulmonary Dysplasia. *N. Engl. J. Med.* **2022**, *386*, 1121–1131. [CrossRef] [PubMed]
33. Zayat, N.; Truffert, P.; Drumez, E.; Duhamel, A.; Labreuche, J.; Zemlin, M.; Milligan, D.; Maier, R.F.; Jarreau, P.H.; Torchin, H.; et al. EPICEResearch Group Systemic Steroids in Preventing Bronchopulmonary Dysplasia (BPD): Neurodevelopmental Outcome According to the Risk of BPDin the EPICE Cohort. *Int. J. Environ. Res. Public Health* **2022**, *19*, 5600. [CrossRef] [PubMed]
34. Van de Loo, M.; van Kaam, A.; Offringa, M.; Doyle, L.W.; Cooper, C.; Onland, W. Corticosteroids for the prevention and treatment of bronchopulmonary dysplasia: An overview of systemic reviews. *Cochrane Database Syst. Rev.* **2024**, *2024*, CD013271. [CrossRef]
35. Doyle, L.W.; Mainzer, R.; Cheong, J.L.Y. Systemic Postnatal Corticosteroids, Bronchopulmonary Dysplasia, and Survival Free of Cerebral Palsy. *JAMA Pediatr.* **2025**, *179*, 65–72. [CrossRef] [PubMed]
36. Zhou, J.; Yu, Z.; Chen, C. Hydrocortisone for Preventing Mortality and Bronchopulmonary Dysplasia in Preterm Infants with or without Chorioamnionitis Exposure: A Meta-Analysis of Randomized Trials. *Am. J. Perinatol.* **2021**, *38*, 662–668. [CrossRef] [PubMed]
37. Bhandari, A.; Schramm, C.M.; Kimble, C.; Pappagallo, M.; Hussain, N. Effect of a short course of prednisolone in infants with oxygen- dependent bronchopulmonary dysplasia. *Pediatrics* **2008**, *121*, e344–e349. [CrossRef] [PubMed]
38. Linafelter, A.; Cuna, A.; Liu, C.; Quigley, A.; Truog, W.E.; Sampath, V.; Oschman, A. Extended course of prednisolone in infants with severe bronchopulmonary dysplasia. *Early Hum. Dev.* **2019**, *136*, 1–6. [CrossRef] [PubMed]
39. Duijts, L.; van Meel, E.R.; Moschino, L.; Baraldi, E.; Barnhoorn, M.; Bramer, W.M.; Bolton, C.E.; Boyd, J.; Buchvald, F.; Del Cerro, M.J.; et al. European Respiratory Society guideline on long-term management of children with bronchopulmonary dysplasia. *Eur. Respir. J.* **2020**, *55*, 1900788. [CrossRef] [PubMed]
40. Bassler, D.; Plavka, R.; Shinwell, E.S.; Hallman, M.; Jarreau, P.H.; Carnielli, V.; Van den Anker, J.N.; Meisner, C.; Engel, C.; Schwab, M.; et al. Early inhaled budesonide for the prevention of bronchopulmonary dysplasia. *N. Engl. J. Med.* **2015**, *373*, 1497–1506. [CrossRef] [PubMed]
41. Bassler, D. Inhaled budesonide for the prevention of bronchopulmonary dysplasia. *J. Matern. Fetal Neonatal Med.* **2017**, *30*, 2372–2374. [CrossRef] [PubMed]
42. Delara, M.; Chauhan, B.F.; Le, M.-L.; Abou-Setta, A.M.; Zarychanski, R.; 'tJong, G.W. Efficacy and safety of pulmonary application of corticosteroids in preterm infants with respiratory distress syndrome: A systematic review and meta-analysis. *Arch. Dis. Child Fetal Neonatal Ed.* **2019**, *104*, F137–F144. [CrossRef] [PubMed]
43. Yeh, T.F.; Chen, C.M.; Wu, S.Y.; Husan, Z.; Li, T.C.; Hsieh, W.S.; Tsai, C.H.; Lin, H.C. Intratracheal administration of budesonide/surfactant to prevent bronchopulmonary dysplasia. *Am. J. Respir. Crit. Care Med.* **2016**, *193*, 86–95. [CrossRef] [PubMed]
44. Venkataraman, R.; Kamaluddeen, M.; Hasan, S.U.; Robertson, H.L.; Lodha, A. Intratracheal Administration of Budesonide-Surfactant in Prevention of Bronchopulmonary Dysplasia in Very Low Birth Weight Infants: A Systematic Review and Meta-Analysis. *Pediatr. Pulmonol.* **2017**, *52*, 968–975. [CrossRef] [PubMed]
45. Liu, M.M.; Ji, L.; Dong, M.Y.; Zhu, X.F.; Wang, H.J. Efficacy and safety of intratracheal administration of budesonide combined with pulmonary surfactant in preventing bronchopulmonary dysplasia: A prospective randomized controlled trial. *Zhongguo Dang Dai Er Ke Za Zhi* **2022**, *24*, 78–84. [PubMed]
46. Manley, B.J.; Kamlin, C.O.F.; Donath, S.M.; Francis, K.L.; Cheong, J.L.Y.; Dargaville, P.A.; Dawson, J.A.; Jacobs, S.E.; Birch, P.; Resnick, S.M.; et al. Intratracheal Budesonide Mixed With Surfactant for Extremely Preterm Infants: A Randomized Clinical Trial. *JAMA* **2024**, *332*, 1889–1899. [CrossRef] [PubMed]
47. Zheng, Y.; Xiu, W.; Lin, Y.; Ren, Y.; Zhang, B.; Yang, C. Long-term effects of the intratracheal administration of corticosteroids for the prevention of bronchopulmonary dysplasia: A meta-analysis. *Pediatr. Pulmonol.* **2019**, *54*, 1722–1734. [CrossRef] [PubMed]
48. Zhang, M.; Zhang, W.; Liao, H. Efficacy and safety of different inhaled corticosteroids for bronchopulmonary dysplasia prevention in preterm infants: A systematic review and meta-analysis. *Respir. Med. Res.* **2024**, *85*, 101096. [CrossRef] [PubMed]
49. Onland, W.; Offringa, M.; van Kaam, A. Late ($\geq$7 days) inhaled corticosteroids to reduce bronchopulmonary dysplasia in preterm infants. *Cochrane Database Syst. Rev.* **2022**, *12*, CD002311. [CrossRef] [PubMed]
50. Shah, S.S.; Ohlsson, A.; Halliday, H.L.; Shah, V.S. Inhaled versus systemic corticosteroids for the treatment of bronchopulmonary dysplasia in ventilated very low birth weight preterm infants. *Cochrane Database Syst. Rev.* **2017**, *10*, CD002057. [CrossRef] [PubMed]

antioxidants

MDPI

Review

The Destructive Cycle in Bronchopulmonary Dysplasia: The Rationale for Systems Pharmacology Therapeutics

Mia Teng [1], Tzong-Jin Wu [2], Kirkwood A. Pritchard, Jr. [3,4], Billy W. Day [4], Stephen Naylor [4] and Ru-Jeng Teng [2,*]

1 Department of Integrative Biology, University of Wisconsin-Madison, 145 Noland Hall, 250 N. Mills St., Madison, WI 53706, USA; mteng7@wisc.edu

2 Department of Pediatrics, Medical College of Wisconsin, Suite C410, Children's Corporate Center, 999 N. 92nd Street, Milwaukee, WI 53226, USA; twu@mcw.edu

3 Department of Surgery, Medical College of Wisconsin, 8701 Watertown Plank Rd., Milwaukee, WI 53226, USA; kpritch@mcw.edu

4 ReNeuroGen LLC, 2160 San Fernando Dr, Elm Grove, WI 53122, USA; billy.day@rngen.com (B.W.D.); snaylor@rngen.com (S.N.)

* Correspondence: rteng@mcw.edu; Tel.: +1-414-9552337

Abstract: Bronchopulmonary dysplasia (BPD) remains a significant complication of premature birth and neonatal intensive care. While much is known about the drivers of lung injury, few studies have addressed the interrelationships between oxidative stress, inflammation, and downstream events, such as endoplasmic reticulum (ER) stress. In this review, we explore the concept of a "destructive cycle" in which these drivers self-amplify to push the lung into a state of maladaptive repair. Animal models, primarily the hyperoxic rat pup model, support a sequential progression from the generation of reactive oxygen species (ROS) and inflammation to endoplasmic reticulum (ER) stress and mitochondrial injury. We highlight how these intersecting pathways offer not just therapeutic targets but also opportunities for interventions that reprogram system-wide responses. Accordingly, we explore the potential of systems pharmacology therapeutics (SPTs) to address the multifactorial nature of BPD. As a prototype SPT, we describe the development of N-acetyl-L-lysyl-L-tyrosyl-L-cysteine amide (KYC), a systems chemico-pharmacology drug (SCPD), which is selectively activated in inflamed tissues and modulates key nodal targets such as high-mobility group box-1 (HMGB1) and Kelch-like ECH-associated protein-1 (Keap1). Collectively, the data suggest that future therapies may require a coordinated, network-level approach to break the destructive cycle and enable proper regeneration rather than partial repair.

Keywords: bronchopulmonary dysplasia; systems pharmacology therapeutic; unfolded protein response; mitochondria; inflammation; cellular senescence; endothelial nitric oxide synthase

1. Introduction

Bronchopulmonary dysplasia (BPD) is a multifactorial lung disorder frequently seen in premature infants. Classical BPD, first described by Northway et al. in 1967 [1], is a severe phenotype that was characterized by moderate prematurity (>30 weeks), severe lung damage, inflammation, and fibrosis resulting from prolonged exposure to high-concentration oxygen and prolonged mechanical ventilation. With the introduction of exogenous surfactant and the advancement of neonatal care, extremely premature neonates (<28 weeks) now have a much higher survival rate. Still, many require prolonged oxygen treatment and develop what is now termed "new BPD." The pathologic changes of "new BPD" include alveolar simplification, large alveolar sacs with less extensive lung fibrosis and damage [2]. It is believed that the mechanism behind the new BPD is growth arrest [3] secondary to the impaired angiogenesis [4].

Respiratory distress syndrome resulting from a lack of surfactant formation used to be the primary cause of death in premature neonates. The introduction of exogenous surfactant [5] and advancements in neonatal care [6] have successfully increased the survival rate of premature neonates, including periviable (22–25 weeks) premature neonates [7]. Still, as a result, the prevalence of BPD has increased. Clinicians commonly define BPD as oxygen requirement at the postconceptional age of 36 weeks [8]. About half of the premature neonates born less at than 29 weeks of gestation are expected to develop BPD [9]. It is estimated that there are 10,000 to 15,000 new BPD cases diagnosed in the United States each year [10]. Postnatal steroid treatment used to be the standard treatment to prevent BPD under the belief that inflammation is the key contributor. However, the increased odds of neurodevelopmental impairment with early initiation [11] have resulted in a debate about its routine use in premature neonates [12].

Prolonged supplemental oxygen use, aggressive nutritional support, restricted fluid intake, systemic or airway steroids, and bronchodilators are commonly given to infants with established BPD. Although controversial, diuretics are also commonly included in the therapeutic regimen. With the increased work of breathing and concerns about aspiration, when and how to offer oral feeding to BPD infants remains a debated topic among neonatologists, dietitians, and speech therapists [13,14]. Each neonatal unit should establish its own feeding guidelines through a multidisciplinary discussion and drawing on best practices.

Postnatal vitamin A [15] and early postnatal caffeine treatment [16] are clinically proven beneficial treatments that can prevent premature neonates from developing BPD. Each treatment reduced BPD by about 10% in the randomized controlled study. Clinicians have not widely adopted the three-times-a-week intramuscular megadose vitamin A injection due to concerns about the lack of muscle mass in extremely premature neonates. The implementation of early postnatal caffeine and vitamin A treatments in premature neonates has not changed the prevalence of BPD [17]. Part of the reason for the unwavering prevalence of BPD is the improved survival rate in extremely premature neonates. This suggests that there is still a lack of clarity in our ability to treat such patients effectively.

Premature neonates frequently require supplemental oxygen to maintain adequate tissue oxygenation. However, the abrupt transition from a hypoxic (32–35 mmHg) intrauterine environment [18] to the ambient 21% oxygen (~150 mmHg) environment generates oxidative stress (OS) in their premature lungs. This development is especially challenging for extremely premature neonates, as they lack a fully developed antioxidative capacity and the capability to generate surfactant [19]. Although exogenous surfactant treatment improves oxygenation, surfactant does not contain an antioxidative property. Mechanical ventilation [20], antenatal inflammation [21], and postnatal infection [22] also contribute to the development of OS in premature lungs. Innate inflammatory cells, such as macrophages and neutrophils, recruited to the lung further exacerbate OS by releasing hypochlorous acid (HOCl), generated by myeloperoxidase (MPO) [23]. HOCl is a potent oxidant capable of damaging most biological molecules and cells [24].

OS and inflammation are often described as distinct pathophysiological components of BPD. However, they are functionally interconnected. Reactive oxygen species (ROS) not only damage lipids, DNA, and proteins but also activate pro-inflammatory transcription factors such as NF-κB and AP-1 [25,26]. In turn, inflammatory cytokines stimulate enzymatic ROS production through NADPH oxidase and myeloperoxidase, amplifying oxidative injury [27]. This bidirectional feedback loop drives a self-sustaining cycle of damage in the immature lung, complicating both mechanistic understanding and therapeutic targeting. As we recently reviewed in the context of BPD [28], the mechanistic interdependence between oxidative stress and inflammation undermines the efficacy of monotherapies that target only one of these pathways. Therapeutic approaches that isolate either oxidative stress or inflammation fail to interrupt the self-amplifying cycle that drives lung injury in BPD. [28].

The OS-injured cells release several endogenous molecules known as damage-associated molecular patterns (DAMPs). DAMPs bind to various pattern recognition receptors (PRRs) and exacerbate inflammation by activating the NF-κB transcription factor [29]. The mutual interaction between the OS and inflammation creates a complex, self-perpetuating, and destructive cycle (Figure 1) in the lungs of BPD patients [28]. The process is accompanied by increased endoplasmic reticulum (ER) stress, also referred to as the unfolded protein response (UPR), and mitochondrial dysfunction [27]. The protein refolding process or electron transport chain (ETC) uncoupling generates more reactive oxygen species (ROS), thereby further contributing to OS. After OS-induced lung injuries, the oxygen extraction capacity of the neonatal lung decreases due to fibrosis, alveolar simplification, and decreased alveolar blood vessel formation. Type 2 alveolar (AT2) epithelial cells, considered the resident progenitor cells [30], are reduced in the alveoli, resulting in an impaired growth trajectory [31].

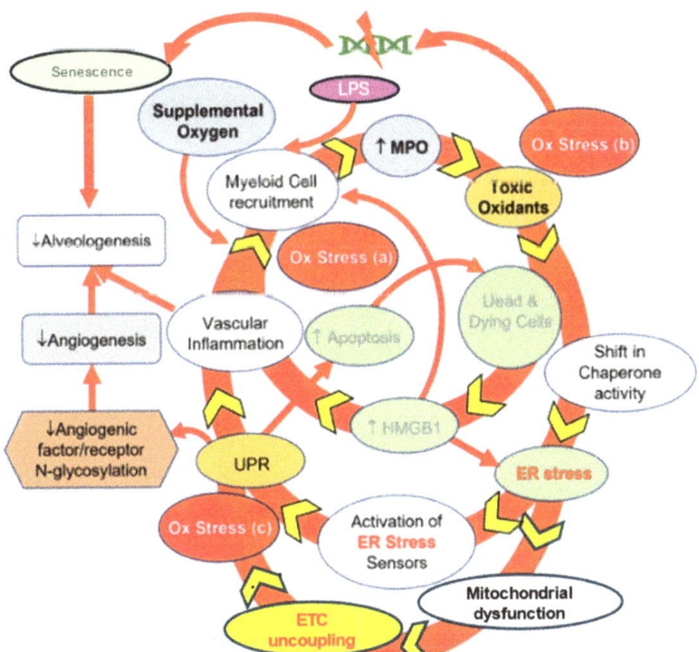

Figure 1. The proposed "destructive cycle" in the neonatal lung after exposure to supplemental oxygen. This figure focuses mainly on inflammation and oxidative stress (OS). Modified from [31].

The complex and self-reinforcing nature of these interconnected, pathological pathways requires a paradigm shift in designing effective therapeutic interventions for BPD. We propose that, instead of targeting isolated, individual pathways, the use of systems pharmacology therapeutics (SPTs) is a more appropriate approach. These SPTs act on multiple interrelated molecular targets within dysregulated biological networks. They differ from conventional multidrug regimens in that they can provide restoration of systems-level homeostasis through a single agent with multi-functional, network-directed activity [32,33]. For example, some SPDs are designed to target "nodes" or "motifs" within cellular circuits to rebalance signaling output [34]. The biological and computational logic behind this approach has been outlined in pioneering studies using disease networks and perturbation theory [35].

We have captured and visualized this conceptual framework in Figure 2. The figure contrasts the traditional single-target drug model with the SPT paradigm. In the classical model, a drug binds to a single receptor or enzyme, blocking or activating its function, often disregarding the redundant, compensatory, or feedback-rich nature of biological systems. In contrast, SPTs engage multiple nodes, which are usually low-affinity or context-specific, and exert their effects through partial modulation of key network motifs, feedback loops, and signaling pathways. This distributed action can achieve greater therapeutic efficacy with reduced off-target toxicity. Importantly, SPT design does not necessarily require targeting multiple independent proteins but may instead involve multi-site engagement within a single regulatory network, restoring systems-level homeostasis [33]. This paradigm has been validated in complex adult disease conditions such as cancer and neurodegeneration, but has not been utilized in neonatology. We argue here that BPD is a quintessential systems disorder and SPT therapies are needed for efficacious treatment. We propose that this paradigm is particularly applicable to BPD and outline its potential in later sections of this review.

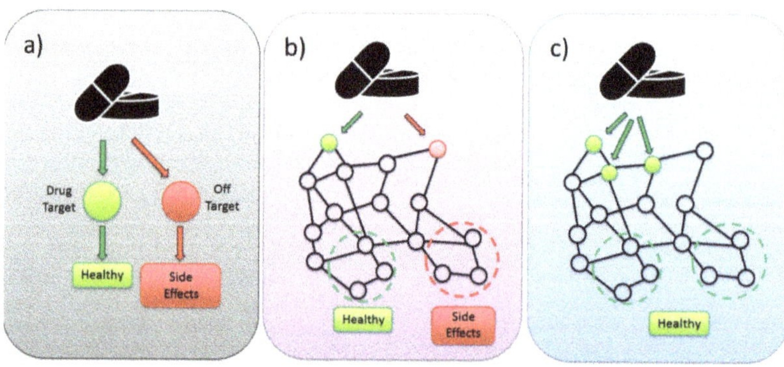

Figure 2. Comparison between classical drug design and systems pharmacology therapeutics (SPTs). (**a**) Representation of the one drug–one target model. (**b**) Representation of a systems biology perspective of therapeutic drug effects operating through a network. (**c**) Representation of the SPT one drug-multi-target-pathway/network approach of systems pharmacology. Green arrows: intended results; Red arrows: unwanted effects; dashed circles: final treatment outcomes.

This review does not aim to provide an exhaustive inventory of mechanisms or clinical trials in BPD. Instead, it presents a focused synthesis of the complex and evolving pathobiology of BPD, integrating clinical insight with biological plausibility. The destructive cycle of oxidative stress and inflammation (Figure 1) provides a unifying framework for understanding disease onset, progression, and heterogeneity. Within this context, we introduce the concept of systems pharmacology therapeutics (SPTs) as a novel and integrative approach to drug development. By highlighting both the mechanistic rationale and translational challenges, we propose that SPTs may be particularly well-suited to disrupt the multifactorial cycles that drive BPD pathogenesis. As an illustrative example, we describe the emerging therapeutic profile of N-acetyl-L-lysyl-L-tyrosyl-L-cysteine amide (KYC), a redox-modulating peptide currently under investigation.

2. Signaling Pathways That Amplify the OS and Inflammation in BPD Lungs

Inflammation, OS, and mechanical injury are the three central contributors to the development and progression of BPD. Limited animal studies have been conducted on the mechanistic role of mechanical injury in BPD. In most animal studies, hyperoxia-induced OS and lipopolysaccharide (LPS)-induced inflammation are employed. However, the temporal relationship between OS and inflammation makes it difficult to discuss them separately [28].

2.1. *Primary Mechanisms of Inflammation*

There are two kinds of innate inflammation involved in BPD. The first is the inflammatory reaction to chorioamnionitis or nosocomial infection, probably in response to pathogens [28]. The second mechanism is through OS-activated residential macrophages, which attract neutrophils from the circulation to infiltrate the damaged site, resulting in sterile inflammation [36,37]. Regardless of the mechanism, neutrophils release molecules to remove pathogens or dead or dying cells after arriving in the alveoli. The activated macrophages and neutrophils generate superoxide ($O_2^{\bullet-}$) or hydrogen peroxide (H_2O_2) through NADPH-oxidases (NOXs). $O_2^{\bullet-}$ interacts non-enzymatically with nitric oxide (•NO) released by the inducible NO synthase (iNOS or NOS2) to form a potent reactive nitrogen species—peroxynitrite ($ONOO^-$) [38]. H_2O_2 interacts with the nearby chloride anion to generate a potent ROS—HOCl—enzymatically by the MPO released from the activated myeloid cells [39]. Other halides and pseudohalides can also form ROS by the exact mechanism. Both $ONOO^-$ and HOCl are highly reactive oxidants that contribute significantly to oxidative injury and cellular dysfunction in the neonatal lung.

The sterile and pathogen-driven inflammatory mechanisms described above are functionally intertwined with OS via a shared signaling architecture involving MPO, NOXs, and cytokine-driven feedback. From a systems pharmacology perspective, this suggests that modulating a single node, such as MPO or NF-κB, may lead to partial or transient benefits unless broader network regulation is achieved. SPTs are designed to simultaneously target multiple redox and inflammatory regulators, exemplifying this systems-level intervention (see Section 4.2 below).

2.1.1. DAMPs and PRRs

The activated neutrophils release their content to facilitate the removal of pathogens or injured cells, mainly through forming free radicals or releasing proteases. The wounded, dying, or dead cells, including neutrophils, release their contents as damage-associated molecular patterns (DAMPs) [40]. The most extensively studied DAMP molecule is high-mobility group box 1 (HMGB1). HMGB1 is a nuclear protein that helps maintain the structure and function of chromosomes [41]. After being released from the nucleus, HMGB1 undergoes post-translational modifications, including oxidation, phosphorylation, *S*-nitrosylation, lactylation, and acetylation [42]. Both lactylation and acetylation promote the release of HMGB1 into the cytosol [43], which is the first step in its extracellular release.

An increased level of HMGB1 has been reported in tracheal aspirates [44] and serum [45] of premature infants who later developed BPD. In a mouse BPD model, the administration of an HMGB1 fragment that competitively inhibits HMGB1 successfully ameliorated the severity of BPD [46]. HMGB1 contains three thiol groups; their redox state determines the biological function of HMGB1. The fully reduced form of HMGB1 is pro-inflammatory but may accelerate tissue regeneration by activating the quiescent stem cells [47]. The fully oxidized form of HMGB1 is generally considered anti-inflammatory and does not promote chemotaxis [48]. Modulating the redox state of HMGB1 has thus been thought to have therapeutic potential in modulating inflammation [49].

Extracellular HMGB1 can bind to several pattern recognition receptors (PRRs) on the cell membrane, including Toll-like receptors 2 and 4 (TLR2 and TLR4), advanced glycosylation end-product specific receptor (AGER), triggering receptor expressed on myeloid cells 1 (TREM1), C-X-C motif chemokine receptor 4 (CXCR4), cluster of differentiation 24 (CD24), and hepatitis A virus cellular receptor 2 (HAVCR2) [50]. The binding between HMGB1 and TLRs, AGER, or TREM1 activates the NF-kB signaling pathway and induces the formation of inflammatory cytokines. HMGB1 also facilitates the binding between CXCL12 and CXCR4, promoting the chemotaxis of inflammatory cells (Figure 3).

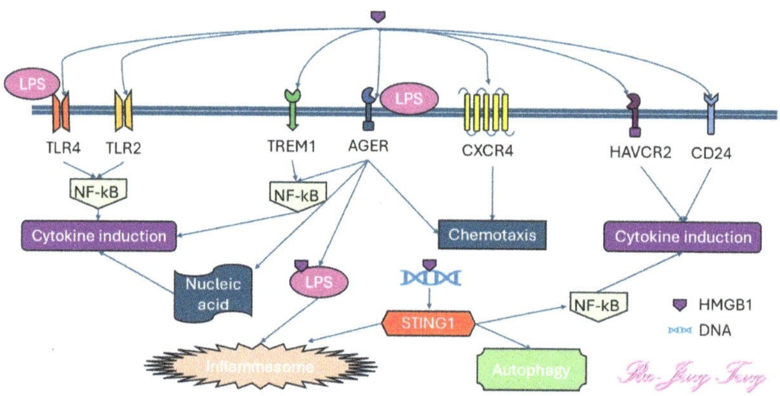

Figure 3. High-mobility group box-1 (HMGB1) is the most potent damage-associated molecular pattern (DAMP) family molecule. The binding between HMGB1 and pattern recognition receptors (PRRs) activates several downstream signaling pathways, exacerbating inflammation by enhancing inflammatory cell chemotaxis, cytokine release, and inflammasome formation. NF-kB is the primary mediator for cytokine induction. STING1 (Stimulator of Interferon Response cGAMP Interactor 1 or Stimulator of Interferon Genes 1) is an adaptor protein that promotes inflammasome formation.

Other DAMP molecules, including extracellular DNA [51], IL1α [52], IL33 [53], cyclophilin A [54], HSP70 [55], and S100 [56] have also been implicated in the development of BPD. However, their mechanistic roles remain incompletely explored.

2.1.2. Inflammasome Pathway

Inflammasomes are intracellular protein complexes comprising caspases, sensors, and adaptors (Figure 4). Inflammasomes cleave pro-inflammatory cytokines into active forms [57]. The caspases, especially caspase-1, induce apoptosis or pyroptosis, which is critical to releasing inflammatory mediators [58]. Caspases involved in the inflammasome formation include the canonical caspase-1, the non-canonical caspase-11 (also known as caspase-4 or caspase-5 in humans), or caspase-8. PYCARD, or ASC, is the adaptor protein for most inflammasomes. The primary sensors are the NOD-like receptors (NLRs), including NLRP3, NLRP1, NLRC4, and AIM1. Pyrin also serves as a sensor for the inflammasome [59,60]. The NLRP3 inflammasome is crucial for sterile inflammation [61]. After detecting specific DAMPs or pathogen-associated molecular patterns (PAMPs), the sensor will either oligomerize or interact with other proteins to form the inflammasome.

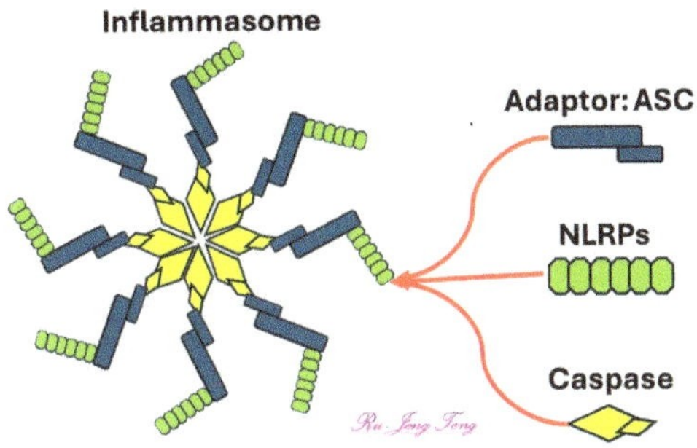

Figure 4. Structure of the inflammasome. NLRP (receptor), caspase (effector), and adaptor are the three main components of the inflammasomes. Some of the inflammasomes do not contain the adaptor.

The inflammasome pathway and DAMPs are closely related. In response to DAMPs, DNA from damaged cells is converted into cyclic guanosine monophosphate–adenosine monophosphate (cGAMP) by the cyclic GMP-AMP synthase (CGAS). Stimulator of interferon genes 1 (STING1) acts as a receptor for cGAMP, producing type I interferons and subsequent inflammasome formation [62]. DAMPs can also activate the inflammasome directly, especially the NLRP3 inflammasome, releasing IL-1β, IL-18, and other pro-inflammatory cytokines [63]. Gasdermin D (GSDMD) is the key executor, which is cleaved by caspase-1/-4/-5/-11, upon inflammasome activation, into the N-terminal fragment (GSDMD-NT) [64]. GSDMD-NT oligomerizes to form a pore in the cell membrane, releasing inflammatory cytokines and causing pyroptosis [65]. The released cytokines then augment inflammation by recruiting more inflammatory cells.

The inflammasome represents a central "hub" within the inflammatory signaling network, influenced by both oxidative inputs and DAMP signaling (Figure 4). Systems pharmacology therapeutics that dampen upstream triggers (e.g., MPO-generated HOCl) and modulate redox-sensitive proteins (e.g., thiolated HMGB1) could disrupt inflammasome activation at multiple control points. This reinforces the logic of targeting interconnected pathways rather than isolated effectors.

2.2. Uncoupled Endothelial Nitric Oxide Synthase (eNOS)

NO generated by the endothelial nitric oxide synthase (eNOS or NOS3) plays a critical role in vascular endothelial cell growth factor (VEGF)-induced angiogenesis [66]. •NO also modulates mitochondrial biogenesis [67], which is essential for endothelial cell function. Proper eNOS activity requires phosphorylation, dimerization, interaction with heat-shock protein 90, and many cofactors, including tetrahydrobiopterin (BH4) [68]. The absence of any cofactor can lead to eNOS uncoupling, generating superoxide instead of •NO [69] and mitochondrial dysfunction [70]. GTP-cyclohydrolase 1 (GCH1) is the critical enzyme in the BH4 biosynthesis. We have reported decreased GCH1 enzyme activity and BH4 levels in BPD rat lungs [71]. The uncoupled eNOS and subsequent mitochondrial dysfunction contribute to the increased OS in BPD lungs. Superoxide generated by the uncoupled eNOS interacts with the limited amount of •NO at a rate constant of $6.7 \pm 0.9 \times 10^9$/mol/s to form the potent $ONOO^-$ [72] that oxidizes almost all biological molecules.

eNOS also regulates neutrophil functions through •NO formation [73]. The basal level of •NO limits neutrophil chemotaxis by reducing the CXCR2 expression [74], prevents the circulating neutrophils from adhering to the activated endothelial cells, suppresses neutrophil phagocytic activity, modulates cellular 5-lipoxygenase activity and leukotriene synthesis, and enhances the cell death mechanism of the infiltrated neutrophils [75]. Thus, disrupted basal eNOS activity may facilitate neutrophil-mediated tissue damage.

2.3. Endoplasmic Reticulum (ER) Stress or Unfolded Protein Response (UPR)

The ER is the subcellular organelle where proteins are sequentially folded and modified to reach their full function. Post-translational protein glycosylation is either accomplished or initiated in the ER. It is also a site for lipid and cholesterol synthesis, calcium storage, and detoxification [27]. The ER also intimately interacts with other organelles, especially the neighboring mitochondria, to support their function. The neonatal lung responds to hyperoxia-induced OS acutely with an ER stress response as a coping mechanism [23]. This response reserves critical biomolecules to increase the chance of cell survival. However, ER stress can also cause translational attenuation, inflammation, autophagy, apoptosis, mRNA and protein degradation, and increase OS when the exogenous stress is insurmountable [27]. Although ER stress resolves quickly after OS is removed [23], its persistence in the early postnatal life still impairs the lung growth trajectory [76].

In the endoplasmic reticulum (ER), there is a group of protein chaperone ER stress sensors, including IRE1α, PERK, and ATF6. These endogenous chaperones include BiP (also known as GRP78), GRP94, CRT, and PDI. BiP is the most extensively investigated chaperone protein. Any exogenous stress that disrupts protein folding (proteostasis perturbation or proteotoxicity) can encourage GRP78 to abandon the ER stress sensors, allowing them to be activated either through phosphorylation and dimer formation (IRE1α and PERK), or be translocated to the Golgi apparatus for activation through enzymatic cleavage (ATF6) (Figure 5). The liberated GRP78 can thus help nascent proteins to be refolded appropriately. Protein folding is a redox process that relies on PDI and ERO1. H_2O_2 is a byproduct of the protein folding in the ER that requires glutathione peroxidases and peroxiredoxin 4 to remove it. The excessive protein refolding will thus exacerbate ER stress, leading to increased OS. The disrupted ER-mitochondria interaction during ER stress also causes mitochondrial dysfunction [77], with an uncoupled ETC [78], contributing to OS.

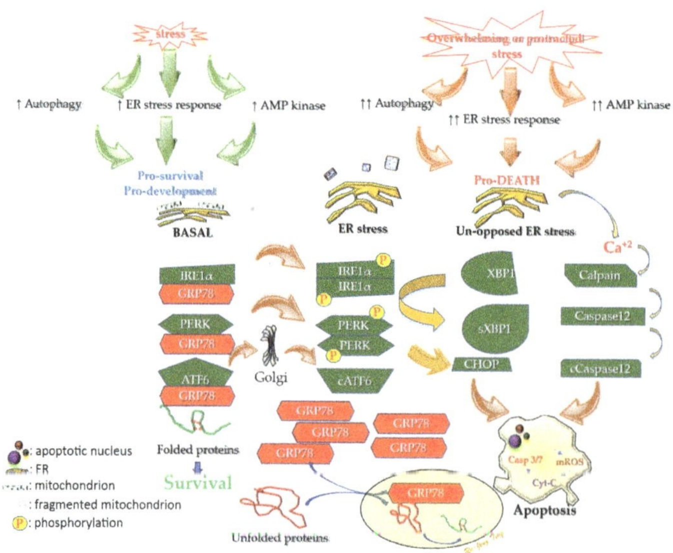

Figure 5. Schematic summary of endoplasmic reticulum (ER) stress or unfolded protein response (UPR). Stress distorting correct protein folding in the ER can elicit the ER stress response. Basal ER stress

upregulates the synthesis of the endogenous chaperone (BiP/GRP78) that assists protein refolding. Basal ER stress also inhibits the synthesis of non-essential proteins or degrades them so that raw material can be generated to synthesize essential proteins. Basal ER stress response is a survival mechanism for cells (**left panel**). If, however, the stress is overwhelming, BiP/GRP78 will all leave the ER stress sensors (IRE1α, PERK, and ATF6) to cope with the unfolded proteins. The protein refolding process generates ROS that exacerbate OS and lead to cell death. AMP: adenosine monophosphate; BiP: binding immunoglobulin protein; GRP78: glucose-regulated protein 78; IRE1α: inositol-requiring enzyme 1α; PERK: protein kinase R-like ER kinase; ATF6: activating transcription factor 6; cATF6: cleaved ATF6; mROS: ROS from mitochondria; Cyt-C: cytochrome C; Casp 3/7: caspase 3 and 7; CHOP: C/EBP homologous protein; XBP1: X-box binding protein 1 (XBP1); sXBP1: split XBP1. (The figure is reproduced from [27] under the Creative Commons CC BY 4.0 license).

All three ER stress sensors participate in the inflammatory response [79]. IRE1α and PERK activation lead to NF-κB-induced inflammatory response [80]. ATF6 cleavage activates NF-κB through the PI3K-Akt pathway [81]. HMGB1 is released from autophagic flux-induced cell death under ER stress, contributing to another mechanism for sterile inflammation [82]. Many signaling pathways are involved in the ER stress-induced inflammation. The phosphorylation and dimerization of IRE1α and PERK activate downstream NF-kB [83] and MAPK [84] signaling pathways, subsequently producing inflammatory cytokines. ER stress also activates the NLRP3 inflammasome, releasing IL-1α to elicit sterile inflammation [85]. Through NF-kB, JNK, and IRF3 pathways, ER stress hyperactivates neutrophils by forming neutrophil extracellular traps (NETosis) [86,87]. Undoubtedly, ER stress-mediated inflammation also contributes to OS. The reciprocal relationship between ER stress and oxidative damage highlights the importance of targeting systems at a higher level. The restoration of ER homeostasis by SPTs, such as through NRF2 activation and modulation of redox-regulated chaperones, may afford dual benefits in reducing inflammation and preserving alveolar development.

2.4. Cellular Senescence

The unopposed OS encountered by neonatal lungs, especially premature lungs, can damage the DNA of lung cells. To avoid faulty replication, tumor suppressor genes are activated to arrest the cell cycle. When cells enter irreversible growth arrest, it is referred to as cellular senescence [88]. However, it is now clear that cellular senescence might not be an irreversible change [89]. There are multiple types of cellular senescence, such as replicative senescence, oncogene-induced senescence, therapy-induced senescence, and developmental senescence [90]. Senescent cells are usually larger and metabolically active, but prone to using glycolysis instead of oxidative phosphorylation in ATP formation, even in a well-oxygenated environment [91]. Their peculiar metabolic behavior inhibits the growth of surrounding cells by depriving nutrients and energy. Senescent cells exhibit a characteristic senescence-associated secretory phenotype (SASP), actively secreting cytokines, growth factors, and proteases [92]. SASP has a paracrine effect that can transform neighboring cells into a senescent state [93].

Developmental senescence plays a critical yet essential role during fetal organ development [94]. This explains why the percentage of senescent cells is high in the mouse lung immediately after birth and decreases to the nadir at the beginning of the alveolar stage [94]. Acute senescence is recognized as a critical biological adaptation in wound healing, as it releases growth factors [95]. To resolve acute senescence, phagocytes remove senescent cells. If the senescent change is too extensive or becomes chronic, it will impair the function and growth of the organ. This is especially detrimental to premature organs, as the growth trajectory will be impaired. We have reported increased cellular senescence in human and rat BPD lungs. The senescent change in AT2 and endothelial cells indicates its contribution to BPD [31].

Senescent cells often exhibit senescence-associated mitochondrial dysfunction (SAMD), which includes increased NOX activity, impaired antioxidant defenses, and uncoupled ETC activity that increases ROS formation [96]. No specific cytokine is consistently present in the SASP. However, IL-1, IL-6, IL-8, IL-18, TNFα [97], and HMGB1 [98] are frequently described in the literature. These pro-inflammatory mediators support a chronic inflammatory milieu and promote fibrosis. Through persistent cytokine release, SASP factors create a pro-inflammatory microenvironment that reinforces tissue injury, impairs repair, and promotes fibrotic remodeling—principal features of the BPD phenotype.

2.5. Metabolic Disruption

Metabolic dysregulation has been observed in BPD [99]. The disturbed metabolism includes glucose, free fatty acids, and amino acids. Ratner et al. were the first group to describe an association between impaired mitochondrial complex I activity and alveolar simplification in mouse pups after a short exposure to hyperoxia, suggesting an essential role of oxidative phosphorylation in alveolar formation [100]. Gong et al. observed an increased glycolysis and pentose phosphate pathway (PPP) in hyperoxia-exposed mouse pups and lung endothelial cells. In their report, the increased PPP, not glycolysis, results in abnormal endothelial cell proliferation, dysmorphic angiogenesis, and alveolar simplification [101]. In the rat BPD model, hyperoxia exposure resulted in an enrichment of glycolysis, oxidative phosphorylation, and PPPs by Affymetrix transcriptomic analysis, but a paradoxical decrease in oxidative phosphorylation (manuscript submitted), probably due to ER stress and mitochondrial fragmentation [76]. Targeted metabolomics in the same study revealed an increased glucose transport into the BPD lungs, without a corresponding increase in ATP generation, suggesting impaired oxidative phosphorylation. The changes in acylcarnitines and amino acids also suggest an uncoupled ETC.

3. Signaling Pathways That Impair Angiogenesis in BPD Lungs

Impaired angiogenesis is believed to play a central role in BPD development and progression [4]. Endothelial cells can be classified into tip and stalk cells during sprouting angiogenesis [102]. However, this classification does not apply to alveolar formation. Alveolar capillary endothelial cells are classified into aerocyte (aCap) and general capillary endothelial cell (gCap) [103]. Some gCap cells are in the proliferating stage and can be classified as "proliferating gCap." Some consider proliferating gCap cells the local stem-like cells that orchestrate microvascular repair after injury [104]. Although these lung endothelial cells rely mainly on glycolysis, impaired oxidative phosphorylation can impair the proliferation, resulting in alveolar simplification. We describe a few signaling pathways that have been studied.

3.1. eNOS Uncoupling

The eNOS-generated NO is crucial for successful angiogenesis [105] as it serves as the second messenger of VEGF and modulates the mitochondrial function of the endothelial cells [106]. It was demonstrated that eNOS is uncoupled in the rat BPD model, partly due to the BH4 deficiency [71]. The uncoupled eNOS and mitochondrial dysfunction thus contribute to OS and impair angiogenesis. The insufficient •NO formation after eNOS uncoupling cannot adequately mediate the VEGF signaling. The •NO deficiency will not be able to remove the superoxide generated by the complex I or to support mitochondrial biogenesis. eNOS uncoupling illustrates how a single dysfunctional enzymatic process can simultaneously impair angiogenesis, increase oxidative stress, and alter the behavior of immune cells. Therapeutics that restore eNOS coupling via cofactor preservation, redox modulation, or upstream antioxidant signaling could address multiple axes of BPD pathology. This is a hallmark of systems pharmacology strategies.

3.2. ER Stress

ER stress impairs angiogenesis by at least two mechanisms. Multiple growth factors, secretory and transmembrane proteins, acquire various post-translational modifications for different biological purposes. N-Glycosylation is one of the post-translational modifications that is initiated in the ER. A characteristic example is the VEGF receptor 2 (VEGFR2) in the ER [107]. Increased ER stress has been demonstrated in the hyperoxia rat BPD models [23, 76], and decreased VEGFR2 glycosylation was also seen in this model. Treatment with tunicamycin, a characteristic ER stress inducer, also supports the contribution of ER stress to BPD, which resulted in a BPD phenotype with alveolar simplification and decreased blood vessel count. The ability of tauroursodeoxycholic acid, a known chemical chaperone that reduces ER stress, to attenuate BPD severity induced by hyperoxia and tunicamycin undoubtedly provides further evidence of the role of ER stress in BPD [76]. The intimate interaction between the ER and mitochondria is another mechanism by which ER stress impacts angiogenesis. The uncoupled electron transport chain resulting from the ER stress can lead to pathological angiogenesis [108] and impair tip cell function during the sprouting angiogenesis [109] as described in the next section.

3.3. Mitochondrial Dysfunction

The lungs generate a significant amount of lactic acid, primarily due to the high percentage of endothelial cells [110]. Although endothelial cells rely mainly on aerobic glycolysis to provide energy [109], they require both glycolysis and oxidative phosphorylation for proliferation [111]. The impaired complex I activity in hyperoxia-exposed neonatal lungs impairs angiogenesis and increases OS. Gong et al. reported that hyperoxia causes dysmorphic angiogenesis [101], which is associated with disrupted mitochondrial metabolism of glucose, free fatty acids, and amino acids. We did not observe dysmorphic angiogenesis, but we did notice a decrease in the percentages of aerocytes and proliferating general capillary endothelial cells, indicating a reduced ability to form new blood vessels (manuscript submitted).

3.4. Cellular Senescence

The relationship between cellular senescence and angiogenesis is a complex one. This complexity comes from different types of senescence and the SASP. In certain situations, senescent cells can promote angiogenesis. This can be observed in the pathologic neovascularization associated with diabetic retinopathy [112]. However, cellular senescence can disrupt the function of endothelial cells, especially progenitor cells, and the response to angiogenic stimuli [113]. Decreased VEGF expression could be the reason behind radiation-induced endothelial cell senescence [114], whereas YAP1 signaling has been considered the cause behind aging-related endothelial cell senescence [115].

We have reported increased senescent changes in endothelial and AT2 cells in rat BPD lungs [31]. Multiple putative signaling pathways are involved in this process, including ER stress, autophagy, TGFβ/SMAD, PTEN/PI3K, AMPK, PGC1α, etc. [94]. These signaling pathways interact with each other and ultimately lead to impaired angiogenesis. The improved alveolar complexity and blood vessel count by foxo4-dri, a synthetic senolytic agent [116], provide further evidence that cellular senescence contributes to the impaired angiogenesis in BPD.

Given the convergence of the multiple pathways in BPD, such as oxidative stress, ER stress, senescence, and impaired angiogenesis, on common molecular regulators, a therapeutic capable of modulating several of these simultaneously could provide superior clinical benefit. The concept of systems pharmacology is particularly compelling in this context: by partially correcting imbalances across a network. SPTs should help restore developmental trajectories more effectively than monotherapies.

4. Therapies for BPD: From One-Target Drugs to Systems Pharmacology

4.1. Historical and Current One Target–One Drug Paradigm

Drug discovery and development in neonatology remain limited and poorly aligned with the unique physiological and maturational trajectories of premature infants [117,118]. In particular, over the past three decades, multiple therapeutic interventions have been tested to reduce the incidence and severity of BPD. Many of these therapies were not initially developed or rigorously tested for neonates, but rather repurposed from adult medicine and used off-label. Most target individual aspects of disease pathophysiology, such as inflammation, oxidative stress, or respiratory drive, reflecting the traditional "one drug–one target" paradigm [28]. While some have demonstrated short-term physiological benefits, their overall impact on long-term pulmonary outcomes has been limited. Moreover, none of these interventions consistently disrupt the self-reinforcing pathological cycle that drives disease progression. In the following sections, we highlight several major therapeutic strategies, including corticosteroids, caffeine, vitamin A, and stem cell therapies and evaluate their benefits, limitations, and implications for future treatment approaches.

4.1.1. Steroids

Although the American Academy of Pediatrics does not recommend it due to the potential complication of neurodevelopmental deficit, some clinicians still use early systemic steroids to prevent or treat BPD in a selective group of premature neonates at high risk [119]. Steroids' main biological effect stems from their non-selective anti-inflammatory properties. Steroids do not have antioxidant activity themselves. The antioxidant property of steroids is believed to be a secondary effect through upregulating certain antioxidant enzymes.

4.1.2. Vitamin A

Vitamin A has anti-inflammatory and antioxidant properties. The hydrophobic chain of polyene units of vitamin A can quench singlet oxygen, neutralize thiyl radicals, and stabilize peroxyl radicals [120]. However, this antioxidant property only occurs in a low-oxygen tension environment. Vitamin A will auto-oxidize when the oxygen tension is high. The anti-inflammatory property has been demonstrated in several randomized controlled studies, using inflammatory markers as the primary endpoint [121]. So far, only intramuscular high-dose injections have been beneficial. Meta-analysis of the four available randomized controlled studies using enteral low-dose vitamin A failed to show efficacy [122]. The inhaled vitamin A study is presently ongoing.

4.1.3. Caffeine

Caffeine is probably one of the most extensively studied medications in neonatology. The primary hypothesis of the CAP study was that a better neurological outcome would be achieved by reducing apnea of prematurity. Interestingly, the successful reduction in the rate of BPD, one of the secondary outcomes, has led to almost universal implementation among clinicians [16]. Our group has demonstrated in the rat BPD model that caffeine increases BH4 levels and eNOS phosphorylation [71], decreasing neutrophil infiltration, apoptosis, and ER stress [23], which may explain the improved angiogenesis and alveolar formation. BH4 is also an antioxidant. Endesfelder et al. demonstrated the antioxidative effect of caffeine by showing decreased H_2O_2, malondialdehyde, and 8-hydroxy-deoxyguanosine with increased expression of superoxide dismutases in the hyperoxia rat BPD model [123]. They also showed caffeine's anti-inflammatory activity by decreasing NF-kB expression using the same animal model [124]. Köroğlu et al. used the rat intra-amniotic lipopolysaccharide injection BPD model to show a reduced inflammatory response with improved lung morphometry and respiratory system resistance by caffeine treatment [125]. Other potential benefits of caffeine treatment include antifibrotic and diuretic effects [126]. Caffeine protects the neonatal lung against the development of BPD through multiple mechanisms.

4.1.4. Stem Cell Therapy

Stem cells [127] and stem cell-derived extracellular vesicles [128] treatments are promising for BPD. Phase I [129,130] and II clinical studies (NCT03601416) are ongoing. Mesenchymal stem cells (MSCs) are the most used source due to their ease of isolation, low immune reaction, anti-inflammatory and antioxidative properties [131,132], and reparative capabilities [127,133]. MSC secretome [134] or exosome [135] are new stem cell-based modalities. The available short-term results of the phase I trials showed that premature neonates tolerated stem cell therapy well without apparent side effects [130]. Still, concerns regarding the difficulty in obtaining the appropriate number of cells and maintaining their quality, as well as the potential for vascular occlusion and tumorigenicity, remain unresolved [136].

These traditional therapeutic strategies for BPD, while biologically grounded and sometimes clinically useful, have not produced consistent or transformative outcomes. This reflects the limitations of targeting isolated disease mechanisms in a condition driven by a complex, interconnected network of oxidative stress, inflammation, and disrupted development. These shortcomings underscore the need for a paradigm shift toward systems-level therapeutics that can address the multifactorial nature of BPD pathogenesis.

4.2. Systems Pharmacology Therapeutics: A New Paradigm for Complex Diseases

4.2.1. Need for a New Therapeutic Framework

The failure of traditional BPD therapies to consistently improve long-term outcomes reflects more than incremental inefficiency. It also underscores a deeper conceptual limitation of the prevailing pharmacologic approach. For decades, drug development has been guided by the "one drug–one target–one disease" paradigm, which has proven effective in treating conditions with discrete, well-defined etiologies, such as bacterial infections or certain forms of cancer [33,35]. However, this reductionist model is poorly suited for treating complex, multifactorial diseases like BPD. As noted above, a myriad of biological processes, such as interdependent OS associated with inflammation, ER stress, senescence, and disrupted lung development, are mechanistically entangled and mutually reinforcing [27,28,137]

The need for a systems-level therapeutic strategy is especially compelling in BPD, where disease drivers exhibit bidirectional feedback, redundancy, and compensation. In such settings, inhibiting a single node often triggers adaptations in adjacent or upstream pathways, leading to limited efficacy or off-target consequences. Thus, a new framework is required that can modulate multiple disease-relevant nodes while accounting for feedback, crosstalk, and nonlinear system responses. This is the conceptual foundation of SPTs. Such therapeutic drugs are an emerging class of interventions designed to address disease as an integrated network rather than a linear sequence of isolated events [138,139].

4.2.2. Definition of an SPT

SPTs are not simply multitarget drugs. Instead, they are intentionally designed or selected to engage multiple nodes within a pathological network in a coordinated, system-aware fashion [33,35,140]. The goal is to restore or reprogram system-level behavior rather than merely suppress an isolated molecular signal.

Historically, the roots of systems pharmacology date back to the 1940s, when early frameworks sought to encompass the complexity and variability of human physiology and pathology. The advent of high-throughput multiomics, network analysis, and computational modeling in the 1990s and 2000s helped crystallize systems biology into a practical foundation for drug design [140]. Talevi later introduced the concept of the "master key" therapeutic capable of unlocking multiple related nodes within a disease network, contrasting with the conventional "lock-and-key" model [141].

Essential features of SPTs include the following:

- Multinodal Engagement: Simultaneous modulation of multiple interrelated targets within the same biological circuit.
- Modulatory (Not Ablative) Action: SPTs often tune rather than eliminate function, favoring homeostatic correction over complete suppression.
- Network-Driven Outcomes: Effects emerge from interaction dynamics across the system, not simple target summation.
- Emergent System Reprogramming: This may include the activation of adaptive responses, such as NRF2, the restoration of mitochondrial balance, or the dampening of DAMP signaling.

To further clarify what constitutes a true systems pharmacology therapeutic, it is useful to examine drugs that, while biologically broad in action, lack the design coherence and network specificity that define SPTs. For example, aspirin is often proposed as an SPT. Indeed, it has been demonstrated to be a pleiotropic agent, binding over two dozen targets including COX-1/2, NF-κB, and numerous acetylation-sensitive proteins [33]. It also exhibits diverse biological effects such as anti-inflammatory, anti-thrombotic, and chemopreventive actions, yet it is not an SPT.

Aspirin's broad actions are the result of biochemical promiscuity rather than rational network targeting. Its effects are neither coordinated nor intentionally designed to modulate disease networks in a systemic, balanced manner. Moreover, its lack of targeted coherence can produce harmful side effects, such as GI bleeding and renal stress. This distinction highlights a fundamental principle: multi-targeting does not equate to systems pharmacology. What defines an SPT is not the number of targets, but their intentional engagement as part of a coherent, disease-relevant network strategy. True SPTs stand at the intersection of systems biology, network pharmacology, and translational medicine, offering a next-generation solution for diseases where linear pharmacology has reached its practical and efficacious limits [139,142].

4.2.3. BPD Complexity and SPTs

BPD epitomizes a systems-level disease. Its progression involves recursive feedback among oxidative stress, inflammation, ER stress, and disrupted alveolar and vascular development [26]. Each of these domains reinforces the others, creating a self-sustaining pathophysiological cycle. Figure 1 illustrates this loop: ROS generation enhances cytokine production, which in turn promotes ER stress and senescence, thereby exacerbating oxidative injury. Furthermore, standard interventions such as corticosteroids or vitamin A act on individual elements of this system and thus do not break this destructive cycle. Redundancy in cytokine signaling, compensatory metabolic flux, and nonlinear amplification all conspire to make one-target therapies ineffective or transient in their benefits.

An ideal BPD-directed SPT should

- Scavenge ROS while modulating MPO activity.
- Activate NRF2-driven antioxidant gene expression.
- Reduce pro-inflammatory cytokine signaling in specific immune cell subsets.
- Protect epithelial and endothelial progenitor cell pools critical for lung development.

Such coordinated actions cannot be achieved through drug cocktails or traditional multitarget drugs. Instead, they require system-aware molecules designed to engage the disease network holistically.

4.2.4. Strategic Considerations in Developing SPTs

Developing SPTs presents unique challenges but holds the potential to transform care for complex disorders like BPD. Critical development principles include the following:

- Network-Informed Targeting: Use of transcriptomic/proteomic data to map disease-relevant subnetworks.
- Phenotypic Screening: Prioritize functional outcomes in disease models over single-target readouts.
- Pharmacodynamic Complexity: Recognize that efficacy may reflect delayed or emergent system changes, requiring longitudinal biomarker strategies [143].
- Safety Profiling at the Network Level: Systems toxicology is essential to predict effects on unintended but connected pathways [138].

Despite the conceptual maturity of systems pharmacology in adult diseases, neonatology remains largely untouched by these frameworks. No published clinical or preclinical studies, to our knowledge, have purposefully applied SPT principles to BPD. This highlights both the novelty and the opportunity of doing so. Our group has been developing an end-capped tripeptide that exemplifies the SPT model, targeting MPO catalysis to generate thiyl radicals that modulate protein networks, and activating NRF2 while reducing HMGB1 release and senescence markers. It demonstrates how intentional molecular design can generate systems-level therapeutic outcomes, and this is discussed in more detail in Section 4.3 below.

4.3. KYC as an SPT

4.3.1. What Is KYC?

KYC is a synthetic end-capped tripeptide originally developed to intercept inflammatory oxidants in vivo. It is now recognized as a prototype of a novel drug class known as systems chemico-pharmacology drugs (SCPDs). SCPDs are not defined solely by their target profile, but rather by their chemical reactivity, context-dependent activation, and network-level pharmacodynamics. As described in our patent [144], an SCPD interacts with a "first target", which is often a catalytically active enzyme such as myeloperoxidase (MPO). The latter is then modified chemically, such as via oxidation, to form the parent compound (KYC). This activation step creates a reactive intermediate (e.g., a thiyl radical) capable of selectively engaging with other nearby biomolecules, which are referred to as the "second target(s)" within the pathological microenvironment. Thus, an SCPD uses a two-step, context-locked mechanism: the first interaction activates the parent molecule, and the second extends its pharmacological sphere of influence across the surrounding system.

This molecular choreography provides both spatial specificity (as activation occurs only in disease-associated biochemical zones) and network coherence, enabling the SCPD to reshape aberrant biological signaling in a systemic yet controlled fashion. KYC embodies this principle, since it becomes pharmacologically active only within inflamed tissues rich in MPO, and its subsequent thiylation of key signaling proteins promotes a homeostatic rebalancing rather than toxic ablation. As such, this SCPD offers a next-generation strategy for treating complex, nonlinear diseases like BPD.

4.3.2. Mechanism of Action: From MPO Activation to Site-Specific Network Modulation

KYC operates through a unique multi-step redox mechanism that defines its systems-level pharmacology:

- Myeloperoxidase (MPO) Activation: KYC is selectively activated at sites of inflammation by MPO, a heme-containing enzyme released by neutrophils. MPO catalyzes the oxidation of the tyrosine residue in KYC, forming an initial tyrosyl radical and then, via intramolecular electron transfer, converting the cysteine thiol into a thiyl radical. This activation is contingent on MPO's catalytic cycle and thus restricted to inflammatory microzones.
- Sphere of Influence and Local Specificity: Once formed, the thiyl radical exits the MPO catalytic site and engages proximal protein thiols or oxidized amino acids, enabling reversible, site-specific covalent modification. We have termed this concept the "sphere of influence," which confers selectivity by limiting KYC's reactive behavior to a defined microenvironment shaped by inflammatory enzyme release, pH, and redox potential. In addition, the short-lived thiyl radical ensures both spatial precision and temporal constraint, limiting off-target effects. Unlike classical inhibitors, KYC does not irreversibly bind to targets but instead modifies them through thiolation, enabling the reprogramming of key signaling hubs, such as HMGB1, GSNOR, Keap1, and NF-κB regulators.
- Emergent Network Effects: These covalent modifications reset cellular redox tone, activate NRF2 signaling, dampen DAMP-associated inflammation, and restore mitochondrial homeostasis. KYC thus engages multiple, interlinked targets as part of a broader strategy to reestablish physiological equilibrium.

This chemical mechanism of KYC satisfies several defining criteria for SPTs outlined in Section 4.2, including multinodal engagement, homeostatic tuning, and emergent reprogramming of complex inflammatory networks. Taken together, these findings not only validate KYC's pharmacodynamic profile but also provide biological confirmation of its system-level engagement in neonatal lung pathology.

4.3.3. Systems Biological Effects of KYC in BPD

KYC has been shown to reversibly inhibit MPO activity [145]. It inhibits the peroxidase activity of MPO by entering the ferric protoporphyrin IX (heme) center to inhibit the binding of halides. The reaction between MPO and KYC transforms MPO into quasi-catalase and prevents HOCl formation. The tyrosine is an electron acceptor and becomes a tyrosyl radical after oxidation. The cysteine thiol group accepts the electron from the tyrosyl radical to become a thiyl radical. The thiyl radical reacts with thiol-containing molecules to form disulfide bonds. The quasi-catalase activity and inhibition of HOCl formation are not the only antioxidant properties of KYC. In the rat BPD model, we have shown that KYC also facilitates the expression of NRF2-induced antioxidant enzymes through thiylation and glutathionylation of KEAP1, followed by the stabilization of NRF2 [146]. The attenuation of ER stress is the third mechanism of the antioxidant activity [74]. These can explain why KYC decreases the chlorinated tyrosine and 8-hydroxy-deoxy-guanosine in BPD lungs.

The anti-inflammatory activities of KYC include decreasing neutrophil infiltration and MPO expression in the rat BPD lungs [146], reducing the expression of cyclooxygenases, TLR4, AGER, and extracellular HMGB1, oxidizing HMGB1, decreasing HMGB1 binding to PRRs [146], and attenuating ER stress [76], autophagy, and cellular senescence [31]. Fully oxidized HMGB1 loses its pro-inflammatory activity, and the decreased HMGB1 binding to TLR4, and AGER prevents the downstream NF-kB signaling. Attenuating autophagy, ER stress, and cellular senescence contribute to KYC's anti-inflammatory properties.

In addition to its antioxidant and anti-inflammatory activities, KYC also increases the AT2 cell count in rat BPD lungs [31]. As AT2 behaves as a resident progenitor epithelial cell, the increased count by KYC indicates a better lung growth potential. Affymetrix studies show enrichment in the E2F, WNT/Catenin, and Notch signaling, with a decrease in TNFα/NF-kB, IL6/JAK/STAT3, P53/apoptosis, and epithelial–mesenchymal transition signaling that are awaiting further investigations. Mouse sickle cell models further demonstrate that KYC improves endothelial cell-mediated vasodilation, indicating an improved eNOS coupling [147]. The evidence suggests that KYC inhibits MPO, protects neonatal lungs against OS and inflammation, and offers more extensive benefits than vitamin A and caffeine. Thus, KYC operates not only as an anti-inflammatory and antioxidant agent but as a prototype for a next-generation class of therapeutics designed to operate within and reprogram the network dynamics of disease. Taken together, these findings not only validate KYC's pharmacodynamic profile but also provide biological confirmation of its system-level engagement in neonatal lung pathology.

4.3.4. Aligning KYC with the SPT Framework

Based on the characteristics and experimental findings discussed above, KYC satisfies multiple criteria that define a well-characterized SPT (Table 1). This is exemplified in the Table below as we map SPT attributes (as defined in Section 4.2) against KYC's mechanisms and preclinical findings:

Table 1. Aligning KYC with the SPT framework.

SPT Attribute	KYC Alignment
Multinodal Network Engagement	Modifies multiple proximal targets (e.g., HMGB1, Keap1) through a shared chemical mechanism (thiyl radical formation).
Context-Dependent Activation	Selectively activated by MPO in inflamed tissues, limiting off-target effects and enhancing safety.
Systemic Rebalancing	Reprograms redox and inflammatory signaling pathways (e.g., NRF2 activation, DAMP suppression).
Emergent Properties	Results in improved alveolarization and lung function not attributable to single-target inhibition.
Mechanism-Guided Design	Engineered with chemical features (thiol, tyrosine) for specific redox activation and network modulation.
Translational Scalability	Mechanistic pathway—MPO activation and downstream NRF2/DAMP effects—relevant to multiple diseases beyond BPD.

The development and characterization of KYC as an SCPD provides a tangible example of how SPT principles can be applied in neonatal disease contexts such as BPD. Its capacity to interface with the biochemical architecture of inflamed tissues and to exert controlled, multilayered network effects points toward a new therapeutic horizon—one where complexity is not bypassed, but instead embraced, in drug design. As a first-in-class SCPD, KYC represents not just a therapeutic candidate for BPD but a model for rationally engineered network-aware drugs designed to restore homeostasis in complex disease systems.

5. Discussion

The dramatic advancement in neonatal care has successfully improved the survival rate of extremely premature neonates and contributed to the unchanged prevalence of BPD.

Clinicians are facing an increasing number of survivors of premature neonates born at the late canalicular stage (22–26 weeks) of lung development [6,7]. Unlike diseases affecting other age groups, premature organs are still developing and require maintenance of their growth potential. The most significant contributors to BPD are OS, inflammation, and the use of a positive-pressure mechanical ventilator. Other contributors include fluid overload, patent ductus arteriosus, malnutrition, genetic factors, intrauterine growth restriction, and maternal smoking, among others. [148]. Although animal studies have helped us understand the mechanisms involved in the development and progression of BPD, it is not possible to study all risk factors, such as fluid overload, patent ductus arteriosus, and prolonged mechanical ventilation, in humans. Fortunately, all risk factors are associated with OS and inflammation.

Rodents are commonly used as small animal models for BPD [149,150]. Although rodent pups are born at the saccular stage, they are imperfect as they have no surfactant deficiency with mature antioxidant capacity. However, their large litter size, small body size, and relatively short life span make them the most convenient animal models. Researchers use LPS or hyperoxia (greater than 70% O_2) to simulate inflammation and OS, respectively, to generate the BPD phenotype. We used outbred rat pups to study BPD, assuming they can better mirror the genetic diversity in the human population. Combining data from our experiments and those in the literature, we have constructed a theoretical "destructive cycle for BPD." (Figure 1), which exhibits a self-perpetuating nature. The complexity of this cycle suggests that any therapy that only targets one mechanism will be insufficient in BPD treatment.

Available evidence suggests that caffeine and vitamin A have antioxidant and anti-inflammatory properties. The clinical success of early postnatal caffeine and high-dose vitamin A treatment in decreasing BPD incidence supports the importance of a multi-prong approach. Recent insights from systems pharmacology have highlighted the limitations of single-target drugs in addressing complex, nonlinear diseases, such as BPD. The concept of SPTs, which leverages network-aware, multinodal engagement, offers a promising strategy to overcome these limitations. In this context, our lead candidate, KYC, a first-in-class SCPD, represents a new therapeutic class capable of reprogramming disease networks. The deliberate mapping of KYC's mechanism onto features of SPTs that includes modular engagement, context-specific activation, and emergent system-level effects supports its potential as a paradigm-shifting intervention in BPD.

6. Conclusions

BPD is a complex, multifactorial, and incompletely understood complication of RDS. As prematurity is not physiologic, decreasing OS and inflammation and preserving growth potential make BPD treatment different from the disease in other age groups. We propose the concept of utilizing SPTs to treat premature birth-associated disorders. In conclusion, integrating a systems pharmacology framework into the therapeutic landscape of BPD enables a more comprehensive and mechanistically coherent approach to treatment. Our work introduces KYC as a prototype SCPD, a molecule that conforms to the principles of an SPT by engaging disease-relevant networks in a spatially and biochemically restricted manner. The convergence of experimental evidence, mechanistic rationale, and systems-level design reinforces the promise of KYC not merely as a drug candidate but as a conceptual advance in how we approach pharmacotherapy for complex neonatal diseases.

7. Patents

1. SYSTEMS CHEMICO–PHARMACOLOGY DRUGS AND METHODS OF USE. Kirkwood Arthur Pritchard, Jr.; Dustin Paul Martin; Ru-Jeng Teng; Billy W. Day; and Stephen Naylor. Pub. No.: US 2022/0409692 A1 (NON-PROVISIONAL FULL APPLICATION).

2. DUAL ROLE COMPOUNDS WITH PRODRUG AND SYSTEMS CHEMICO-PHARMACOLOGY DRUG PROPERTIES. Billy W. Day; Kirkwood Arthur Pritchard, Jr.; Ru-Jeng Teng; and Stephen Naylor. FILED 04/21/2025 US PROVISIONAL PATENT APPLICATION 63/791,992.
3. SYSTEMS CHEMICO-BIOPROBES DERIVED FROM SYSTEMS CHEMICO-PHARMACOLOGY DRUGS FOR TARGET DISCOVERY AND COMPANION DIAGNOSTIC APPLICATIONS. Kirkwood Arthur Pritchard, Jr.; Billy W. Day; Ru-Jeng Teng; and Stephen Naylor. FILED 04/21/2025 US PROVISIONAL PATENT APPLICATION 63/792,015.

Author Contributions: Conceptualization, writing, and visualization: M.T., T.-J.W., S.N. and R.-J.T.; organization and supervision: K.A.P.J., B.W.D., S.N. and R.-J.T.; funding acquisition: R.-J.T., B.W.D., K.A.P.J. and S.N. All authors have read and agreed to the published version of the manuscript.

Funding: This work was supported by NHLBI R44HL166018 (X.J., B.W.D., K.A.P.Jr., S.N., and R.-J.T.) and NHLBI R01HL128371 (R.-J.T. and K.A.P.Jr.).

Data Availability Statement: The original contributions presented in the study are included in the article; further inquiries can be directed to the corresponding author.

Acknowledgments: This is a summary project for Mia Teng.

Conflicts of Interest: K.A.P., B.W.D., and S.N. own ReNeuroGen LLC, which is developing the drug candidate K.Y.C. as a potential therapeutic intervention in B.P.D. At present, K.Y.C. is in the preclinical phase of development. The remaining authors declare that the research was conducted in the absence of any commercial or financial relationships that could be construed as a potential conflict of interest.

Abbreviations

The following abbreviations are used in this manuscript:

AGER	Advanced glycosylation end-product receptor
AT2	Type 2 alveolar epithelial cell
ATF6	Activating Transcription Factor 6
BiP	Binding immunoglobulin protein
BPD	Bronchopulmonary dysplasia
CD24	Cluster of differentiation 24
cGAMP	Cyclic guanosine monophosphate-adenosine monophosphate
CGAS	Cyclic GMP-AMP synthase
CHOP	C/EBP homologous protein
CXCR4	C-X-C motif chemokine receptor 4
DAMP	Damage-associated molecular pattern
eNOS	Endothelial nitric oxide synthase
ER	Endoplasmic reticulum
ETC	Electron transport chain
GCH1	GTP-cyclohydrolase 1
GRP78	Glucose-regulated protein 78
GSDMD	Gasdermin D
HAVCR2	Hepatitis A virus cellular receptor 2
HMGB1	High-mobility group box 1
HOCl	Hypochlorous acid
HSP	Heat-shock protein
iNOS	Inducible nitric oxide synthase
IRE1α	Inositol-requiring enzyme 1 α
JAK	Janus kinase
KEAP1	Kelch-like ECH-associated protein 1

KYC	N-acetyl-lysyltyrosylcysteine-amide
LPS	Lipopolysaccharide
MPO	Myeloperoxidase
mROS	ROS from mitochondria
MSC	Mesenchymal stem cell
NF-kB	Nuclear factor-kappa B
NLR	NOD-like receptors
NOX	NADPH oxidase
NRF2	Nuclear factor erythroid 2-related factor 2
$ONOO^-$	Peroxynitrite
OS	Oxidative stress
PAMP	Pathogen-associated molecular patterns
PERK	PRKR-Like Endoplasmic Reticulum Kinase
PI3K	Phosphoinositide 3-Kinase
PPP	Pentose phosphate pathway
PRR	Pattern recognition receptor
PTEN	Phosphatase and tensin homolog
ROS	Reactive oxygen species
SAMD	Senescence-associated mitochondrial dysfunction
SASP	Senescence-associated secretory phenotype
SCPD	Systems chemico-pharmacology drug
SMAD	Suppressor of mothers against decapentaplegic.
SPT	Systems pharmacology therapeutics
STAT	Signal transducer and activator of transcription
STING1	Stimulator of interferon genes 1
TGFβ	Transforming growth factor β
TLR	Toll-like receptor
TNFα	Tumor necrosis factor α
TREM1	Triggering receptor expressed on myeloid cells 1
VEGF	Vascular endothelial cell growth factor
VEGFR2	Type 2 VEGF receptor
UPR	Unfolded protein response
WNT	Wingless/Integrated
XBP1	X-box binding protein 1

References

1. Northway, W.H., Jr.; Rosan, R.C.; Porter, D.Y. Pulmonary disease following respirator therapy of hyaline-membrane disease. Bronchopulmonary dysplasia. *N. Engl. J. Med.* **1967**, *276*, 357–368. [CrossRef] [PubMed]
2. Philip, A.G. Bronchopulmonary dysplasia: Then and now. *Neonatology* **2012**, *102*, 1–8. [CrossRef] [PubMed]
3. Jobe, A.J. The new BPD: An arrest of lung development. *Pediatr. Res.* **1999**, *46*, 641–643. [CrossRef] [PubMed]
4. Baker, C.D.; Abman, S.H. Impaired pulmonary vascular development in bronchopulmonary dysplasia. *Neonatology* **2015**, *107*, 344–351. [CrossRef] [PubMed] [PubMed Central]
5. Long, W.; Corbet, A.; Cotton, R.; Courtney, S.; McGuiness, G.; Walter, D.; Watts, J.; Smyth, J.; Bard, H.; Chernick, V. A controlled trial of synthetic surfactant in infants weighing 1250 g or more with respiratory distress syndrome. The American Exosurf Neonatal Study Group, I; and the Canadian Exosurf Neonatal Study Group. *N. Engl. J. Med.* **1991**, *325*, 1696–1703. [CrossRef] [PubMed]
6. Edwards, E.M.; Ehret, D.E.Y.; Soll, R.F.; Horbar, J.D. Survival of Infants Born at 22 to 25 Weeks' Gestation Receiving Care in the NICU: 2020–2022. *Pediatrics* **2024**, *154*, e2024065963. [CrossRef] [PubMed]
7. Qattea, I.; Farghaly, M.A.A.; Kattea, M.O.; Abdula, N.; Mohamed, M.A.; Aly, H. Survival of infants born at periviable gestation: The US national database. *Lancet Reg. Health Am.* **2022**, *14*, 100330. [CrossRef] [PubMed] [PubMed Central]
8. Jobe, A.H.; Bancalari, E. Bronchopulmonary dysplasia. *Am. J. Respir. Crit. Care Med.* **2001**, *163*, 1723–1729. [CrossRef] [PubMed]

9. Stoll, B.J.; Hansen, N.I.; Bell, E.F.; Shankaran, S.; Laptook, A.R.; Walsh, M.C.; Hale, E.C.; Newman, N.S.; Schibler, K.; Carlo, W.A.; et al. Neonatal outcomes of extremely preterm infants from the NICHD Neonatal Research Network. *Pediatrics* **2010**, *126*, 443–456. [CrossRef] [PubMed] [PubMed Central]
10. Jensen, E.A.; Schmidt, B. Epidemiology of bronchopulmonary dysplasia. *Birth Defects Res. A Clin. Mol. Teratol.* **2014**, *100*, 145–157. [CrossRef] [PubMed] [PubMed Central]
11. Yeh, T.F.; Lin, Y.J.; Lin, H.C.; Huang, C.C.; Hsieh, W.S.; Lin, C.H.; Tsai, C.H. Outcomes at school age after postnatal dexamethasone therapy for lung disease of prematurity. *N. Engl. J. Med.* **2004**, *350*, 1304–1313. [CrossRef] [PubMed]
12. Cummings, J.J.; Pramanik, A.K.; Committee on Fetus and Newborn. Postnatal Corticosteroids to Prevent or Treat Chronic Lung Disease Following Preterm Birth. *Pediatrics* **2022**, *149*, e2022057530. [CrossRef] [PubMed]
13. Hanin, M.; Nuthakki, S.; Malkar, M.B.; Jadcherla, S.R. Safety and Efficacy of Oral Feeding in Infants with BPD on Nasal CPAP. *Dysphagia* **2015**, *30*, 121–127. [CrossRef] [PubMed] [PubMed Central]
14. Reynolds, E.W.; Grider, D.; Caldwell, R.; Capilouto, G.; Patwardhan, A.; Charnigo, R. Effects of Bronchopulmonary Dysplasia on Swallow: Breath Interaction and Phase of Respiration with Swallow During Non-nutritive Suck. *J. Nat. Sci.* **2018**, *4*, e531. [PubMed] [PubMed Central]
15. Tyson, J.E.; Wright, L.L.; Oh, W.; Kennedy, K.A.; Mele, L.; Ehrenkranz, R.A.; Stoll, B.J.; Lemons, J.A.; Stevenson, D.K.; Bauer, C.R.; et al. Vitamin A supplementation for extremely-low-birth-weight infants. National Institute of Child Health and Human Development Neonatal Research Network. *N. Engl. J. Med.* **1999**, *340*, 1962–1968. [CrossRef] [PubMed]
16. Schmidt, B.; Roberts, R.S.; Davis, P.; Doyle, L.W.; Barrington, K.J.; Ohlsson, A.; Solimano, A.; Tin, W.; Caffeine for Apnea of Prematurity Trial Group. Caffeine therapy for apnea of prematurity. *N. Engl. J. Med.* **2006**, *354*, 2112–2121. [CrossRef] [PubMed]
17. Lee, S.M.; Sie, L.; Liu, J.; Profit, J.; Lee, H.C. Evaluation of Trends in Bronchopulmonary Dysplasia and Respiratory Support Practice for Very Low Birth Weight Infants: A Population-Based Cohort Study. *J. Pediatr.* **2022**, *243*, 47–52.e2. [CrossRef] [PubMed] [PubMed Central]
18. Vali, P.; Lakshminrusimha, S. The Fetus Can Teach Us: Oxygen and the Pulmonary Vasculature. *Children* **2017**, *4*, 67. [CrossRef] [PubMed] [PubMed Central]
19. Martini, S.; Aceti, A.; Della Gatta, A.N.; Beghetti, I.; Marsico, C.; Pilu, G.; Corvaglia, L. Antenatal and Postnatal Sequelae of Oxidative Stress in Preterm Infants: A Narrative Review Targeting Pathophysiological Mechanisms. *Antioxidants* **2023**, *12*, 422. [CrossRef] [PubMed] [PubMed Central]
20. Zergeroglu, M.A.; McKenzie, M.J.; Shanely, R.A.; Van Gammeren, D.; DeRuisseau, K.C.; Powers, S.K. Mechanical ventilation-induced oxidative stress in the diaphragm. *J. Appl. Physiol.* **2003**, *95*, 1116–1124. [CrossRef] [PubMed]
21. Cheah, F.C.; Jobe, A.H.; Moss, T.J.; Newnham, J.P.; Kallapur, S.G. Oxidative stress in fetal lambs exposed to intra-amniotic endotoxin in a chorioamnionitis model. *Pediatr. Res.* **2008**, *63*, 274–279. [CrossRef] [PubMed]
22. Lapcharoensap, W.; Kan, P.; Powers, R.J.; Shaw, G.M.; Stevenson, D.K.; Gould, J.B.; Wirtschafter, D.D.; Lee, H.C. The Relationship of Nosocomial Infection Reduction to Changes in Neonatal Intensive Care Unit Rates of Bronchopulmonary Dysplasia. *J. Pediatr.* **2017**, *180*, 105–109.e1. [CrossRef] [PubMed]
23. Teng, R.J.; Jing, X.; Michalkiewicz, T.; Afolayan, A.J.; Wu, T.J.; Konduri, G.G. Attenuation of endoplasmic reticulum stress by caffeine ameliorates hyperoxia-induced lung injury. *Am. J. Physiol. Lung Cell Mol. Physiol.* **2017**, *312*, L586–L598. [CrossRef] [PubMed] [PubMed Central]
24. Block, M.S.; Rowan, B.G. Hypochlorous Acid: A Review. *J. Oral. Maxillofac. Surg.* **2020**, *78*, 1461–1466. [CrossRef] [PubMed] [PubMed Central]
25. Saugstad, O.D. Oxidative stress in the newborn–a 30-year perspective. *Biol. Neonate* **2005**, *88*, 228–236. [CrossRef]
26. Alzahrani, A.S.; Smith, J.L.; García, M.; Chen, X.; Tanaka, H.; Müller, F.; Ivanova, K.; Kim, Y.; Silva, R.; Dubois, P. Oxidative stress and inflammation: The vicious cycle in BPD pathogenesis. *Front. Pediatr.* **2022**, *10*, 857234.
27. Wu, T.J.; Teng, M.; Jing, X.; Pritchard, K.A., Jr.; Day, B.W.; Naylor, S.; Teng, R.J. Endoplasmic Reticulum Stress in Bronchopulmonary Dysplasia: Contributor or Consequence? *Cells* **2024**, *13*, 1774. [CrossRef] [PubMed] [PubMed Central]
28. Teng, M.; Wu, T.J.; Jing, X.; Day, B.W.; Pritchard, K.A., Jr.; Naylor, S.; Teng, R.J. Temporal Dynamics of Oxidative Stress and Inflammation in Bronchopulmonary Dysplasia. *Int. J. Mol. Sci.* **2024**, *25*, 10145. [CrossRef] [PubMed] [PubMed Central]
29. Liu, T.; Zhang, L.; Joo, D.; Sun, S.C. NF-κB signaling in inflammation. *Signal Transduct. Target. Ther.* **2017**, *2*, 17023. [CrossRef] [PubMed] [PubMed Central]
30. Vaughan, A.E.; Chapman, H.A. Failure of Alveolar Type 2 Cell Maintenance Links Neonatal Distress with Adult Lung Disease. *Am. J. Respir. Cell Mol. Biol.* **2017**, *56*, 415–416. [CrossRef] [PubMed]
31. Jing, X.; Jia, S.; Teng, M.; Day, B.W.; Afolayan, A.J.; Jarzembowski, J.A.; Lin, C.W.; Hessner, M.J.; Pritchard, K.A., Jr.; Naylor, S.; et al. Cellular Senescence Contributes to the Progression of Hyperoxic Bronchopulmonary Dysplasia. *Am. J. Respir. Cell Mol. Biol.* **2024**, *70*, 94–109. [CrossRef] [PubMed] [PubMed Central]

32. Anighoro, A.; Bajorath, J.; Rastelli, G. Polypharmacology: Challenges and opportunities in drug discovery. *J. Med. Chem.* **2014**, *57*, 7874–7887. [CrossRef] [PubMed]
33. Pritchard, K.A., Jr.; Martin, D.P.; Naylor, S. Perspective on Systems Pharmacology: When Multi-Targeting Is Advantageous. Drug Discovery World, Winter 2018. Available online: https://www.ddw-online.com/systems-pharmacology-when-multi-targeting-is-advantageous-503-201812/ (accessed on 14 June 2025).
34. Ramsay, R.R.; Popovic-Nikolic, M.R.; Nikolic, K.; Uliassi, E.; Bolognesi, M.L. A perspective on multi-target drug discovery and design for complex diseases. *Clin. Transl. Med.* **2018**, *7*, 3. [CrossRef] [PubMed] [PubMed Central]
35. Hopkins, A.L. Network pharmacology: The next paradigm in drug discovery. *Nat. Chem. Biol.* **2008**, *4*, 682–690. [CrossRef] [PubMed]
36. Chen, G.Y.; Nuñez, G. Sterile inflammation: Sensing and reacting to damage. *Nat. Rev. Immunol.* **2010**, *10*, 826–837. [CrossRef]
37. Murch, S.H.; Costeloe, K.; Klein, N.J.; MacDonald, T.T. Early production of macrophage inflammatory protein-1 alpha occurs in respiratory distress syndrome and is associated with poor outcome. *Pediatr. Res.* **1996**, *40*, 490–497. [CrossRef]
38. Beckman, J.S.; Viera, L.; Estévez, A.G.; Teng, R. Nitric oxide and peroxynitrite in the perinatal period. *Semin. Perinatol.* **2000**, *24*, 37–41. [CrossRef]
39. Pattison, D.I.; Davies, M.J.; Hawkins, C.L. Reactions and reactivity of myeloperoxidase-derived oxidants: Differential biological effects of hypochlorous and hypothiocyanous acids. *Free Radic. Res.* **2012**, *46*, 975–995. [CrossRef]
40. Roh, J.S.; Sohn, D.H. Damage-Associated Molecular Patterns in Inflammatory Diseases. *Immune Netw.* **2018**, *18*, e27. [CrossRef] [PubMed]
41. Zhong, H.; Li, X.; Zhou, S.; Jiang, P.; Liu, X.; Ouyang, M.; Nie, Y.; Chen, X.; Zhang, L.; Liu, Y.; et al. Interplay between RAGE and TLR4 Regulates HMGB1-Induced Inflammation by Promoting Cell Surface Expression of RAGE and TLR4. *J Immunol.* **2020**, *205*, 767–775. [CrossRef]
42. Chen, R.; Kang, R.; Tang, D. The mechanism of HMGB1 secretion and release. *Exp. Mol. Med.* **2022**, *54*, 91–102. [CrossRef] [PubMed] [PubMed Central]
43. Yang, K.; Fan, M.; Wang, X.; Xu, J.; Wang, Y.; Tu, F.; Gill, P.S.; Ha, T.; Liu, L.; Williams, D.L.; et al. Lactate promotes macrophage HMGB1 lactylation, acetylation, and exosomal release in polymicrobial sepsis. *Cell Death Differ.* **2022**, *29*, 133–146. [CrossRef] [PubMed] [PubMed Central]
44. Aghai, Z.H.; Saslow, J.G.; Meniru, C.; Porter, C.; Eydelman, R.; Bhat, V.; Stahl, G.; Sannoh, S.; Pyon, K.; Hewitt, C.; et al. High-mobility group box-1 protein in tracheal aspirates from premature infants: Relationship with bronchopulmonary dysplasia and steroid therapy. *J. Perinatol.* **2010**, *30*, 610–615. [CrossRef] [PubMed]
45. Shima, Y.; Kumasaka, S.; Negishi, Y. Sustained sterile inflammation is related to pulmonary morbidities in premature infants. *J. Matern. Fetal Neonatal Med.* **2022**, *35*, 6928–6932. [CrossRef] [PubMed]
46. Hara, T.; Shimbo, T.; Masuda, T.; Kitayama, T.; Fujii, M.; Hanawa, M.; Yokota, K.; Endo, M.; Tomimatsu, T.; Kimura, T.; et al. High-mobility group box-1 peptide ameliorates bronchopulmonary dysplasia by suppressing inflammation and fibrosis in a mouse model. *Biochem. Biophys. Res. Commun.* **2023**, *671*, 357–365. [CrossRef] [PubMed]
47. Lee, G.; Espirito Santo, A.I.; Zwingenberger, S.; Cai, L.; Vogl, T.; Feldmann, M.; Horwood, N.J.; Chan, J.K.; Nanchahal, J. Fully reduced HMGB1 accelerates the regeneration of multiple tissues by transitioning stem cells to GAlert. *Proc. Natl. Acad. Sci. USA* **2018**, *115*, E4463–E4472. [CrossRef] [PubMed] [PubMed Central]
48. Kwak, M.S.; Kim, H.S.; Lee, B.; Kim, Y.H.; Son, M.; Shin, J.S. Immunological Significance of HMGB1 Post-Translational Modification and Redox Biology. *Front. Immunol.* **2020**, *11*, 1189. [CrossRef] [PubMed] [PubMed Central]
49. Ferrara, M.; Chialli, G.; Ferreira, L.M.; Ruggieri, E.; Careccia, G.; Preti, A.; Piccirillo, R.; Bianchi, M.E.; Sitia, G.; Venereau, E. Oxidation of HMGB1 Is a Dynamically Regulated Process in Physiological and Pathological Conditions. *Front. Immunol.* **2020**, *11*, 1122. [CrossRef] [PubMed] [PubMed Central]
50. Tang, D.; Kang, R.; Zeh, H.J.; Lotze, M.T. The multifunctional protein HMGB1: 50 years of discovery. *Nat. Rev. Immunol.* **2023**, *23*, 824–841. [CrossRef] [PubMed]
51. Chen, C.; Weng, H.; Zhang, X.; Wang, S.; Lu, C.; Jin, H.; Chen, S.; Liu, Y.; Sheng, A.; Sun, Y. Low-Dose Vitamin D Protects Hyperoxia-Induced Bronchopulmonary Dysplasia by Inhibiting Neutrophil Extracellular Traps. *Front. Pediatr.* **2020**, *8*, 335. [CrossRef] [PubMed] [PubMed Central]
52. Tao, X.; Mo, L.; Zeng, L. Hyperoxia Induced Bronchopulmonary Dysplasia-Like Inflammation via miR34a-TNIP2-IL-1β Pathway. *Front. Pediatr.* **2022**, *10*, 805860. [CrossRef] [PubMed] [PubMed Central]
53. Jin, R.; Xu, J.; Gao, Q.; Mao, X.; Yin, J.; Lu, K.; Guo, Y.; Zhang, M.; Cheng, R. IL-33-induced neutrophil extracellular traps degrade fibronectin in a murine model of bronchopulmonary dysplasia. *Cell Death Discov.* **2020**, *6*, 33. [CrossRef] [PubMed] [PubMed Central]

54. Nigro, P.; Pompilio, G.; Capogrossi, M.C. Cyclophilin A: A key player for human disease. *Cell Death Dis.* **2013**, *4*, e888. [CrossRef] [PubMed] [PubMed Central]
55. Hsiao, C.C.; Lee, C.H.; Yang, R.C.; Chen, J.Y.; Su, T.C.; Chang, Y.J.; Lin, C.Y.; Tsai, Y.G. Heat Shock Protein-70 Levels Are Associated with a State of Oxidative Damage in the Development of Bronchopulmonary Dysplasia. *Front. Pediatr.* **2021**, *9*, 616452. [CrossRef] [PubMed] [PubMed Central]
56. Loughran-Fowlds, A.; Leach, S.; Lin, J.; Oei, J.; Henry, R.; Day, A.S.; Lui, K. Respiratory disease and early serum S100A12 changes in very premature infants. *Acta Paediatr.* **2011**, *100*, 1538–1543. [CrossRef] [PubMed]
57. Martinon, F.; Burns, K.; Tschopp, J. The inflammasome: A molecular platform triggering activation of inflammatory caspases and processing of proIL-beta. *Mol. Cell* **2002**, *10*, 417–426. [CrossRef] [PubMed]
58. Latz, E.; Xiao, T.S.; Stutz, A. Activation and regulation of the inflammasomes. *Nat. Rev. Immunol.* **2013**, *13*, 397–411. [CrossRef] [PubMed] [PubMed Central]
59. Malik, A.; Kanneganti, T.D. Inflammasome activation and assembly at a glance. *J. Cell Sci.* **2017**, *130*, 3955–3963. [CrossRef] [PubMed] [PubMed Central]
60. Zheng, D.; Liwinski, T.; Elinav, E. Inflammasome activation and regulation: Toward a better understanding of complex mechanisms. *Cell Discov.* **2020**, *6*, 36. [CrossRef] [PubMed] [PubMed Central]
61. Cho, S.; Ying, F.; Sweeney, G. Sterile inflammation and the NLRP3 inflammasome in cardiometabolic disease. *Biomed. J.* **2023**, *46*, 100624. [CrossRef] [PubMed] [PubMed Central]
62. Zhang, R.; Kang, R.; Tang, D. The STING1 network regulates autophagy and cell death. *Signal Transduct. Target. Ther.* **2021**, *6*, 208. [CrossRef] [PubMed] [PubMed Central]
63. Schaefer, L. Complexity of danger: The diverse nature of damage-associated molecular patterns. *J. Biol. Chem.* **2014**, *289*, 35237–35245. [CrossRef] [PubMed] [PubMed Central]
64. He, W.T.; Wan, H.; Hu, L.; Chen, P.; Wang, X.; Huang, Z.; Yang, Z.H.; Zhong, C.Q.; Han, J. Gasdermin D is an executor of pyroptosis and required for interleukin-1β secretion. *Cell Res.* **2015**, *25*, 1285–1298. [CrossRef] [PubMed] [PubMed Central]
65. Kamajaya, L.J.; Boucher, D. Gasdermin D Cleavage Assay Following Inflammasome Activation. *Methods Mol. Biol.* **2022**, *2459*, 39–49. [CrossRef] [PubMed]
66. Fukumura, D.; Gohongi, T.; Kadambi, A.; Izumi, Y.; Ang, J.; Yun, C.O.; Buerk, D.G.; Huang, P.L.; Jain, R.K. Predominant role of endothelial nitric oxide synthase in vascular endothelial growth factor-induced angiogenesis and vascular permeability. *Proc. Natl. Acad. Sci. USA* **2001**, *98*, 2604–2609. [CrossRef] [PubMed] [PubMed Central]
67. Nisoli, E.; Clementi, E.; Paolucci, C.; Cozzi, V.; Tonello, C.; Sciorati, C.; Bracale, R.; Valerio, A.; Francolini, M.; Moncada, S.; et al. Mitochondrial biogenesis in mammals: The role of endogenous nitric oxide. *Science* **2003**, *299*, 896–899. [CrossRef] [PubMed]
68. Teng, R.J.; Du, J.; Xu, H.; Bakhutashvili, I.; Eis, A.; Shi, Y.; Pritchard, K.A., Jr.; Konduri, G.G. Sepiapterin improves angiogenesis of pulmonary artery endothelial cells with in utero pulmonary hypertension by recoupling endothelial nitric oxide synthase. *Am. J. Physiol. Lung Cell Mol. Physiol.* **2011**, *301*, L334–L345. [CrossRef] [PubMed] [PubMed Central]
69. Teng, R.J.; Wu, T.J.; Bisig, C.G.; Eis, A.; Pritchard, K.A.; Konduri, G.G. Nitrotyrosine impairs angiogenesis and uncouples eNOS activity of pulmonary artery endothelial cells isolated from developing sheep lungs. *Pediatr. Res.* **2011**, *69*, 112–117. [CrossRef] [PubMed] [PubMed Central]
70. Teng, R.J.; Wu, T.J.; Afolayan, A.J.; Konduri, G.G. Nitrotyrosine impairs mitochondrial function in fetal lamb pulmonary artery endothelial. *Cells Am. J. Physiol. Cell Physiol.* **2016**, *310*, C80–C88. [CrossRef] [PubMed] [PubMed Central]
71. Jing, X.; Huang, Y.W.; Jarzembowski, J.; Shi, Y.; Konduri, G.G.; Teng, R.J. Caffeine ameliorates hyperoxia-induced lung injury by protecting GCH1 function in neonatal rat pups. *Pediatr. Res.* **2017**, *82*, 483–489. [CrossRef] [PubMed] [PubMed Central]
72. Huie, R.E.; Padmaja, S. The reaction of no with superoxide. *Free Radic Res Commun.* **1993**, *18*, 195–199. [CrossRef] [PubMed]
73. Kumar, S.; Dikshit, M. Nitric Oxide: Regulation and Function in Neutrophil Immune Responses. *Antioxid Redox Signal.* **2024**, *40*, 998–1024. [CrossRef] [PubMed]
74. Chishti, A.D.; Shenton, B.K.; Kirby, J.A.; Baudouin, S.V. Neutrophil chemotaxis and receptor expression in clinical septic shock. *Intensive Care Med.* **2004**, *30*, 605–611. [CrossRef] [PubMed]
75. Galkina, S.I.; Golenkina, E.A.; Viryasova, G.M.; Romanova, Y.M.; Sud'ina, G.F. Nitric Oxide in Life and Death of Neutrophils. *Curr. Med. Chem.* **2019**, *26*, 5764–5780. [CrossRef] [PubMed]
76. Pritchard, K.A., Jr.; Jing, X.; Teng, M.; Wells, C.; Jia, S.; Afolayan, A.J.; Jarzembowski, J.; Day, B.W.; Naylor, S.; Hessner, M.J.; et al. Role of endoplasmic reticulum stress in impaired neonatal lung growth and bronchopulmonary dysplasia. *PLoS ONE* **2022**, *17*, e0269564. [CrossRef] [PubMed] [PubMed Central]
77. Degechisa, S.T.; Dabi, Y.T.; Gizaw, S.T. The mitochondrial associated endoplasmic reticulum membranes: A platform for the pathogenesis of inflammation-mediated metabolic diseases. *Immun. Inflamm. Dis.* **2022**, *10*, e647. [CrossRef] [PubMed] [PubMed Central]

78. Malhotra, J.D.; Kaufman, R.J. ER stress and its functional link to mitochondria: Role in cell survival and death. *Cold Spring Harb. Perspect. Biol.* **2011**, *3*, a004424. [CrossRef] [PubMed] [PubMed Central]
79. Garg, A.D.; Kaczmarek, A.; Krysko, O.; Vandenabeele, P.; Krysko, D.V.; Agostinis, P. ER stress-induced inflammation: Does it aid or impede disease progression? *Trends Mol. Med.* **2012**, *18*, 589–598. [CrossRef] [PubMed]
80. Tam, A.B.; Mercado, E.L.; Hoffmann, A.; Niwa, M. ER stress activates NF-κB by integrating functions of basal IKK activity, IRE1 and PERK. *PLoS ONE* **2012**, *7*, e45078. [CrossRef] [PubMed] [PubMed Central]
81. Kitamura, M. Biphasic, bidirectional regulation of NF-kappaB by endoplasmic reticulum stress. *Antioxid. Redox Signal.* **2009**, *11*, 2353–2364. [CrossRef] [PubMed]
82. Thorburn, J.; Horita, H.; Redzic, J.; Hansen, K.; Frankel, A.E.; Thorburn, A. Autophagy regulates selective HMGB1 release in tumor cells that are destined to die. *Cell Death Differ.* **2009**, *16*, 175–183. [CrossRef] [PubMed] [PubMed Central]
83. Deng, J.; Lu, P.D.; Zhang, Y.; Scheuner, D.; Kaufman, R.J.; Sonenberg, N.; Harding, H.P.; Ron, D. Translational repression mediates activation of nuclear factor kappa B by phosphorylated translation initiation factor 2. *Mol. Cell Biol.* **2004**, *24*, 10161–10168. [CrossRef] [PubMed] [PubMed Central]
84. Zhang, K.; Kaufman, R.J. From endoplasmic-reticulum stress to the inflammatory response. *Nature* **2008**, *454*, 455–462. [CrossRef] [PubMed] [PubMed Central]
85. Lerner, A.G.; Upton, J.P.; Praveen, P.V.; Ghosh, R.; Nakagawa, Y.; Igbaria, A.; Shen, S.; Nguyen, V.; Backes, B.J.; Heiman, M.; et al. IRE1α induces thioredoxin-interacting protein to activate the NLRP3 inflammasome and promote programmed cell death under irremediable ER stress. *Cell Metab.* **2012**, *16*, 250–264. [CrossRef] [PubMed] [PubMed Central]
86. Injarabian, L.; Skerniskyte, J.; Giai Gianetto, Q.; Witko-Sarsat, V.; Marteyn, B.S. Reducing neutrophil exposure to oxygen allows their basal state maintenance. *Immunol. Cell Biol.* **2021**, *99*, 782–789. [CrossRef] [PubMed] [PubMed Central]
87. Janssens, S.; Pulendran, B.; Lambrecht, B.N. Emerging functions of the unfolded protein response in immunity. *Nat. Immunol.* **2014**, *15*, 910–919. [CrossRef] [PubMed] [PubMed Central]
88. Huang, W.; Hickson, L.J.; Eirin, A.; Kirkland, J.L.; Lerman, L.O. Cellular senescence: The good, the bad and the unknown. *Nat. Rev. Nephrol.* **2022**, *18*, 611–627. [CrossRef] [PubMed] [PubMed Central]
89. Reimann, M.; Lee, S.; Schmitt, C.A. Cellular senescence: Neither irreversible nor reversible. *J. Exp. Med.* **2024**, *221*, e20232136. [CrossRef] [PubMed] [PubMed Central]
90. Regulski, M.J. Cellular Senescence: What, Why, and How. *Wounds* **2017**, *29*, 168–174. [PubMed]
91. Sabbatinelli, J.; Prattichizzo, F.; Olivieri, F.; Procopio, A.D.; Rippo, M.R.; Giuliani, A. Where Metabolism Meets Senescence: Focus on Endothelial Cells. *Front. Physiol.* **2019**, *10*, 1523. [CrossRef] [PubMed] [PubMed Central]
92. Childs, B.G.; Gluscevic, M.; Baker, D.J.; Laberge, R.M.; Marquess, D.; Dananberg, J.; van Deursen, J.M. Senescent cells: An emerging target for diseases of ageing. *Nat. Rev. Drug Discov.* **2017**, *16*, 718–735. [CrossRef] [PubMed] [PubMed Central]
93. Gonzalez-Meljem, J.M.; Apps, J.R.; Fraser, H.C.; Martinez-Barbera, J.P. Paracrine roles of cellular senescence in promoting tumourigenesis. *Br. J. Cancer.* **2018**, *118*, 1283–1288. [CrossRef] [PubMed] [PubMed Central]
94. Dennery, P.A.; Yao, H. Emerging role of cellular senescence in normal lung development and perinatal lung injury. *Chin. Med. J. Pulm. Crit. Care Med.* **2024**, *2*, 10–16. [CrossRef] [PubMed] [PubMed Central]
95. Demaria, M.; Ohtani, N.; Youssef, S.A.; Rodier, F.; Toussaint, W.; Mitchell, J.R.; Laberge, R.M.; Vijg, J.; Van Steeg, H.; Dollé, M.E.; et al. An essential role for senescent cells in optimal wound healing through secretion of PDGF-AA. *Dev. Cell* **2014**, *31*, 722–733. [CrossRef] [PubMed] [PubMed Central]
96. von Zglinicki, T. Oxidative stress and cell senescence as drivers of ageing: Chicken and egg. *Ageing Res. Rev.* **2024**, *102*, 102558. [CrossRef] [PubMed]
97. Han, Z.; Wang, K.; Ding, S.; Zhang, M. Cross-talk of inflammation and cellular senescence: A new insight into the occurrence and progression of osteoarthritis. *Bone Res.* **2024**, *12*, 69. [CrossRef] [PubMed] [PubMed Central]
98. Sofiadis, K.; Josipovic, N.; Nikolic, M.; Kargapolova, Y.; Übelmesser, N.; Varamogianni-Mamatsi, V.; Zirkel, A.; Papadionysiou, I.; Loughran, G.; Keane, J.; et al. HMGB1 coordinates SASP-related chromatin folding and RNA homeostasis on the path to senescence. *Mol. Syst. Biol.* **2021**, *17*, e9760. [CrossRef] [PubMed] [PubMed Central]
99. Yue, L.; Lu, X.; Dennery, P.A.; Yao, H. Metabolic dysregulation in bronchopulmonary dysplasia: Implications for identification of biomarkers and therapeutic approaches. *Redox Biol.* **2021**, *48*, 102104. [CrossRef] [PubMed] [PubMed Central]
100. Ratner, V.; Starkov, A.; Matsiukevich, D.; Polin, R.A.; Ten, V.S. Mitochondrial dysfunction contributes to alveolar developmental arrest in hyperoxia-exposed mice. *Am. J. Respir. Cell Mol. Biol.* **2009**, *40*, 511–518. [CrossRef] [PubMed] [PubMed Central]
101. Gong, J.; Feng, Z.; Peterson, A.L.; Carr, J.F.; Lu, X.; Zhao, H.; Ji, X.; Zhao, Y.Y.; De Paepe, M.E.; Dennery, P.A.; et al. The pentose phosphate pathway mediates hyperoxia induced lung vascular dysgenesis and alveolar simplification in neonates. *JCI Insight.* **2021**, *6*, e137594. [CrossRef] [PubMed] [PubMed Central]

102. Herbert, S.P.; Cheung, J.Y.; Stainier, D.Y. Determination of endothelial stalk versus tip cell potential during angiogenesis by H2.0-like homeobox-1. *Curr. Biol.* **2012**, *22*, 1789–1794. [CrossRef] [PubMed] [PubMed Central]
103. Gillich, A.; Zhang, F.; Farmer, C.G.; Travaglini, K.J.; Tan, S.Y.; Gu, M.; Zhou, B.; Feinstein, J.A.; Krasnow, M.A.; Metzger, R.J. Capillary cell-type specialization in the alveolus. *Nature* **2020**, *586*, 785–789. [CrossRef] [PubMed] [PubMed Central]
104. Godoy, R.S.; Cober, N.D.; Cook, D.P.; McCourt, E.; Deng, Y.; Wang, L.; Schlosser, K.; Rowe, K.; Stewart, D.J. Single-cell transcriptomic atlas of lung microvascular regeneration after targeted endothelial cell ablation. *eLife* **2023**, *12*, e80900. [CrossRef] [PubMed] [PubMed Central]
105. Cooke, J.P.; Losordo, D.W. Nitric oxide and angiogenesis. *Circulation* **2002**, *105*, 2133–2135. [CrossRef] [PubMed]
106. Citrin, K.M.; Chaube, B.; Fernández-Hernando, C.; Suárez, Y. Intracellular endothelial cell metabolism in vascular function and dysfunction. *Trends Endocrinol. Metab.* **2024**. [CrossRef] [PubMed]
107. Chandler, K.B.; Leon, D.R.; Kuang, J.; Meyer, R.D.; Rahimi, N.; Costello, C.E. N-Glycosylation regulates ligand-dependent activation and signaling of vascular endothelial growth factor receptor 2 (VEGFR2). *J. Biol. Chem.* **2019**, *294*, 13117–13130. [CrossRef] [PubMed] [PubMed Central]
108. Bayona-Bafaluy, M.P.; Esteban, O.; Ascaso, J.; Montoya, J.; Ruiz-Pesini, E. Oxidative phosphorylation inducers fight pathological angiogenesis. *Drug Discov. Today* **2019**, *24*, 1731–1734. [CrossRef] [PubMed]
109. Yetkin-Arik, B.; Vogels, I.M.C.; Neyazi, N.; van Duinen, V.; Houtkooper, R.H.; van Noorden, C.J.F.; Klaassen, I.; Schlingemann, R.O. Endothelial tip cells in vitro are less glycolytic and have a more flexible response to metabolic stress than non-tip. *Cells Sci. Rep.* **2019**, *9*, 10414. [CrossRef] [PubMed] [PubMed Central]
110. Crapo, J.D.; Barry, B.E.; Gehr, P.; Bachofen, M.; Weibel, E.R. Cell number and cell characteristics of the normal human lung. *Am. Rev. Respir. Dis.* **1982**, *126*, 332–337.
111. Dumas, S.J.; García-Caballero, M.; Carmeliet, P. Metabolic Signatures of Distinct Endothelial Phenotypes. *Trends Endocrinol. Metab.* **2020**, *31*, 580–595. [CrossRef]
112. Crespo-Garcia, S.; Tsuruda, P.R.; Dejda, A.; Ryan, R.D.; Fournier, F.; Chaney, S.Y.; Pilon, F.; Dogan, T.; Cagnone, G.; Patel, P.; et al. Pathological angiogenesis in retinopathy engages cellular senescence and is amenable to therapeutic elimination via BCL-xL inhibition. *Cell Metab.* **2021**, *33*, 818–832.e7. [CrossRef] [PubMed]
113. Han, Y.; Kim, S.Y. Endothelial senescence in vascular diseases: Current understanding and future opportunities in senotherapeutics. *Exp. Mol. Med.* **2023**, *55*, 1–12. [CrossRef] [PubMed] [PubMed Central]
114. Chang, H.; Rha, S.Y.; Jeung, H.C.; Park, K.H.; Kim, T.S.; Kim, Y.B.; Chung, H.C. Telomerase- and angiogenesis-related gene responses to irradiation in human umbilical vein endothelial. *Cells Int. J. Mol. Med.* **2013**, *31*, 1202–1208. [CrossRef] [PubMed]
115. Mammoto, T.; Torisawa, Y.S.; Muyleart, M.; Hendee, K.; Anugwom, C.; Gutterman, D.; Mammoto, A. Effects of age-dependent changes in cell size on endothelial cell proliferation and senescence through YAP1. *Aging* **2019**, *11*, 7051–7069. [CrossRef] [PubMed] [PubMed Central]
116. Baar, M.P.; Brandt, R.M.C.; Putavet, D.A.; Klein, J.D.D.; Derks, K.W.J.; Bourgeois, B.R.M.; Stryeck, S.; Rijksen, Y.; van Willigenburg, H.; Feijtel, D.A.; et al. Targeted Apoptosis of Senescent Cells Restores Tissue Homeostasis in Response to Chemotoxicity and Aging. *Cell* **2017**, *169*, 132–147.e16. [CrossRef] [PubMed] [PubMed Central]
117. Laughon, M.M.; Benjamin, D.K., Jr.; Capparelli, E.V.; Kearns, G.L.; Berezny, K.; Paul, I.M.; Wade, K.; Barrett, J.; Smith, P.B.; Cohen-Wolkowiez, M. Innovative clinical trial design for pediatric therapeutics. *Expert. Rev. Clin. Pharmacol.* **2011**, *4*, 643–652. [CrossRef] [PubMed] [PubMed Central]
118. Ward, R.M.; Benjamin, D.K., Jr.; Davis, J.M.; Gorman, R.L.; Kauffman, R.; Kearns, G.L.; Murphy, M.D.; Sherwin, C.M.T. The Need for Pediatric Drug Development. *J. Pediatr.* **2018**, *192*, 13–21. [CrossRef] [PubMed]
119. Jensen, E.A.; Wiener, L.E.; Rysavy, M.A.; Dysart, K.C.; Gantz, M.G.; Eichenwald, E.C.; Greenberg, R.G.; Harmon, H.M.; Laughon, M.M.; Watterberg, K.L.; et al. Assessment of Corticosteroid Therapy and Death or Disability According to Pretreatment Risk of Death or Bronchopulmonary Dysplasia in Extremely Preterm Infants. *JAMA Netw. Open.* **2023**, *6*, e2312277. [CrossRef]
120. Palace, V.P.; Khaper, N.; Qin, Q.; Singal, P.K. Antioxidant potentials of vitamin A and carotenoids and their relevance to heart disease. *Free Radic. Biol. Med.* **1999**, *26*, 746–761. [CrossRef]
121. Gholizadeh, M.; Basafa Roodi, P.; Abaj, F.; Shab-Bidar, S.; Saedisomeolia, A.; Asbaghi, O.; Lak, M. Influence of Vitamin A supplementation on inflammatory biomarkers in adults: A systematic review and meta-analysis of randomized clinical trials. *Sci. Rep.* **2022**, *12*, 21384. [CrossRef]
122. Manapurath, R.M.; Kumar, M.; Pathak, B.G.; Chowdhury, R.; Sinha, B.; Choudhary, T.; Chandola, N.; Mazumdar, S.; Taneja, S.; Bhandari, N.; et al. Enteral Low-Dose Vitamin A Supplementation in Preterm or Low Birth Weight Infants to Prevent Morbidity and Mortality: A Systematic Review and Meta-analysis. *Pediatrics* **2022**, *150* (Suppl. S1), e2022057092L. [CrossRef]
123. Endesfelder, S.; Strauß, E.; Scheuer, T.; Schmitz, T.; Bührer, C. Antioxidative effects of caffeine in a hyperoxia-based rat model of bronchopulmonary dysplasia. *Respir. Res.* **2019**, *20*, 88. [CrossRef]

124. Endesfelder, S.; Strauß, E.; Bendix, I.; Schmitz, T.; Bührer, C. Prevention of Oxygen-Induced Inflammatory Lung Injury by Caffeine in Neonatal Rats. *Oxid Med. Cell. Longev.* **2020**, *2020*, 3840124. [CrossRef] [PubMed]
125. Köroğlu, O.A.; MacFarlane, P.M.; Balan, K.V.; Zenebe, W.J.; Jafri, A.; Martin, R.J.; Kc, P. Anti-inflammatory effect of caffeine is associated with improved lung function after lipopolysaccharide-induced amnionitis. *Neonatology* **2014**, *106*, 235–240. [CrossRef] [PubMed]
126. Yuan, Y.; Yang, Y.; Lei, X.; Dong, W. Caffeine and bronchopulmonary dysplasia: Clinical benefits and the mechanisms involved. *Pediatr. Pulmonol.* **2022**, *57*, 1392–1400. [CrossRef] [PubMed]
127. Omar, S.A.; Abdul-Hafez, A.; Ibrahim, S.; Pillai, N.; Abdulmageed, M.; Thiruvenkataramani, R.P.; Mohamed, T.; Madhukar, B.V.; Uhal, B.D. Stem-Cell Therapy for Bronchopulmonary Dysplasia (BPD) in Newborns. *Cells* **2022**, *11*, 1275. [CrossRef] [PubMed] [PubMed Central]
128. Saneh, H.; Wanczyk, H.; Walker, J.; Finck, C. Stem cell-derived extracellular vesicles: A potential intervention for Bronchopulmonary Dysplasia. *Pediatr. Res.* **2025**, *97*, 497–509. [CrossRef] [PubMed] [PubMed Central]
129. Wu, X.; Xia, Y.; Zhou, O.; Song, Y.; Zhang, X.; Tian, D.; Li, Q.; Shu, C.; Liu, E.; Yuan, X.; et al. Allogeneic human umbilical cord-derived mesenchymal stem cells for severe bronchopulmonary dysplasia in children: Study protocol for a randomized controlled trial (MSC-BPD trial). *Trials* **2020**, *21*, 125. [CrossRef] [PubMed] [PubMed Central]
130. Cerro Marín, M.J.D.; Ormazábal, I.G.; Gimeno-Navarro, A.; Álvarez-Fuente, M.; López-Ortego, P.; Avila-Alvarez, A.; Arruza Gómez, L.; González-Menchen, C.; Labrandero de Lera, C.; Lozano Balseiro, M.; et al. Repeated intravenous doses of human umbilical cord-derived mesenchymal stromal cells for bronchopulmonary dysplasia: Results of a phase 1 clinical trial with 2-year follow-up. *Cytotherapy* **2024**, *26*, 632–640. [CrossRef] [PubMed]
131. Luo, Q.; Xian, P.; Wang, T.; Wu, S.; Sun, T.; Wang, W.; Wang, B.; Yang, H.; Yang, Y.; Wang, H.; et al. Antioxidant activity of mesenchymal stem cell-derived extracellular vesicles restores hippocampal neurons following seizure damage. *Theranostics* **2021**, *11*, 5986–6005. [CrossRef]
132. Stavely, R.; Nurgali, K. The emerging antioxidant paradigm of mesenchymal stem cell therapy. *Stem Cells Transl. Med.* **2020**, *9*, 985–1006. [CrossRef] [PubMed]
133. Zhang, S.; Mulder, C.; Riddle, S.; Song, R.; Yue, D. Mesenchymal stromal/stem cells and bronchopulmonary dysplasia. *Front. Cell Dev. Biol.* **2023**, *11*, 1247339. [CrossRef] [PubMed]
134. Kumar, P.; Kandoi, S.; Misra, R.; Vijayalakshmi, S.; Rajagopal, K.; Verma, R.S. The mesenchymal stem cell secretome: A new paradigm towards cell-free therapeutic mode in regenerative medicine. *Cytokine Growth Factor. Rev.* **2019**, *46*, 1–9.
135. Mendt, M.; Rezvani, K.; Shpall, E. Mesenchymal stem cell-derived exosomes for clinical use. *Bone Marrow Transplant.* **2019**, *54* (Suppl. S2), 789–792. [CrossRef] [PubMed]
136. Simones, A.A.; Beisang, D.J.; Panoskaltsis-Mortari, A.; Roberts, K.D. Mesenchymal stem cells in the pathogenesis and treatment of bronchopulmonary dysplasia: A clinical review. *Pediatr. Res.* **2018**, *83*, 308–317. [CrossRef] [PubMed]
137. Wu, T.J.; Jing, X.; Teng, M.; Pritchard, K.A., Jr.; Day, B.W.; Naylor, S.; Teng, R.J. Role of Myeloperoxidase, Oxidative Stress, and Inflammation in Bronchopulmonary Dysplasia. *Antioxidants* **2024**, *13*, 889. [CrossRef] [PubMed] [PubMed Central]
138. Berger, S.I.; Iyengar, R. Role of systems pharmacology in understanding drug adverse events. *Wiley Interdiscip. Rev. Syst. Biol. Med.* **2011**, *3*, 129–135. [CrossRef] [PubMed]
139. Goel, R.; Singh, P.K.; Katiyar, A. Systems pharmacology in the spotlight: Trending technologies in network biology and drug discovery. *Front. Pharmacol.* **2024**, *15*, 1298712. [CrossRef] [PubMed]
140. van der Greef, J.; McBurney, R.N. Rescuing drug discovery: In vivo systems pathology and systems pharmacology. *Nat. Rev. Drug Discov.* **2006**, *4*, 961–967. [CrossRef] [PubMed]
141. Talevi, A. Multi-target pharmacology: Possibilities and limitations of the "skeleton key approach" from a medicinal chemist perspective. *Front. Pharmacol.* **2015**, *6*, 205. [CrossRef] [PubMed]
142. Li, Z.; Wei, Y.; Zhao, H. Network pharmacology-driven therapeutic interventions for interstitial lung diseases. *Pharmacol. Res.* **2024**, *198*, 106274.
143. Ramasubbu, M.K.; Paleja, B.; Srinivasan, A.; Maiti, R.; Kumar, R. Applying Quantitative and Systems Pharmacology to Drug Development and Beyond: An Introduction for Clinical Pharmacologists. *Indian. J. Pharmacol.* **2024**, *56*, 268–276. [CrossRef] [PubMed]
144. Pritchard, K.A.; Martin, D.P.; Naylor, S. US20220409692A1. Available online: https://www.patentguru.com/assignee/medical-college-of-wisconsin?page=24 (accessed on 14 June 2025).
145. Zhang, H.; Jing, X.; Shi, Y.; Xu, H.; Du, J.; Guan, T.; Weihrauch, D.; Jones, D.W.; Wang, W.; Gourlay, D.; et al. N-acetyl lysyltyrosylcysteine amide inhibits myeloperoxidase, a novel tripeptide inhibitor. *J. Lipid Res.* **2013**, *54*, 3016–3029. [CrossRef] [PubMed] [PubMed Central]

146. Teng, R.J.; Jing, X.; Martin, D.P.; Hogg, N.; Haefke, A.; Konduri, G.G.; Day, B.W.; Naylor, S.; Pritchard, K.A., Jr. N-acetyl-lysyltyrosylcysteine amide, a novel systems pharmacology agent, reduces bronchopulmonary dysplasia in hyperoxic neonatal rat pups. *Free Radic. Biol. Med.* **2021**, *166*, 73–89. [CrossRef]
147. Rarick, K.R.; Li, K.; Teng, R.J.; Jing, X.; Martin, D.P.; Xu, H.; Jones, D.W.; Hogg, N.; Hillery, C.A.; Garcia, G.; et al. Sterile inflammation induces vasculopathy and chronic lung injury in murine sickle cell disease. *Free Radic. Biol. Med.* **2024**, *215*, 112–126. [CrossRef] [PubMed]
148. Ito, M.; Kato, S.; Saito, M.; Miyahara, N.; Arai, H.; Namba, F.; Ota, E.; Nakanishi, H. Bronchopulmonary Dysplasia in Extremely Premature Infants: A Scoping Review for Identifying Risk Factors. *Biomedicines* **2023**, *11*, 553. [CrossRef] [PubMed] [PubMed Central]
149. Berger, J.; Bhandari, V. Animal models of bronchopulmonary dysplasia. The term mouse models. *Am. J. Physiol. Lung Cell Mol. Physiol.* **2014**, *307*, L936–L947. [CrossRef] [PubMed] [PubMed Central]
150. O'Reilly, M.; Thébaud, B. Animal models of bronchopulmonary dysplasia. The term rat models. *Am. J. Physiol. Lung Cell Mol. Physiol.* **2014**, *307*, L948–L958. [CrossRef] [PubMed]

antioxidants

MDPI

Review

Impact of E-Cigarettes on Fetal and Neonatal Lung Development: The Influence of Oxidative Stress and Inflammation

Antonella Gambadauro [1,*,†], Francesca Galletta [1,†], Beatrice Andrenacci [2], Simone Foti Randazzese [1,*], Maria Francesca Patria [2] and Sara Manti [1]

[1] Pediatric Unit, Department of Human Pathology in Adult and Developmental Age "Gaetano Barresi", University of Messina, 98124 Messina, Italy; francygall.92@gmail.com (F.G.); sara.manti@unime.it (S.M.)

[2] S.C. Pneumoinfettivologia Pediatrica, Fondazione IRCCS Ca' Granda Ospedale Maggiore Policlinico, 20122 Milan, Italy; beatrice.andrenacci@policlinico.mi.it (B.A.); francesca.patria@policlinico.mi.it (M.F.P.)

* Correspondence: gambadauroa92@gmail.com (A.G.); simone.foti.92@gmail.com (S.F.R.)

† These authors contributed equally to the work.

Abstract: Electronic cigarettes (e-cigs) recently increased their popularity as "safer" alternatives to traditional tobacco smoking, including among pregnant women. However, the effect of e-cig exposure on fetal and neonatal developing lungs remains poorly investigated. In this review, we analysed the impact of e-cig aerosol components (e.g., nicotine, solvents, and flavouring agents) on respiratory system development. We particularly emphasized the role of e-cig-related oxidative stress and inflammation on lung impairment. Nicotine contained in e-cigs can impair lung development at anatomical and molecular levels. Solvents and flavours induce inflammation and oxidative stress and contribute to compromising neonatal lung function. Studies suggest that prenatal e-cig aerosol exposure may increase the risk of future development of respiratory diseases in offspring, such as asthma and chronic obstructive pulmonary disease (COPD). Preventive strategies, such as smoking cessation programs and antioxidant supplementation, may be essential for safeguarding respiratory health. There is an urgent need to explore the safety profile and potential risks of e-cigs, especially considering the limited studies in humans. This review highlights the necessity of regulating e-cig use during pregnancy and promoting awareness of its potential consequences on fetal and neonatal development.

Keywords: air pollution; asthma; COPD; e-cigarette; lung development; neonates; oxidative stress; pregnancy

1. Introduction

In the last decade, researchers warned the global population about indoor and outdoor pollution's harmful effects on respiratory health in adults and children [1]. Despite the increased knowledge about the deleterious impact of nicotine and other additives on the lungs, new electronic nicotine delivery systems (ENDSs) became popular among consumers and often suggested for smoking cessation and reported as safer than tobacco cigarettes [2,3]. Electronic cigarettes (e-cigs) are one of the most popular ENDSs in commerce. These devices may also not contain nicotine, and, in this case, they can be classified as electronic non-nicotine delivery systems (ENNDSs) [4]. E-cigs are non-combustible systems comprising a battery, a microprocessor, a start button, a heating part, and a liquid reservoir [5]. The solution in the reservoir is vaporized during use, producing an aerosol (or vapour) of highly concentrated droplets mainly formed by sub-micrometre particles [6]. In addition to the potential presence of nicotine, the liquid solution may contain a wide variety of flavours, solvents (e.g., propylene glycol (PG), vegetable glycerine (VG), glycerol), and preservatives [7,8]. Trace metals and additional impurities (e.g., polyaromatic hydrocarbons, aldehydes, and acrolein) have also been detected in the aerosol [9–11].

ENDS use during pregnancy was reported in percentage rates ranging from 0.6% to 15% in different analyses [12–15]. An online national survey conducted in the United States (USA) on 445 pregnant women reported that 6% and 7% used only tobacco cigarettes or only e-cigs, respectively, while 9% used both tobacco cigarettes and e-cigs [16]. These findings must be read considering the general perception that e-cigs are a healthier alternative to traditional cigarettes and may assist with smoking cessation during pregnancy [17].

The fetal lungs are organs sensitive to toxins, which can alter their normal neonatal development and predispose them to respiratory diseases later in life [18]. Few data are available on the short- and long-term influences of e-cigs on respiratory development during the fetal and neonatal period due to the lack of safety studies [19]. The latest evidence suggests that nicotine and chemical products generated from e-cigs during the perinatal period could promote the activation of inflammation and the production of oxidative compounds, altering lung development and increasing the incidence of respiratory diseases in offspring [20,21].

In this review, we summarized and critically analysed the existing literature about the impact of e-cigs on fetal and neonatal lung development. There is an urgent need to determine if these electronic devices could be safe for offspring or potentially related to functional and structural lung impairment. By combining the key terms "e-cigarettes" OR "e-cigs" AND "lung development" OR "airway" OR "lung" AND "pregnancy" OR "neonates" in a computerized search of PubMed limited to the last 10 years, we provided a comprehensive overview of the literature, based on a critical evaluation without standardized methodologies or statistical analyses.

2. Stages of Normal Lung Development

Lung development is a sequential, multistep process that begins prenatally and continues postnatally, reaching completion in early adulthood [22,23]. This extended timeline makes the respiratory system vulnerable to prenatal and postnatal influences, including prematurity, low birth weight, maternal and personal nutrition, exposure to smoking, environmental pollutants, and early-life infections. These factors interact with individual genetic and epigenetic backgrounds to shape respiratory function [22]. Lung function at birth is a key determinant of lung health across the lifespan, and supporting optimal lung development during early life is crucial to mitigate the natural decline in respiratory function with ageing [22].

Prenatal lung development is classically divided into five stages from 3 to 36 weeks of gestation: the embryonic, pseudo-glandular, canalicular, saccular, and alveolar stages (Figure 1). Each phase represents a critical window during which specific structural and functional components of the lungs are established [23].

During the embryonic stage (0–7 weeks), the first rudiment of the respiratory system, which is called the laryngotracheal groove, appears on the ventral wall of the anterior foregut, extending caudally into the respiratory diverticulum [22]. The trachea separates from the oesophagus via tracheoesophageal septa, and the "branching morphogenesis" begins [22,24]. This process first generates the two main bronchi and progresses to lobar and segmental bronchi until the development of 18 major lobules, laying the groundwork for the airway tree. Simultaneously, angiogenesis begins around the developing respiratory structures, initiating the formation of the pulmonary vasculature [22].

In the pseudo-glandular stage (7–17 weeks), vasculogenesis and branching morphogenesis advance rapidly, forming terminal bronchioles by 14 weeks of gestation. By the end of this stage, approximately 70% of the airway tree is complete [22]. Airway epithelial cells also begin to differentiate, with columnar cells appearing in the proximal airways and cuboidal cells in the distal regions [22]. Fetal breathing movements and airway contractions play a pivotal role in stimulating epithelial proliferation, differentiation, and early surfactant production. These mechanical stimuli stretch the lung tissue, promoting cellular development through pathways like serotonin upregulation [25,26].

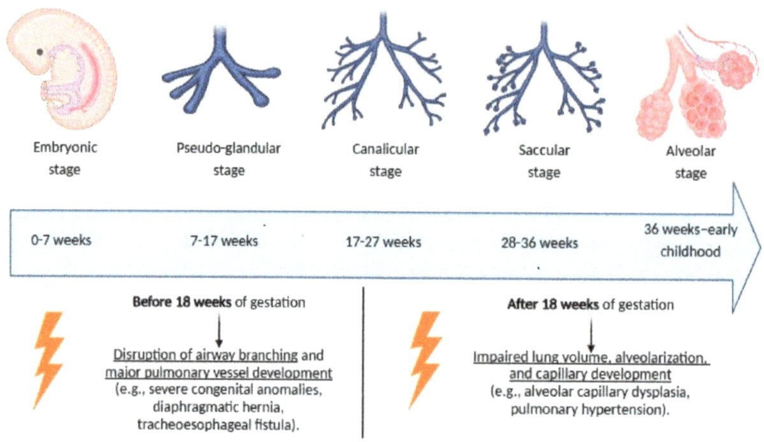

Figure 1. Stages of normal lung development and impact of adverse exposures.

The canalicular stage (17–27 weeks) is characterized by the development of respiratory bronchioles, alveolar ducts, and primitive alveoli, establishing the functional units of the lungs known as "pulmonary acini" [24]. During this stage, type I and type II pneumocytes differentiate, with type II cells beginning to produce surfactant precursors, known as lamellar bodies, by 24 weeks of gestation [24]. Early surfactant production reduces surface tension within the lungs, stabilizes alveoli during inflation, and facilitates effective breathing [24]. At the end of this stage, the formation of the alveolar–capillary barrier and the onset of gas exchange capability makes it possible for a preterm newborn to survive [24].

Branching morphogenesis concludes in the saccular stage (28–36 weeks of gestational age), with saccules developing from pulmonary acini, accompanied by a significant increase in gas exchange potential and surfactant production [24].

The alveolar stage (36 weeks–early childhood) begins shortly before birth, and it involves the development of new alveoli through secondary septation of terminal saccules [22]. The alveolar number increases rapidly during the first 2–4 years of life, reaching an estimated 300–800 million at around 8 years of age; then, alveolarization slows, though alveolar size continues to grow until adolescence, with interindividual variability and slight gender differences [27,28]. Recent advances in imaging techniques, such as high-resolution tomography and helium-3 magnetic resonance imaging, suggest that alveolarization may persist into adulthood, though the exact timeline and molecular mechanisms remain under investigation [29–31]. During this period airways expand proportionally in diameter and length, maintaining a constant length-to-diameter ratio [32]. Concurrently, the pulmonary microvasculature undergoes remodelling, with the double-layered network fusing into a single-layered structure through capillary fusion and thinning of the alveolar septa. The resulting alveolar–capillary membrane, composed of alveolar epithelial cells, capillary endothelial cells, and their shared basement membrane, becomes the key site for efficient respiratory gas exchange [22,24].

The respiratory system remains highly sensitive to environmental and genetic factors throughout its development, with long-term outcomes depending on the timing and nature of exposures during critical windows of susceptibility (Figure 1) [33].

Adverse exposures occurring before 18 weeks of gestation may disrupt airway branching and major pulmonary vessel development, resulting in severe congenital anomalies, such as pulmonary agenesis or aplasia, which are often incompatible with life, or conditions like diaphragmatic hernia with pulmonary hypoplasia and tracheoesophageal fistula, which carry significant morbidity and mortality [24].

Instead, later exposures, occurring after 18 weeks of gestation, generally affect lung volume, alveolarization, and capillary development. For example, alveolar capillary dysplasia (ACD) may result from insults during this stage, characterized by reduced capillary density, misaligned pulmonary veins, and thickened alveolar septa, leading to pulmonary hypertension [34]. Additionally, disturbances in the normal pattern of angiogenesis and alveolar maturation can impair gas exchange efficiency and lung compliance, often leading to chronic respiratory complications [35].

Finally, adverse exposures after birth, such as air pollution, second-hand smoke, or recurrent respiratory infections, might impair lung growth and the development of pulmonary microvasculature [22]. Young children are particularly vulnerable to inhaled pollutants due to their relatively higher airway flow rates and greater deposition of small particles than adults [36]. These vulnerabilities highlight the importance of protective measures and early interventions to preserve respiratory health and promote optimal lung development.

3. Are E-Cigarettes Safe?

3.1. Impact on Lung Development

In recent years, especially among young people, the interest in e-cigs has increased. Furthermore, their use is intensifying in pregnant women due to the perception of their safety compared to tobacco cigarettes [17,37]. The reservoir of these devices contains e-liquid, which is converted into aerosols by a heating process. E-liquid is a mixture of water, different nicotine levels, solvents, flavours, preservatives, and other potential additives [7,8] (Table 1).

Nearly 40% of pregnant women believe that e-cigs do not contain nicotine [12]. Nicotine is a dinitrogen alkaloid extracted from tobacco plants. It is a parasympathomimetic stimulant and a dependence-forming constituent of e-cigs and tobacco cigarettes [38]. In pregnant women, nicotine crosses the placenta and reaches high concentrations in fetal blood and amniotic fluid [39,40]. During breastfeeding, nicotine levels are higher in breast milk than in maternal plasma [41,42]. No amount of nicotine is known to be safe during pregnancy [43]. ENDSs contain diverse nicotine concentrations, and previous studies conducted on animals and humans reported similar or higher nicotine levels in e-cigs compared to tobacco cigarettes [44,45]. In a study conducted on rodents, nicotine concentrations were eight times higher after e-cig exposure (JUUL type) compared to tobacco cigarette exposure [44]. A study conducted on 13 healthy adult e-cig users who took 15 puffs from their usual e-cigs revealed that nicotine levels were comparable to or higher than classic tobacco cigarettes [45]. These results are relevant for understanding the potential impact of e-cigs during pregnancy. However, the most important limitation of these findings is the different nicotine levels in the diverse formulations of e-cigs studied.

Few data are reported in the literature about the effects of nicotine contained in e-cigs on human lung development. Most studies have been conducted on animal models and not exclusively during the in utero period. However, they can assist in comprehending the effect of maternal e-cig use and exposure during the critical first few weeks of life on the lung development of offspring. Ozekin et al. tested the impact of maternal vaping on fetal lung development in mice by low–moderate daily exposure to 2.4% nicotine vapour during the entire gestational period [46]. They specifically focused on the mouse E18.5 timepoint, considered as the last day of the embryonic stage, and reported that maternally vaped wildtype mouse lungs have smaller airspaces compared to room-air-exposed mice [46]. This finding was consistent with the disruption of the saccular stage of lung development in the same model [46]. A separate study analysed the impact of the daily inhalation of 36 mg/mL of nicotine cinnamon-flavoured e-cig aerosols during the preconception period (12 days before mating) and during gestation in a murine model [47]. Lung morphometry evaluations of preconception e-cig-exposed offspring reported a notably increased tissue fraction at birth related to a delayed thinning of the lung epithelium during the saccular stage [47]. Both of these studies showed wide gene expression changes in maternally vaped embryos dominated by the downregulation of developmental signalling pathways such as Notch and Wnt [46,47]. Notch and Wnt are pivotal regulators of lung development. Notch regulates ciliated versus secretory cell expression in the developing airways, and ciliated cell loss associated with downregulated Notch expression was reported in smokers and in patients with chronic obstructive pulmonary disease (COPD) [48,49]. The Wnt signalling pathway is essential in the saccular stage, promoting the proliferation and differentiation of alveolar epithelial cells, stem cell maintenance, and lung branching morphogenesis [50]. These findings confirm, at both the anatomical and molecular levels, that the nicotine contained in e-cigs impairs fetal lung development. Another study on mice compared e-cig aerosol exposure during pregnancy with and without nicotine to room-air controls at gestational day (GD) 21, at postnatal day (PND) 4, and at PND 10 [19]. The group exposed to e-cig aerosol vaping containing nicotine reported decreased lung size, increased free space within the lung parenchyma, and decreased alveolar septation compared to the other two groups at PND 4 [19]. In humans, the equivalent developmental stage of murine PND 4 occurs in the alveolar stage. These results suggest the specific role of nicotine contained in e-cig vapours (without the confounding factor of flavours) in determining an emphysematous phenotype with larger and fewer distal air spaces in neonatal mice [19]. Postnatal lung growth and weight gain were also reduced in neonatal mice after exposure to e-cigs containing 1.8% nicotine and PG for the first ten days of life compared to room air controls [51]. Based on these results, we can conclude that nicotine contained in e-cigs can affect lung development by altering airway growth, especially during the embryonic, saccular, and alveolar stages, inducing a final emphysematous phenotype and low lung size at birth.

Mucociliary clearance is essential to defend against the organisms that cause respiratory infections or the inhalation of irritants [52]. A bullfrog palate paradigm (a well-established model used to study mucociliary clearance) confirmed a modest dampening effect of e-cig aerosol on mucociliary clearance [53]. The impact on the mucociliary function seems principally related to the nicotine contained in e-cigs. In a study conducted on a murine model, chronic (3 weeks), daily, 20 min-exposure to nicotine and PG (e-cigs) slowed mucociliary clearance in murine lungs. Conversely, exposure to PG alone did not impact mucociliary clearance [54]. A study conducted in both in vitro human bronchial epithelial cells obtained from never-smoking individuals and in vivo sheep bronchial epithelial cells showed that nicotine-containing e-cig vapour induced mucociliary dysfunction by acting on TRPA1 (transient receptor potential ankyrin 1) [55]. TRPA1 is a nicotine-sensitive receptor that belongs to the TRP ion channel family, and it is extensively distributed in the cellular membranes of humans and mammals [56]. The impairment in mucociliary clearance, by altering mucus viscoelasticity, increasing mucus/mucin concentrations, and damaging mucociliary transport, may affect the protection of the airway and increase infectious diseases in neonates and infants born to vaping women. Studies on the neonatal period are needed to confirm this hypothesis.

The two solvents most frequently included in e-cigs are PG and VG. Both PG and VG were initially considered innocuous components of e-cigs. However, recent studies have analysed their impact on lung development. During vaping, the thermal decomposition of PG generates acetone, acetaldehyde, and formaldehyde, while the thermal decomposition of VG produces principally acrolein and formaldehyde [57]. Among these substances, acetaldehyde and formaldehyde are classified as carcinogens, while acrolein is considered a human carcinogen able to irritate the upper respiratory tract after inhalation [58]. An in vivo study on human bronchial epithelial cells (HBECs) obtained from never-smoking individuals revealed that a one-week exposure to VG and PG/VG-containing e-cigs (in the absence of nicotine and flavours) decreased the function of cystic fibrosis transmembrane conductance regulator (CFTR), an anion channel important for mucus hydration [59]. In volunteers exposed to VG-containing e-cigs, the reduction in CFTR function was associated with high inflammatory biomarkers, such as Interleukin-6 (IL-6), IL-8, matrix metalloproteinase-9 (MMP9) mRNAs, MMP-9 activity, and mucin 5AC (MUC5AC) expression levels [59]. These modifications induced a reduction in ciliary beating and increased mucus concentration as a possible consequence of goblet cell hyperplasia [60]. Exposure during pregnancy to PG/VG-containing e-cig vapour with and without nicotine was tested in murine neonates at 5 months of age [61]. Female mice exposed to PG/VG-containing e-cig vapour with and without nicotine reported goblet cell hyperplasia. The group exposed only to solvents (without nicotine) showed increased lung and alveolar stiffness (both genders) and decreased lung compliance (only female mice) [61]. This last finding could be the result of increased extracellular matrix (ECM) deposition, which was reported as a cause of PG/VG-containing e-cig aerosol exposure (without nicotine) during pregnancy [62].

Few studies consider the impact of flavours contained in e-cigs during pregnancy, and most of them consider the preferences of pregnant women. Among flavouring agents, fruit and mint are the most used in this population [8]. However, the impact of specific flavours on lung development in the fetal and neonatal periods was not critically analysed in previous research. Each flavouring agent may potentially have different inflammatory and toxic effects on lung morphogenesis and function. Moreover, e-cigs containing several flavours are more toxic and induce a higher inflammatory response than single flavours [63]. Maltol, ethyl maltol, ethyl vanillin, vanillin, and furaneol have been reported as the most cytotoxic flavouring agents in e-cigs [64]. By considering the respiratory system, cinnamaldehyde, 2-methoxycinnamaldehyde, O-vanillin, and pentanedione seem to have cytotoxic effects on airways [65]. Further studies are needed to analyse the role of different flavours on the developmental lungs.

Table 1. An overview of the main harmful effects of e-cigarette compounds on lung development.

Chemical Compound	Harmful Effect	References
Nicotine	Crosses the placenta, accumulates in fetal blood and amniotic fluid, disrupts lung development (embryonic, saccular, and alveolar stages), reduces lung size, induces the emphysematous phenotype, impairs mucociliary clearance.	[19,39,40,46–51,53–55]
PG/VG	Their thermal decomposition generates carcinogens and irritants (PG produces acetone, acetaldehyde, and formaldehyde, while VG produces acrolein and formaldehyde); PG/VG exposure reduces CFTR function and increases inflammatory biomarkers, leading to lower ciliary beating and higher mucus concentration; PG/VG exposure decreases lung compliance, increases alveolar stiffness, and induces extracellular matrix deposition.	[57–62]
Flavouring agents	Some flavours (e.g., cinnamaldehyde, O-vanillin, pentanedione) have cytotoxic and inflammatory effects on airways.	[63–65]
$PM_{2.5}$ and UFPs	Oxidative stress by generating ROS; activation of inflammatory pathways.	[66]

CFTR = cystic fibrosis transmembrane conductance regulator; PG = propylene glycol; VG = vegetable glycerine; UFPs = ultrafine particles; $PM_{2.5}$ = particulate matter 2.5.

3.2. *The Role of Oxidative Stress*

Oxidative stress has a central role in the development and maintenance of numerous lung diseases [67]. E-cigs can directly or indirectly induce oxidative damage to human lungs due to the presence of reactive chemical substances in their aerosols, the activation of pro-inflammatory compounds, and the modulation of intracellular pro-inflammatory pathways. E-cig aerosols contain diverse toxic compounds (e.g., formaldehyde, acetaldehyde, and acrolein), reactive oxygen species (ROS), heavy metals, flavours, and propylene oxide, which is derived from PG heating [68]. It was reported that e-cig vapour contains 7×10^{11} free radicals per puff, eliciting an important increase in oxidative stress [69]. Moreover, several studies have described the presence of significant amounts of fine ($PM_{2.5}$) and ultrafine particles (UFPs) in e-cig aerosol [66] (Table 1). Nicotine, PG, and VG play a pivotal role in determining the cytotoxic and pro-inflammatory effects of these aerosols. PG and VG induce cytotoxic effects depending on puff number, while nicotine seems protective against cellular toxicity [66]. Lung inflammation is secondary to the activation of the pro-inflammatory NF-kB pathway, which induces IL-1β and tumour necrosis factor (TNF)-α production [66]. Interestingly, a study on murine models reported an abnormal inflammatory environment in the airways of both mothers and offspring after e-cig exposure during pregnancy [70]. In that research, TNF-α was increased in the lungs of both the mothers and offspring; and IL-1β was increased only in the mothers [70]. An in vivo study conducted on human bronchial airway epithelial cells (H292) and human fetal lung fibroblasts (HFL1) showed that the inflammatory response and the stress phenotype depended on nicotine content and flavour type [71]. For example, cinnamon-flavoured e-liquid produced a higher IL-8 response than cigarette smoke extract [71]. The exposure of wild-type mice to e-cig aerosols decreased lung glutathione levels, increasing oxidative airway damage [71]. A recent study on rat lung tissues confirmed that exposure to e-cig aerosols reduced the activity of antioxidant compounds, such as superoxide dismutase, catalase, and glutathione peroxidase [72]. Pulmonary microvascular endothelial cells can also be the target of oxidative stress damage. Exposure to e-cig vapour contributes to vascular endothelial dysfunction in the cardio-respiratory system, predisposing to hypertension and atherosclerosis [73,74]. E-cig vapour exposure (with or without nicotine) increases the expression of NADPH oxidase (NOX), an enzymatic source for ROS during inflammation [75]. Furthermore, it induces endothelial nitric oxide synthase (eNOS) uncoupling, increasing superoxide generation and peroxynitrite production, as well as tetrahydrobiopterin (BH4) reduction [75]. All these changes decrease nitric oxide (NO) synthesis and availability, inducing endothelial dysfunction and long-term cardio-respiratory system impairment [75]. These findings are useful in improving our understanding of the potential oxidative harm produced by e-cigs on the airways (Figure 2). However, studies on fetal and neonatal lungs may be helpful in understanding the effects of e-cig-induced oxidative stress during pregnancy.

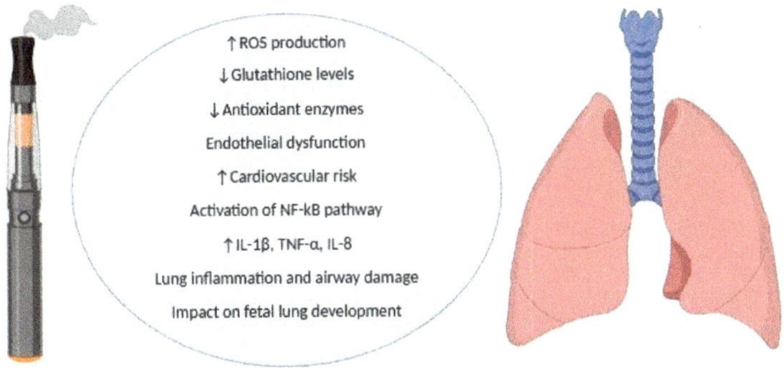

Figure 2. The role of e-cig-induced oxidative stress on the airways.

3.3. Long-Term Effects on Respiratory Health

Maternal smoking during pregnancy is related to a twofold increase in the risk of wheezing and asthma in offspring [76,77]. Several studies reported that this increased risk seems to be associated with nicotine alone during perinatal smoke exposure [18]. During gestation, the use of devices containing tobacco, such as heated tobacco products (HTPs), was associated with an increased risk of onset of allergic diseases (e.g., asthma, allergic rhino/conjunctivitis, atopic dermatitis) in offspring [78]. However, few studies have been conducted on e-cigs, and most of the literature is based on the effect of tobacco cigarettes. Nicotine is the primary agent responsible for impaired lung development. Maternal smoking exposure is associated with reduced forced expiratory flows, decreased passive expiratory compliance, higher hospitalization rates for respiratory infections, and increased prevalence of childhood wheezing and asthma [20]. A study conducted in a murine model analysed e-cig aerosol exposure (with and without nicotine) during pregnancy [19]. The findings of that research revealed increased respiratory resistance and reduced compliance in mice exposed to nicotine-containing e-cigs compared to controls [19]. Infants and children exposed to passive smoking during pregnancy showed higher airway resistance and increased rates of persistent wheezing [79,80]. Adolescents using e-cigs are more likely to have chronic cough, bronchitis, and asthma [81]. Asthmatic children and adolescents exposed to second-hand e-cig smoke at home have more daily respiratory symptoms compared to unexposed controls [82]. In adults, an association between e-cigs and the development of asthma or COPD was found, especially in non-smokers [83]. E-cig vapour may also suppress the immune response to viral and bacterial infections, increasing the host susceptibility by impairing pulmonary anti-microbial defences [69]. The literature on the long-term effects of e-cigs on respiratory health is still lacking. However, asthma and COPD seem to be the two conditions more associated with e-cig exposure.

4. Prevention Strategies

The adoption of effective prevention strategies represents a pivotal step in mitigating the adverse effects of e-cig use on fetal lung development. These strategies are essential not only for safeguarding the fetus's health but also for enhancing maternal health outcomes [84]. Prevention strategies require a multifaceted approach that integrates smoking cessation programs, legislative efforts, dietary interventions, and the active involvement of healthcare providers [20].

Smoking cessation remains the cornerstone of preventing the harmful effects of e-cigs on both the mother and fetus [84]. Despite the well-documented risks to lung development and overall fetal health, a significant number of pregnant women continue smoking due to addiction, limited access to support, or socioeconomic barriers [85–87]. Successful cessation strategies, therefore, play a critical role in reducing these risks [84]. The effectiveness of various smoking cessation interventions has been extensively evaluated. Behavioural counselling is one of the most effective interventions for smoking cessation during pregnancy. Evidence from systematic reviews demonstrated that intensive psychosocial support, including motivational interviewing and behavioural therapies, can significantly increase cessation rates. Counselling programs tailored to the unique needs of pregnant women are especially effective, particularly when delivered through multiple sessions by trained professionals [85,88]. Physical activity programs have been explored as adjunct strategies to traditional behavioural support [89]. Although their effectiveness in promoting smoking cessation during pregnancy remains uncertain, physical activity contributes to overall health benefits and has been shown to reduce smoking urges in non-pregnant individuals. These findings suggest the potential for further research of these interventions in future studies [89]. Emerging technologies such as Artificial Intelligence (AI) have introduced innovative approaches to smoking cessation [90]. AI-powered tools can provide real-time, personalized support, analyse individual behavioural patterns, and predict relapse risks [90]. For instance, mobile applications can deliver motivational messages, monitor smoking habits, and recommend tailored coping strategies. Virtual support groups facilitated by AI platforms have also proven effective in fostering community engagement [90]. Financial incentives have been identified as another effective tool in promoting smoking cessation among pregnant individuals, particularly those from disadvantaged socioeconomic backgrounds. Studies reveal that offering monetary or voucher-based rewards for achieving abstinence significantly increases cessation rates and reduces the economic burden of smoking-related complications [91–93].

Healthcare providers, particularly midwives and obstetricians, play a critical role in smoking cessation efforts during pregnancy. Routine screening for tobacco use during prenatal visits, combined with timely referrals to cessation resources, is essential. However, challenges such as time limitations and resource restrictions often hinder the consistent implementation of these interventions [88,94–96]. Overall, integrating evidence-based strategies like counselling, financial incentives, and healthcare provider engagement into prenatal care can significantly enhance smoking cessation rates during pregnancy, improving both maternal and fetal outcomes [95]. Additionally, systematic reviews and population-based studies have highlighted both the potential benefits and risks associated with Nicotine Replacement Therapy (NRT) and other pharmacological interventions, such as bupropion and varenicline, for smoking cessation during pregnancy [97–101]. NRT is commonly used to support smoking cessation in the general population and is sometimes prescribed during pregnancy to mitigate nicotine withdrawal symptoms [98]. Studies have demonstrated that NRT can increase the likelihood of smoking cessation during pregnancy when combined with behavioural support. A Cochrane review showed that NRT improves cessation rates by approximately 37% compared to placebos, although with low-certainty evidence [98,99]. Faster-acting formulations, such as nicotine gum or tablets, have been suggested as potentially safer options compared to nicotine patches, which provide continuous nicotine delivery [102]. However, concerns persist about potential risks, including increased rates of infant colic and attention-deficit/hyperactivity disorder

(ADHD) in children exposed to nicotine during gestation [97,103]. Consequently, the use of NRT during pregnancy remains controversial due to its potential effects on fetal development [98]. Bupropion, an atypical antidepressant, and varenicline, a partial nicotine receptor agonist, are other pharmacological options that have been studied for smoking cessation during pregnancy [100]. Evidence suggests that bupropion can be effective without significantly increasing the risk of adverse perinatal outcomes, such as preterm birth or congenital anomalies [100,101]. Varenicline, while effective in the general population, has limited data on its safety in pregnancy, and guidelines generally recommend caution due to insufficient evidence [100,101]. To date, no information is currently available regarding the effects of either bupropion or varenicline on fetal lung development. Overall, while pharmacological interventions such as NRT, bupropion, and varenicline may support smoking cessation during pregnancy, their use should be carefully balanced against potential risks [100]. Healthcare providers should adopt a patient-centred approach, tailoring pharmacological strategies to individual needs and circumstances, and should ensure that such treatments are complemented by behavioural and psychosocial support [99,101,104]. Further research is needed to provide clearer evidence regarding the safety and efficacy of these interventions in pregnancy.

Legislative measures could have a pivotal role in promoting smoking cessation during pregnancy. Policies that restrict the sale of e-cigs prohibit their use in public spaces and mandate clear labelling of nicotine content can discourage use. Additionally, public health campaigns focused on raising awareness about the risks of e-cigs, particularly during pregnancy, can enhance the effectiveness of these legislative actions by educating individuals to make healthier choices [105].

The use of antioxidants during pregnancy for smoking mothers has shown promise in mitigating respiratory issues in offspring. Vitamin C has been extensively studied for its protective effects against oxidative stress induced by maternal smoking [106–108]. A randomized trial demonstrated that daily supplementation with 500 mg of vitamin C during pregnancy improves newborn pulmonary function, reduces wheezing, and prevents DNA methylation changes associated with maternal smoking [109]. These changes suggest that vitamin C mitigates the epigenetic alterations linked to reduced lung development and long-term respiratory risks.

Vitamin E, another powerful antioxidant, has been shown to synergize with vitamin C to contrast oxidative stress in placental tissues, thereby protecting fetal development from the adverse effects of nicotine [110]. Similarly, N-acetylcysteine (NAC), by enhancing glutathione synthesis, has demonstrated protective roles in reducing oxidative stress and supporting placental function [110]. Meanwhile, L-carnitine has shown efficacy in reducing inflammation and oxidative damage in animal studies, improving birth weight and partially normalizing inflammatory markers in offspring exposed to maternal smoke [111].

5. Conclusions

E-cigs are significant modifiable risk factors for short- and long-term respiratory damage in fetal and neonatal airways. Nicotine and other compounds are responsible for anatomical and molecular modifications of the lungs during pregnancy and the perinatal period. Oxidative stress has a pivotal role in the activation of inflammatory damage in the respiratory system. Studies in humans are urgently needed to confirm the sequelae of the exposure on the developing airways.

A comprehensive prevention strategy should integrate multiple interventions to address the diverse factors influencing e-cig use during pregnancy (Figure 3). Legislative measures and counselling are the most relevant strategies for mothers and their children.

The use of antioxidants and/or NRT without stopping the smoking habit can increase the nicotine burden on fetal and neonatal lungs and encourage the mother to continue nicotine dependence and transfer it to the next generations. Community-based programs that involve partners, family members, and social networks can create a supportive environment for behaviour change. Similarly, collaborations between healthcare systems, public health agencies, and advocacy groups can amplify the reach and effectiveness of prevention efforts.

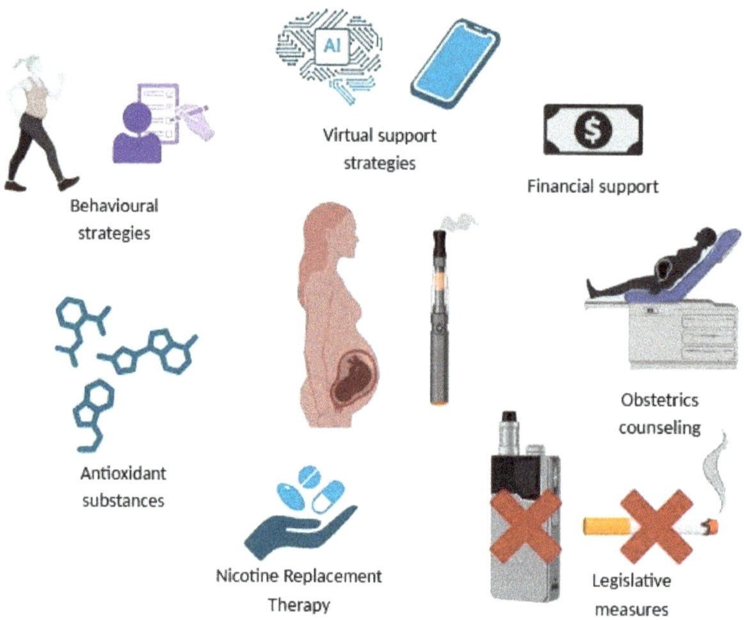

Figure 3. Multiple integrated prevention strategies to influence e-cig use during pregnancy.

Author Contributions: Conceptualization, S.M. and M.F.P.; methodology, A.G. and S.F.R.; data curation, F.G. and B.A.; writing—original draft preparation, A.G., F.G. and B.A.; writing—review and editing, A.G. and S.F.R.; supervision, S.M. and M.F.P. All authors have read and agreed to the published version of the manuscript.

Funding: This research received no external funding.

Conflicts of Interest: The authors declare no conflicts of interest.

References

1. La Grutta, S.; Indinnimeo, L.; di Coste, A.; Ferrante, G.; Landi, M.; Pelosi, U.; Rusconi, F. Environmental Risk Factors and Lung Diseases in Children: From Guidelines to Health Effects. *Early Hum. Dev.* **2013**, *89*, S59–S62. [CrossRef] [PubMed]
2. Palazzolo, D.L. Electronic Cigarettes and Vaping: A New Challenge in Clinical Medicine and Public Health. A Literature Review. *Front. Public Health* **2013**, *1*, 56. [CrossRef] [PubMed]
3. McRobbie, H.; Bullen, C.; Hartmann-Boyce, J.; Hajek, P. Electronic Cigarettes for Smoking Cessation and Reduction. In *Cochrane Database of Systematic Reviews*; McRobbie, H., Ed.; John Wiley & Sons, Ltd.: Chichester, UK, 2014. [CrossRef]
4. Yoong, S.L.; Hall, A.; Turon, H.; Stockings, E.; Leonard, A.; Grady, A.; Tzelepis, F.; Wiggers, J.; Gouda, H.; Fayokun, R.; et al. Association between Electronic Nicotine Delivery Systems and Electronic Non-Nicotine Delivery Systems with Initiation of Tobacco Use in Individuals Aged < 20 Years. A Systematic Review and Meta-Analysis. *PLoS ONE* **2021**, *16*, e0256044. [CrossRef]
5. Breitbarth, A.K.; Morgan, J.; Jones, A.L. E-Cigarettes—An Unintended Illicit Drug Delivery System. *Drug Alcohol. Depend.* **2018**, *192*, 98–111. [CrossRef] [PubMed]

6. Sosnowski, T.R.; Odziomek, M. Particle Size Dynamics: Toward a Better Understanding of Electronic Cigarette Aerosol Interactions With the Respiratory System. *Front. Physiol.* **2018**, *9*, 853. [CrossRef] [PubMed]
7. Stefaniak, A.B.; LeBouf, R.F.; Ranpara, A.C.; Leonard, S.S. Toxicology of Flavoring- and Cannabis-Containing e-Liquids Used in Electronic Delivery Systems. *Pharmacol. Ther.* **2021**, *224*, 107838. [CrossRef] [PubMed]
8. Stroud, L.R.; Papandonatos, G.D.; Borba, K.; Kehoe, T.; Scott-Sheldon, L.A.J. Flavored Electronic Cigarette Use, Preferences, and Perceptions in Pregnant Mothers: A Correspondence Analysis Approach. *Addict. Behav.* **2019**, *91*, 21–29. [CrossRef] [PubMed]
9. Goniewicz, M.L.; Knysak, J.; Gawron, M.; Kosmider, L.; Sobczak, A.; Kurek, J.; Prokopowicz, A.; Jablonska-Czapla, M.; Rosik-Dulewska, C.; Havel, C.; et al. Levels of Selected Carcinogens and Toxicants in Vapour from Electronic Cigarettes. *Tob. Control* **2014**, *23*, 133–139. [CrossRef]
10. Palazzolo, D.L.; Crow, A.P.; Nelson, J.M.; Johnson, R.A. Trace Metals Derived from Electronic Cigarette (ECIG) Generated Aerosol: Potential Problem of ECIG Devices That Contain Nickel. *Front. Physiol.* **2017**, *7*, 663. [CrossRef]
11. Jensen, R.P.; Luo, W.; Pankow, J.F.; Strongin, R.M.; Peyton, D.H. Hidden Formaldehyde in E-Cigarette Aerosols. *N. Engl. J. Med.* **2015**, *372*, 392–394. [CrossRef]
12. Mark, K.S.; Farquhar, B.; Chisolm, M.S.; Coleman-Cowger, V.H.; Terplan, M. Knowledge, Attitudes, and Practice of Electronic Cigarette Use Among Pregnant Women. *J. Addict. Med.* **2015**, *9*, 266–272. [CrossRef]
13. Ashford, K.; Wiggins, A.; Butler, K.; Ickes, M.; Rayens, M.K.; Hahn, E. E-Cigarette Use and Perceived Harm Among Women of Childbearing Age Who Reported Tobacco Use During the Past Year. *Nurs. Res.* **2016**, *65*, 408–414. [CrossRef] [PubMed]
14. Reynolds, C.M.E.; Egan, B.; McKeating, A.; Daly, N.; Sheehan, S.R.; Turner, M.J. Five Year Trends in Maternal Smoking Behaviour Reported at the First Prenatal Appointment. *Ir. J. Med. Sci.* **2017**, *186*, 971–979. [CrossRef]
15. Whittington, J.R.; Simmons, P.M.; Phillips, A.M.; Gammill, S.K.; Cen, R.; Magann, E.F.; Cardenas, V.M. The Use of Electronic Cigarettes in Pregnancy: A Review of the Literature. *Obs. Gynecol. Surv.* **2018**, *73*, 544–549. [CrossRef]
16. Wagner, N.J.; Camerota, M.; Propper, C. Prevalence and Perceptions of Electronic Cigarette Use during Pregnancy. *Matern. Child. Health J.* **2017**, *21*, 1655–1661. [CrossRef]
17. McCubbin, A.; Fallin-Bennett, A.; Barnett, J.; Ashford, K. Perceptions and Use of Electronic Cigarettes in Pregnancy. *Health Educ. Res.* **2017**, *32*, 22–32. [CrossRef] [PubMed]
18. Kuniyoshi, K.M.; Rehan, V.K. The Impact of Perinatal Nicotine Exposure on Fetal Lung Development and Subsequent Respiratory Morbidity. *Birth Defects Res.* **2019**, *111*, 1270–1283. [CrossRef]
19. Orzabal, M.R.; Naik, V.D.; Lee, J.; Hillhouse, A.E.; Brashear, W.A.; Threadgill, D.W.; Ramadoss, J. Impact of E-Cig Aerosol Vaping on Fetal and Neonatal Respiratory Development and Function. *Transl. Res.* **2022**, *246*, 102–114. [CrossRef]
20. McEvoy, C.T.; Spindel, E.R. Pulmonary Effects of Maternal Smoking on the Fetus and Child: Effects on Lung Development, Respiratory Morbidities, and Life Long Lung Health. *Paediatr. Respir. Rev.* **2017**, *21*, 27–33. [CrossRef]
21. Wills, T.A.; Soneji, S.S.; Choi, K.; Jaspers, I.; Tam, E.K. E-Cigarette Use and Respiratory Disorders: An Integrative Review of Converging Evidence from Epidemiological and Laboratory Studies. *Eur. Respir. J.* **2021**, *57*, 1901815. [CrossRef]
22. Calogero, C.; Sly, P.D. Developmental Physiology: Lung Function during Growth and Development from Birth to Old Age. In *Paediatric Lung Function*; European Respiratory Society: Lausanne, Switzerland, 2010. [CrossRef]
23. Rehman, S.; Bacha, D. *Embryology, Pulmonary*; StatPearls Publishing LLC.: Tampa, FL, USA, 2025.
24. Schittny, J.C. Development of the Lung. *Cell Tissue Res.* **2017**, *367*, 427–444. [CrossRef] [PubMed]
25. Koos, B.J.; Rajaee, A. *Fetal Breathing Movements and Changes at Birth*; Springer: New York, NY, USA, 2014; pp. 89–101. [CrossRef]
26. Pan, J.; Copland, I.; Post, M.; Yeger, H.; Cutz, E. Mechanical Stretch-Induced Serotonin Release from Pulmonary Neuroendocrine Cells: Implications for Lung Development. *Am. J. Physiol.-Lung Cell Mol. Physiol.* **2006**, *290*, L185–L193. [CrossRef]
27. Dunnill, M.S. Postnatal Growth of the Lung. *Thorax* **1962**, *17*, 329–333. [CrossRef]
28. Thurlbeck, W.M. Postnatal Human Lung Growth. *Thorax* **1982**, *37*, 564–571. [CrossRef] [PubMed]
29. Lovric, G.; Barré, S.F.; Schittny, J.C.; Roth-Kleiner, M.; Stampanoni, M.; Mokso, R. Dose Optimization Approach to Fast X-Ray Microtomography of the Lung Alveoli. *J. Appl. Crystallogr.* **2013**, *46*, 856–860. [CrossRef] [PubMed]
30. Vasilescu, D.M.; Gao, Z.; Saha, P.K.; Yin, L.; Wang, G.; Haefeli-Bleuer, B.; Ochs, M.; Weibel, E.R.; Hoffman, E.A. Assessment of Morphometry of Pulmonary Acini in Mouse Lungs by Nondestructive Imaging Using Multiscale Microcomputed Tomography. *Proc. Natl. Acad. Sci. USA* **2012**, *109*, 17105–17110. [CrossRef] [PubMed]
31. Yablonskiy, D.A.; Sukstanskii, A.L.; Woods, J.C.; Gierada, D.S.; Quirk, J.D.; Hogg, J.C.; Cooper, J.D.; Conradi, M.S. Quantification of Lung Microstructure with Hyperpolarized ^{3}He Diffusion MRI. *J. Appl. Physiol.* **2009**, *107*, 1258–1265. [CrossRef]
32. Hislop, A.; Muir, D.C.F.; Jacobsen, M.; Simon, G.; Reid, L. Postnatal Growth and Function of the Pre-Acinar Airways. *Thorax* **1972**, *27*, 265–274. [CrossRef] [PubMed]

33. Selevan, S.; Kimmel, C.; Mendola, P. Windows of Susceptibility to Environmental Exposures in Children: (Excerpt from Children's Health and the Environment: A Global Perspective). In *Everyday Environmental Toxins*; Apple Academic Press: Cambridge, MA, USA, 2015; pp. 23–34. [CrossRef]
34. Kovesi, T.; Rubin, S. Long-Term Complications of Congenital Esophageal Atresia and/or Tracheoesophageal Fistula. *Chest* **2004**, *126*, 915–925. [CrossRef] [PubMed]
35. Kool, H.; Mous, D.; Tibboel, D.; de Klein, A.; Rottier, R.J. Pulmonary Vascular Development Goes Awry in Congenital Lung Abnormalities. *Birth Defects Res. Part C Embryo Today* **2014**, *102*, 343–358. [CrossRef] [PubMed]
36. Phalen, R.F.; Oldham, M.J.; Beaucage, C.B.; Crocker, T.T.; Mortensen, J. Postnatal Enlargement of Human Tracheobronchial Airways and Implications for Particle Deposition. *Anat. Rec.* **1985**, *212*, 368–380. [CrossRef]
37. Wen, X.; Liu, L.; Moe, A.A.; Ormond, I.K.; Shuren, C.C.; Scott, I.N.; Ozga, J.E.; Stanton, C.A.; Ruybal, A.L.; Hart, J.L.; et al. Use of E-Cigarettes and Cigarettes During Late Pregnancy Among Adolescents. *JAMA Netw. Open* **2023**, *6*, e2347407. [CrossRef]
38. Mescolo, F.; Ferrante, G.; La Grutta, S. Effects of E-Cigarette Exposure on Prenatal Life and Childhood Respiratory Health: A Review of Current Evidence. *Front. Pediatr.* **2021**, *9*, 711573. [CrossRef]
39. Prochaska, J.J.; Vogel, E.A.; Benowitz, N. Nicotine Delivery and Cigarette Equivalents from Vaping a JUULpod. *Tob. Control* **2022**, *31*, e88–e93. [CrossRef] [PubMed]
40. Luck, W.; Nau, H.; Hansen, R.; Steldinger, R. Extent of Nicotine and Cotinine Transfer to the Human Fetus, Placenta and Amniotic Fluid of Smoking Mothers. *Dev. Pharmacol. Ther.* **1985**, *8*, 384–395. [CrossRef]
41. Napierala, M.; Mazela, J.; Merritt, T.A.; Florek, E. Tobacco Smoking and Breastfeeding: Effect on the Lactation Process, Breast Milk Composition and Infant Development. A Critical Review. *Environ. Res.* **2016**, *151*, 321–338. [CrossRef]
42. Rowe, H.; Baker, T.; Hale, T.W. Maternal Medication, Drug Use, and Breastfeeding. *Child. Adolesc. Psychiatr. Clin. N. Am.* **2015**, *24*, 1–20. [CrossRef] [PubMed]
43. Suter, M.A.; Mastrobattista, J.; Sachs, M.; Aagaard, K. Is There Evidence for Potential Harm of Electronic Cigarette Use in Pregnancy? *Birth Defects Res. A Clin. Mol. Teratol.* **2015**, *103*, 186–195. [CrossRef] [PubMed]
44. Rao, P.; Liu, J.; Springer, M.L. JUUL and Combusted Cigarettes Comparably Impair Endothelial Function. *Tob. Regul. Sci.* **2020**, *6*, 30–37. [CrossRef]
45. St. Helen, G.; Havel, C.; Dempsey, D.A.; Jacob, P.; Benowitz, N.L. Nicotine Delivery, Retention and Pharmacokinetics from Various Electronic Cigarettes. *Addiction* **2016**, *111*, 535–544. [CrossRef] [PubMed]
46. Ozekin, Y.H.; Saal, M.L.; Pineda, R.H.; Moehn, K.; Ordonez-Erives, M.A.; Delgado Figueroa, M.F.; Frazier, C.; Korth, K.M.; Königshoff, M.; Bates, E.A.; et al. Intrauterine Exposure to Nicotine through Maternal Vaping Disrupts Embryonic Lung and Skeletal Development via the Kcnj2 Potassium Channel. *Dev. Biol.* **2023**, *501*, 111–123. [CrossRef] [PubMed]
47. Noël, A.; Hansen, S.; Zaman, A.; Perveen, Z.; Pinkston, R.; Hossain, E.; Xiao, R.; Penn, A. In Utero Exposures to Electronic-Cigarette Aerosols Impair the *Wnt* Signaling during Mouse Lung Development. *Am. J. Physiol.-Lung Cell. Mol. Physiol.* **2020**, *318*, L705–L722. [CrossRef] [PubMed]
48. Beers, M.F.; Morrisey, E.E. The Three R's of Lung Health and Disease: Repair, Remodeling, and Regeneration. *J. Clin. Investig.* **2011**, *121*, 2065–2073. [CrossRef]
49. Tilley, A.E.; Harvey, B.-G.; Heguy, A.; Hackett, N.R.; Wang, R.; O'Connor, T.P.; Crystal, R.G. Down-Regulation of the Notch Pathway in Human Airway Epithelium in Association with Smoking and Chronic Obstructive Pulmonary Disease. *Am. J. Respir. Crit. Care Med.* **2009**, *179*, 457–466. [CrossRef] [PubMed]
50. Hussain, M.; Xu, C.; Lu, M.; Wu, X.; Tang, L.; Wu, X. Wnt/β-Catenin Signaling Links Embryonic Lung Development and Asthmatic Airway Remodeling. *Biochim. Biophys. Acta (BBA)—Mol. Basis Dis.* **2017**, *1863*, 3226–3242. [CrossRef]
51. McGrath-Morrow, S.A.; Hayashi, M.; Aherrera, A.; Lopez, A.; Malinina, A.; Collaco, J.M.; Neptune, E.; Klein, J.D.; Winickoff, J.P.; Breysse, P.; et al. The Effects of Electronic Cigarette Emissions on Systemic Cotinine Levels, Weight and Postnatal Lung Growth in Neonatal Mice. *PLoS ONE* **2015**, *10*, e0118344. [CrossRef]
52. Gambadauro, A.; Galletta, F.; Li Pomi, A.; Manti, S.; Piedimonte, G. Immune Response to Respiratory Viral Infections. *Int. J. Mol. Sci.* **2024**, *25*, 6178. [CrossRef] [PubMed]
53. Palazzolo, D.L.; Nelson, J.M.; Ely, E.A.; Crow, A.P.; Distin, J.; Kunigelis, S.C. The Effects of Electronic Cigarette (ECIG)-Generated Aerosol and Conventional Cigarette Smoke on the Mucociliary Transport Velocity (MTV) Using the Bullfrog (*R. catesbiana*) Palate Paradigm. *Front. Physiol.* **2017**, *8*, 1023. [CrossRef] [PubMed]
54. Laube, B.L.; Afshar Mohajer, N.; Koehler, K.; Chen, G.; Lazarus, P.; Collaco, J.M.; McGrath-Morrow, S.A. Acute and Chronic in Vivo Effects of Exposure to Nicotine and Propylene Glycol from an E-Cigarette on Mucociliary Clearance in a Murine Model. *Inhal. Toxicol.* **2017**, *29*, 197–205. [CrossRef]

55. Chung, S.; Baumlin, N.; Dennis, J.S.; Moore, R.; Salathe, S.F.; Whitney, P.L.; Sabater, J.; Abraham, W.M.; Kim, M.D.; Salathe, M. Electronic Cigarette Vapor with Nicotine Causes Airway Mucociliary Dysfunction Preferentially via TRPA1 Receptors. *Am. J. Respir. Crit. Care Med.* **2019**, *200*, 1134–1145. [CrossRef]
56. Manti, S.; Gambadauro, A.; Galletta, F.; Ruggeri, P.; Piedimonte, G. Update on the Role of B2AR and TRPV1 in Respiratory Diseases. *Int. J. Mol. Sci.* **2024**, *25*, 10234. [CrossRef] [PubMed]
57. Ogunwale, M.A.; Li, M.; Ramakrishnam Raju, M.V.; Chen, Y.; Nantz, M.H.; Conklin, D.J.; Fu, X.-A. Aldehyde Detection in Electronic Cigarette Aerosols. *ACS Omega* **2017**, *2*, 1207–1214. [CrossRef] [PubMed]
58. Sinharoy, P.; McAllister, S.L.; Vasu, M.; Gross, E.R. *Environmental Aldehyde Sources and the Health Implications of Exposure*; Springer: Singapore, 2019; pp. 35–52. [CrossRef]
59. Kim, M.D.; Chung, S.; Dennis, J.S.; Yoshida, M.; Aguiar, C.; Aller, S.P.; Mendes, E.S.; Schmid, A.; Sabater, J.; Baumlin, N.; et al. Vegetable Glycerin E-Cigarette Aerosols Cause Airway Inflammation and Ion Channel Dysfunction. *Front. Pharmacol.* **2022**, *13*, 1012723. [CrossRef] [PubMed]
60. Kim, M.D.; Chung, S.; Baumlin, N.; Qian, J.; Montgomery, R.N.; Sabater, J.; Berkland, C.; Salathe, M. The Combination of Propylene Glycol and Vegetable Glycerin E-Cigarette Aerosols Induces Airway Inflammation and Mucus Hyperconcentration. *Sci. Rep.* **2024**, *14*, 1942. [CrossRef]
61. Aslaner, D.M.; Alghothani, O.; Saldana, T.A.; Ezell, K.G.; Yallourakis, M.D.; MacKenzie, D.M.; Miller, R.A.; Wold, L.E.; Gorr, M.W. E-Cigarette Vapor Exposure in Utero Causes Long-Term Pulmonary Effects in Offspring. *Am. J. Physiol.-Lung Cell. Mol. Physiol.* **2022**, *323*, L676–L682. [CrossRef] [PubMed]
62. Wang, Q.; Sundar, I.K.; Blum, J.L.; Ratner, J.R.; Lucas, J.H.; Chuang, T.-D.; Wang, Y.; Liu, J.; Rehan, V.K.; Zelikoff, J.T.; et al. Prenatal Exposure to Electronic-Cigarette Aerosols Leads to Sex-Dependent Pulmonary Extracellular-Matrix Remodeling and Myogenesis in Offspring Mice. *Am. J. Respir. Cell Mol. Biol.* **2020**, *63*, 794–805. [CrossRef]
63. Lucas, J.H.; Muthumalage, T.; Wang, Q.; Friedman, M.R.; Friedman, A.E.; Rahman, I. E-Liquid Containing a Mixture of Coconut, Vanilla, and Cookie Flavors Causes Cellular Senescence and Dysregulated Repair in Pulmonary Fibroblasts: Implications on Premature Aging. *Front. Physiol.* **2020**, *11*, 924. [CrossRef]
64. Hua, M.; Omaiye, E.E.; Luo, W.; McWhirter, K.J.; Pankow, J.F.; Talbot, P. Identification of Cytotoxic Flavor Chemicals in Top-Selling Electronic Cigarette Refill Fluids. *Sci. Rep.* **2019**, *9*, 2782. [CrossRef]
65. Behar, R.Z.; Davis, B.; Wang, Y.; Bahl, V.; Lin, S.; Talbot, P. Identification of Toxicants in Cinnamon-Flavored Electronic Cigarette Refill Fluids. *Toxicol. Vitr.* **2014**, *28*, 198–208. [CrossRef]
66. Ma, T.; Wang, X.; Li, L.; Sun, B.; Zhu, Y.; Xia, T. Electronic Cigarette Aerosols Induce Oxidative Stress-Dependent Cell Death and NF-KB Mediated Acute Lung Inflammation in Mice. *Arch. Toxicol.* **2021**, *95*, 195–205. [CrossRef]
67. Hecker, L. Mechanisms and Consequences of Oxidative Stress in Lung Disease: Therapeutic Implications for an Aging Populace. *Am. J. Physiol.-Lung Cell Mol. Physiol.* **2018**, *314*, L642–L653. [CrossRef] [PubMed]
68. Emma, R.; Caruso, M.; Campagna, D.; Pulvirenti, R.; Li Volti, G. The Impact of Tobacco Cigarettes, Vaping Products and Tobacco Heating Products on Oxidative Stress. *Antioxidants* **2022**, *11*, 1829. [CrossRef]
69. Sussan, T.E.; Gajghate, S.; Thimmulappa, R.K.; Ma, J.; Kim, J.H.; Sudini, K.; Consolini, N.; Cormier, S.A.; Lomnicki, S.; Hasan, F.; et al. Exposure to Electronic Cigarettes Impairs Pulmonary Anti-Bacterial and Anti-Viral Defenses in a Mouse Model. *PLoS ONE* **2015**, *10*, e0116861. [CrossRef]
70. Chen, H.; Li, G.; Chan, Y.L.; Chapman, D.G.; Sukjamnong, S.; Nguyen, T.; Annissa, T.; McGrath, K.C.; Sharma, P.; Oliver, B.G. Maternal E-Cigarette Exposure in Mice Alters DNA Methylation and Lung Cytokine Expression in Offspring. *Am. J. Respir. Cell Mol. Biol.* **2018**, *58*, 366–377. [CrossRef] [PubMed]
71. Lerner, C.A.; Sundar, I.K.; Yao, H.; Gerloff, J.; Ossip, D.J.; McIntosh, S.; Robinson, R.; Rahman, I. Vapors Produced by Electronic Cigarettes and E-Juices with Flavorings Induce Toxicity, Oxidative Stress, and Inflammatory Response in Lung Epithelial Cells and in Mouse Lung. *PLoS ONE* **2015**, *10*, e0116732. [CrossRef]
72. Abusara, O.H.; Hammad, A.M.; Debas, R.; Al-Shalabi, E.; Waleed, M.; Scott Hall, F. The Inflammation and Oxidative Status of Rat Lung Tissue Following Smoke/Vapor Exposure via E-Cigarette, Cigarette, and Waterpipe. *Gene* **2025**, *935*, 149066. [CrossRef]
73. El-Mahdy, M.A.; Mahgoup, E.M.; Ewees, M.G.; Eid, M.S.; Abdelghany, T.M.; Zweier, J.L. Long-Term Electronic Cigarette Exposure Induces Cardiovascular Dysfunction Similar to Tobacco Cigarettes: Role of Nicotine and Exposure Duration. *Am. J. Physiol.-Heart Circ. Physiol.* **2021**, *320*, H2112–H2129. [CrossRef] [PubMed]
74. Moheimani, R.S.; Bhetraratana, M.; Yin, F.; Peters, K.M.; Gornbein, J.; Araujo, J.A.; Middlekauff, H.R. Increased Cardiac Sympathetic Activity and Oxidative Stress in Habitual Electronic Cigarette Users. *JAMA Cardiol.* **2017**, *2*, 278. [CrossRef] [PubMed]

75. El-Mahdy, M.A.; Ewees, M.G.; Eid, M.S.; Mahgoup, E.M.; Khaleel, S.A.; Zweier, J.L. Electronic Cigarette Exposure Causes Vascular Endothelial Dysfunction Due to NADPH Oxidase Activation and ENOS Uncoupling. *Am. J. Physiol.-Heart Circ. Physiol.* **2022**, *322*, H549–H567. [CrossRef]
76. Neuman, Å.; Hohmann, C.; Orsini, N.; Pershagen, G.; Eller, E.; Kjaer, H.F.; Gehring, U.; Granell, R.; Henderson, J.; Heinrich, J.; et al. Maternal Smoking in Pregnancy and Asthma in Preschool Children. *Am. J. Respir. Crit. Care Med.* **2012**, *186*, 1037–1043. [CrossRef] [PubMed]
77. Burke, H.; Leonardi-Bee, J.; Hashim, A.; Pine-Abata, H.; Chen, Y.; Cook, D.G.; Britton, J.R.; McKeever, T.M. Prenatal and Passive Smoke Exposure and Incidence of Asthma and Wheeze: Systematic Review and Meta-Analysis. *Pediatrics* **2012**, *129*, 735–744. [CrossRef] [PubMed]
78. Zaitsu, M.; Kono, K.; Hosokawa, Y.; Miyamoto, M.; Nanishi, K.; Okawa, S.; Niki, S.; Takahashi, K.; Yoshihara, S.; Kobashi, G.; et al. Maternal Heated Tobacco Product Use during Pregnancy and Allergy in Offspring. *Allergy* **2023**, *78*, 1104–1112. [CrossRef]
79. Martinez, F.D.; Wright, A.L.; Taussig, L.M.; Holberg, C.J.; Halonen, M.; Morgan, W.J. Asthma and Wheezing in the First Six Years of Life. *N. Engl. J. Med.* **1995**, *332*, 133–138. [CrossRef] [PubMed]
80. Lodrup Carlsen, K.; Jaakkola, J.; Nafstad, P.; Carlsen, K. In Utero Exposure to Cigarette Smoking Influences Lung Function at Birth. *Eur. Respir. J.* **1997**, *10*, 1774–1779. [CrossRef]
81. Gotts, J.E.; Jordt, S.-E.; McConnell, R.; Tarran, R. What Are the Respiratory Effects of E-Cigarettes? *BMJ* **2019**, *366*, l5275. [CrossRef] [PubMed]
82. Costantino, S.; Torre, A.; Foti Randazzese, S.; Mollica, S.A.; Motta, F.; Busceti, D.; Ferrante, F.; Caminiti, L.; Crisafulli, G.; Manti, S. Association between Second-Hand Exposure to E-Cigarettes at Home and Exacerbations in Children with Asthma. *Children* **2024**, *11*, 356. [CrossRef] [PubMed]
83. Wills, T.A.; Pagano, I.; Williams, R.J.; Tam, E.K. E-Cigarette Use and Respiratory Disorder in an Adult Sample. *Drug Alcohol. Depend.* **2019**, *194*, 363–370. [CrossRef]
84. Bailey, C.; Medeiros, P.d.B.; Ellwood, D.A.; Middleton, P.; Andrews, C.J.; Flenady, V.J. A Systematic Review of Interventions to Increase the Use of Smoking Cessation Services for Women Who Smoke during Pregnancy. *Aust. N. Z. J. Obstet. Gynaecol.* **2023**, *63*, 737–745. [CrossRef] [PubMed]
85. Diamanti, A.; Papadakis, S.; Schoretsaniti, S.; Rovina, N.; Vivilaki, V.; Gratziou, C.; Katsaounou, P. Smoking Cessation in Pregnancy: An Update for Maternity Care Practitioners. *Tob. Induc. Dis.* **2019**, *17*, 57. [CrossRef] [PubMed]
86. Bérard, A.; Zhao, J.-P.; Sheehy, O. Success of Smoking Cessation Interventions during Pregnancy. *Am. J. Obs. Gynecol.* **2016**, *215*, e1–e611. [CrossRef] [PubMed]
87. Madureira, J.; Camelo, A.; Silva, A.I.; Reis, A.T.; Esteves, F.; Ribeiro, A.I.; Teixeira, J.P.; Costa, C. The Importance of Socioeconomic Position in Smoking, Cessation and Environmental Tobacco Smoke Exposure during Pregnancy. *Sci. Rep.* **2020**, *10*, 15584. [CrossRef] [PubMed]
88. Chamberlain, C.; O'Mara-Eves, A.; Porter, J.; Coleman, T.; Perlen, S.M.; Thomas, J.; McKenzie, J.E. Psychosocial Interventions for Supporting Women to Stop Smoking in Pregnancy. *Cochrane Database Syst. Rev.* **2017**, *2020*, CD001055. [CrossRef]
89. Ussher, M.; Lewis, S.; Aveyard, P.; Manyonda, I.; West, R.; Lewis, B.; Marcus, B.; Riaz, M.; Taylor, A.; Daley, A.; et al. Physical Activity for Smoking Cessation in Pregnancy: Randomised Controlled Trial. *BMJ* **2015**, *350*, h2145. [CrossRef]
90. Georgakopoulou, V.E.; Diamanti, A. Artificial Intelligence for Smoking Cessation in Pregnancy. *Cureus* **2024**, *16*, e63732. [CrossRef] [PubMed]
91. McMeekin, N.; Sinclair, L.; Robinson-Smith, L.; Mitchell, A.; Bauld, L.; Tappin, D.M.; Boyd, K.A. Financial Incentives for Quitting Smoking in Pregnancy: Are They Cost-effective? *Addiction* **2023**, *118*, 1445–1456. [CrossRef] [PubMed]
92. Boyd, K.A.; Briggs, A.H.; Bauld, L.; Sinclair, L.; Tappin, D. Are Financial Incentives Cost-Effective to Support Smoking Cessation during Pregnancy? *Addiction* **2016**, *111*, 360–370. [CrossRef] [PubMed]
93. Tappin, D.; Bauld, L.; Purves, D.; Boyd, K.; Sinclair, L.; MacAskill, S.; McKell, J.; Friel, B.; McConnachie, A.; de Caestecker, L.; et al. Financial Incentives for Smoking Cessation in Pregnancy: Randomised Controlled Trial. *BMJ* **2015**, *350*, h134. [CrossRef]
94. Scherman, A.; Tolosa, J.E.; McEvoy, C. Smoking Cessation in Pregnancy: A Continuing Challenge in the United States. *Ther. Adv. Drug Saf.* **2018**, *9*, 457–474. [CrossRef]
95. Vila-Farinas, A.; Pérez-Rios, M.; Montes-Martinez, A.; Ruano-Ravina, A.; Forray, A.; Rey-Brandariz, J.; Candal-Pedreira, C.; Fernández, E.; Casal-Acción, B.; Varela-Lema, L. Effectiveness of Smoking Cessation Interventions among Pregnant Women: An Updated Systematic Review and Meta-Analysis. *Addict. Behav.* **2024**, *148*, 107854. [CrossRef] [PubMed]
96. Gould, G.S.; Twyman, L.; Stevenson, L.; Gribbin, G.R.; Bonevski, B.; Palazzi, K.; Bar Zeev, Y. What Components of Smoking Cessation Care during Pregnancy Are Implemented by Health Providers? A Systematic Review and Meta-Analysis. *BMJ Open* **2019**, *9*, e026037. [CrossRef]

97. Blanc, J.; Tosello, B.; Ekblad, M.O.; Berlin, I.; Netter, A. Nicotine Replacement Therapy during Pregnancy and Child Health Outcomes: A Systematic Review. *Int. J. Environ. Res. Public Health* **2021**, *18*, 4004. [CrossRef] [PubMed]
98. Hartmann-Boyce, J.; Chepkin, S.C.; Ye, W.; Bullen, C.; Lancaster, T. Nicotine Replacement Therapy versus Control for Smoking Cessation. *Cochrane Database Syst. Rev.* **2018**, *2019*, CD000146. [CrossRef] [PubMed]
99. Claire, R.; Chamberlain, C.; Davey, M.-A.; Cooper, S.E.; Berlin, I.; Leonardi-Bee, J.; Coleman, T. Pharmacological Interventions for Promoting Smoking Cessation during Pregnancy. *Cochrane Database Syst. Rev.* **2020**, *2020*, CD010078. [CrossRef] [PubMed]
100. Robijn, A.L.; Tran, D.T.; Cohen, J.M.; Donald, S.; Cesta, C.E.; Furu, K.; Parkin, L.; Pearson, S.-A.; Reutfors, J.; Zoega, H.; et al. Smoking Cessation Pharmacotherapy Use in Pregnancy. *JAMA Netw. Open* **2024**, *7*, e2419245. [CrossRef]
101. Tran, D.T.; Preen, D.B.; Einarsdottir, K.; Kemp-Casey, A.; Randall, D.; Jorm, L.R.; Choi, S.K.Y.; Havard, A. Use of Smoking Cessation Pharmacotherapies during Pregnancy Is Not Associated with Increased Risk of Adverse Pregnancy Outcomes: A Population-Based Cohort Study. *BMC Med.* **2020**, *18*, 15. [CrossRef]
102. Morales-Suárez-Varela, M.; Puig, B.M.; Kaerlev, L.; Peraita-Costa, I.; Perales-Marín, A. Safety of Nicotine Replacement Therapy during Pregnancy: A Narrative Review. *Int. J. Environ. Res. Public Health* **2022**, *20*, 250. [CrossRef] [PubMed]
103. Gambadauro, A.; Foti Randazzese, S.; Currò, A.; Galletta, F.; Crisafulli, G.; Caminiti, L.; Germanò, E.; Di Rosa, G.; Nicotera, A.G.; Manti, S. Impact of the Allergic Therapeutic Adherence in Children with Allergic Rhinitis and ADHD: A Pilot Study. *J. Pers. Med.* **2023**, *13*, 1346. [CrossRef] [PubMed]
104. Barboza, J. Pharmaceutical Strategies for Smoking Cessation during Pregnancy. *Expert. Opin. Pharmacother.* **2018**, *19*, 2033–2042. [CrossRef]
105. Adams, E.K.; Markowitz, S.; Kannan, V.; Dietz, P.M.; Tong, V.T.; Malarcher, A.M. Reducing Prenatal Smoking. *Am. J. Prev. Med.* **2012**, *43*, 34–40. [CrossRef] [PubMed]
106. McEvoy, C.T.; Schilling, D.; Clay, N.; Jackson, K.; Go, M.D.; Spitale, P.; Bunten, C.; Leiva, M.; Gonzales, D.; Hollister-Smith, J.; et al. Vitamin C Supplementation for Pregnant Smoking Women and Pulmonary Function in Their Newborn Infants. *JAMA* **2014**, *311*, 2074. [CrossRef]
107. Shorey-Kendrick, L.E.; McEvoy, C.T.; Ferguson, B.; Burchard, J.; Park, B.S.; Gao, L.; Vuylsteke, B.H.; Milner, K.F.; Morris, C.D.; Spindel, E.R. Vitamin C Prevents Offspring DNA Methylation Changes Associated with Maternal Smoking in Pregnancy. *Am. J. Respir. Crit. Care Med.* **2017**, *196*, 745–755. [CrossRef] [PubMed]
108. Shorey-Kendrick, L.E.; McEvoy, C.T.; O'Sullivan, S.M.; Milner, K.; Vuylsteke, B.; Tepper, R.S.; Morgan, T.K.; Roberts, V.H.J.; Lo, J.O.; Frias, A.E.; et al. Vitamin C Supplementation Improves Placental Function and Alters Placental Gene Expression in Smokers. *Sci. Rep.* **2024**, *14*, 25486. [CrossRef]
109. McEvoy, C.T.; Shorey-Kendrick, L.E.; Milner, K.; Schilling, D.; Tiller, C.; Vuylsteke, B.; Scherman, A.; Jackson, K.; Haas, D.M.; Harris, J.; et al. Oral Vitamin C (500 Mg/d) to Pregnant Smokers Improves Infant Airway Function at 3 Months (VCSIP). A Randomized Trial. *Am. J. Respir. Crit. Care Med.* **2019**, *199*, 1139–1147. [CrossRef] [PubMed]
110. Gallo, C.; Renzi, P.; Loizzo, S.; Loizzo, A.; Piacente, S.; Festa, M.; Caputo, M.; Tecce, M.F.; Capasso, A. Potential Therapeutic Effects of Vitamin E and C on Placental Oxidative Stress Induced by Nicotine: An In Vitro Evidence. *Open Biochem. J.* **2010**, *4*, 77–82. [CrossRef] [PubMed]
111. Wang, B.; Chan, Y.L.; Zhou, S.; Saad, S.; Chen, H.; Oliver, B.G. Offspring Sex Affects the Susceptibility to Maternal Smoking-Induced Lung Inflammation and the Effect of Maternal Antioxidant Supplementation in Mice. *J. Inflamm.* **2020**, *17*, 24. [CrossRef] [PubMed]

antioxidants

MDPI

Review

Apnea of Prematurity and Oxidative Stress: Potential Implications

Lauren Thompson, Joseph W. Werthammer and David Gozal *

Department of Pediatrics, Joan C. Edwards School of Medicine, Marshall University, Huntington, WV 25701, USA; herman8@marshall.edu (L.T.); werthammer@marshall.edu (J.W.W.)
* Correspondence: gozal@marshall.edu; Tel.: +1-304-691-1700

Abstract: Apnea of prematurity (AOP) occurs in 85% of neonates $\leq$34 weeks of gestational age. AOP is frequently associated with intermittent hypoxia (IH). This narrative review reports on the putative relationship of AOP with IH and the resulting oxidative stress (OS). Preterm infants are susceptible to OS due to an imbalance between oxidant and antioxidant systems with the excessive free radical load leading to serious morbidities that may include retinopathy of prematurity, bronchopulmonary dysplasia, and neurodevelopmental delay. Current therapeutic approaches to minimize the adverse effects of AOP and optimize oxygen delivery include noninvasive ventilation and xanthine inhibitor therapy, but these approaches have only been partially successful in decreasing the incidence of AOP and associated morbidities.

Keywords: apnea of prematurity; intermittent hypoxia; CPAP; oxidative stress; retinopathy of prematurity; bronchopulmonary dysplasia; necrotizing enterocolitis; cognitive development; supplemental oxygen

1. Introduction

Preterm birth, defined as delivery before 37 weeks gestation, is a global public health issue of significant magnitude. In 2020, an estimated 13.4 million neonates were born prematurely, accounting for more than 10% of births worldwide [1]. Advances in medical care have led to improved survival at increasingly earlier gestations, but this technological success has come with a rise in the incidence of chronic morbidities among the surviving high-risk neonates.

Common morbidities associated with premature birth include chronic lung disease, retinopathy of prematurity (ROP), sleep-disordered breathing, neurodevelopmental delay, growth failure, necrotizing enterocolitis (NEC), and cardiovascular dysfunction. Recent emerging evidence has pointed to oxidative stress (OS) as a major contributing factor underlying these morbidities, which occurs when there is an imbalance between oxidant and antioxidant systems [2]. Preterm infants are particularly susceptible to OS due to the underlying scarcity in the expression and activity of antioxidant enzymes and consequently immature antioxidant responses. In addition, preterm infants are also exposed to environmental factors that foster a pro-oxidant state such as the need for oxygen supplementation and parenteral nutrition [3]. Clinically, apnea of prematurity (AOP) and the resultant intermittent hypoxia (IH) that characterizes this condition will potentiate the OS that occurs in preterm neonates and if allowed to persist for extended periods, may promote the risk of multisystem complications.

In this review, we will discuss (1) the magnitude of the problem of AOP and IH, (2) the relationship between IH and OS, (3) the currently identified associations between OS and common morbidities seen in the preterm neonate, and (4) the relatively scarce therapeutic approaches to mitigate IH and OS [4].

2. Apnea of Prematurity—The Magnitude of the Problem

Apnea of prematurity (AOP) is one of the most common pathologies seen in premature infants affecting 85% of neonates born ≤34 weeks gestation [5]. The prevalence of AOP is inversely proportional to gestational age, with virtually all infants born ≤28 weeks being affected compared to 20% of the infants born at 34 weeks gestation [6]. AOP generally resolves spontaneously around 36–43 weeks of gestational age, although the age at which AOP resolution will occur is stringently dependent on the degree of prematurity at birth, such that very extreme premature infants may manifest AOP well into 50 weeks post-conceptional age [6–8]. Clinically significant AOP has been defined as breathing pauses that last >20 s or for >10 s if associated with bradycardia (e.g., <80 beats per minute) or oxyhemoglobin desaturation (e.g., O_2 saturation <80–85%) [9]. AOP, i.e., the cessation of airflow, has traditionally been classified as (a) central, characteristically manifesting as the absence of respiratory efforts, (b) obstructive, as illustrated by the presence of ongoing inspiratory efforts in the context of upper airway obstruction, and (c) mixed, whereby a combination of central apnea is either preceded or followed by upper airway obstruction. (Figure 1). Among the obstructive types of AOP, the site of obstruction in the upper airways is primarily in the pharynx, although it may also occur at the larynx and possibly at both sites [10]. All three subtypes of AOP are related to some degree of brainstem immaturity along with contributions and interactions with respiratory mechanics [11–17]. Mixed apnea events with loss of upper airway patency following a central pause are the most common type in premature infants [18,19].

The pathogenesis of AOP is multifactorial and not fully understood [20–22]. AOP may simply resemble the fetal breathing pattern, where breathing is irregular because its main function is to ensure lung growth rather than promote gas exchange [22]. Central nervous system mechanisms and peripheral reflex pathways have all been implicated in the pathophysiology of AOP [8,10,23] (Table 1).

Table 1. Potential risk factors implicated in the pathogenesis of apnea of prematurity.

Central Factors	Peripheral Factors
Reduced central chemosensitivity	Variable carotid body and peripheral chemosensitivity responses
Augmented hypoxic ventilatory depression	Laryngeal chemoreflex
Upregulated inhibitory neural networks	Hering–Breuer and diving reflexes
Immature vasomotor regulatory function	Laryngeal chemoreflex
Neural connectome immaturity	Autonomic dysmaturity facilitating bradycardia
Sensory integration deficits	Reduced diaphragmatic endurance
	Increased airway collapsibility
	Reduced functional residual capacity of lungs and diminished oxygen reserve

Reduced central chemoreceptor responsiveness and instability of the central respiratory pattern generator along with enhanced peripheral chemoreceptor function have all been proposed as contributors to the initiation of apnea events and also to its continuation until some intervention is implemented [10,21]. The dysmaturity of the brainstem respiratory pattern generator and altered ventilatory responses to hypercarbia and hypoxia are all likely to contribute to AOP [24–26]. Changes in CO_2 are sensed at many different central nervous system regions even though the initial discoveries pointed to three different sites at or near the ventral medullary surface of the brainstem; the current consensus suggests that there is a multiplicity of chemosensitive fields collectively contributing to the integrative cardiorespiratory responses along with the preservation of respiratory drive [27–35]. Preterm infants have long been recognized to exhibit a diminished response to CO_2 when compared to more mature infants [14]. Furthermore, the carotid bodies and their developmentally regulated maturational patterning are critically important to the sensing of changes in oxygen levels as well as serving as the major peripheral chemical driver of respiration at all ages [10,36,37]. Peripheral chemoreceptors undergo a very coordinated developmental trajectory initiated by the transition from the hypoxic fetal environment to the environmental air-oxygenated milieu, a phenomenon commonly referred to as "resetting" [38,39]. In settings such as preterm birth in which ventilatory support is required and supplemental oxygen is administered, the resetting phenomenon may be affected and such blunting of peripheral chemoreceptor responses may be long-lasting and involve prolongation of apneic events [40–42].

As mentioned above, peripheral chemoreceptors primarily in the carotid body are responsible for stimulating breathing in response to hypoxia. The classic hypoxic response in the preterm neonate is triphasic, consisting of an initial increase followed by a decrease in ventilation usually ascribed to centrally mediated inhibitory inputs [14,43]. However, preterm infants <1500 g show a virtually absent stimulatory response and instead manifest an immediate reduction in minute ventilation during hypoxia, which is mainly secondary to a decrease in respiratory rate related to central nervous system inhibition [44].

Apnea of prematurity and the resulting intermittent hypoxia (IH) it causes are exceedingly common in the preterm neonate and have been recognized as such with the advent of technology capable of continuously measuring oxygen saturation levels. Historically, invasive arterial blood gas sampling was the gold standard for measuring oxygen levels in the NICU. However, these sporadic estimates cannot identify the frequent fluctuations in oxygen levels that occur in preterm infants. In contrast, transcutaneous and pulse oximetry monitoring can provide continuous monitoring of an infant's oxygen levels. Transcutaneous monitoring requires periodic calibration and involves heating the membrane positioned on the skin of the infant at the probe site, leading pulse oximetry to become adopted as the standard practice for cardiorespiratory monitoring in the neonatal intensive care unit (NICU). However, pulse oximetry is not without flaws and has limited value in identifying periods of hyperoxemia with the use of supplemental oxygen. Pulse oximeters also have decreased accuracy at lower levels of SpO2. Furthermore, the averaging time used by the embedded algorithm in the instrument can affect accurate characterization of apneic and hypoxic events. While a short averaging time produces the most accurate waveform, a longer averaging time is used in many NICUs to reduce the high frequency of alarms. This can, for example, portray a hypoxic event as an extended event, when in fact it consists of multiple short, self-resolving events occurring in a cluster. This is concerning for two reasons. First, a longer event is more likely to receive supplemental oxygen that may be unnecessary. Second, there is increasing evidence that clustered IH events versus dispersed events are likely associated with different clinical outcomes. Given these challenges, it is difficult to accurately measure and record the true burden of these various types of events and develop informed predictions as to their consequences [3].

Martin et al. [45] tackled this challenge by following a cohort of 79 preterm infants (24–28 weeks gestation) over the first 8 weeks of postnatal life using continuous high-resolution pulse oximetry data (2 s sampling rate and 2 s averaging time). They identified desaturation events $\leq$80% from between 10 s and 3 min and calculated the number of events for each infant for each week of postnatal age. Their results displayed the presence of an abrupt increase in the density of events after the first week of life followed by a progressive increase in weeks 2–4 reaching ~600 events/week and finally a decrease by weeks 6 to 8 of postnatal life. The authors hypothesized that this trend is due to the decline in central chemosensitivity during the first week of postnatal life. They further concluded that the high incidence of IH episodes in this population (50–100 per day) would likely be underestimated using conventional pulse oximetry settings [45]. Thus, IH is an extremely frequent condition in premature infants and can be the consequence of initially delayed maturation and resetting of the peripheral chemoreceptors followed by rebound hypersensitivity of these receptors [34,36].

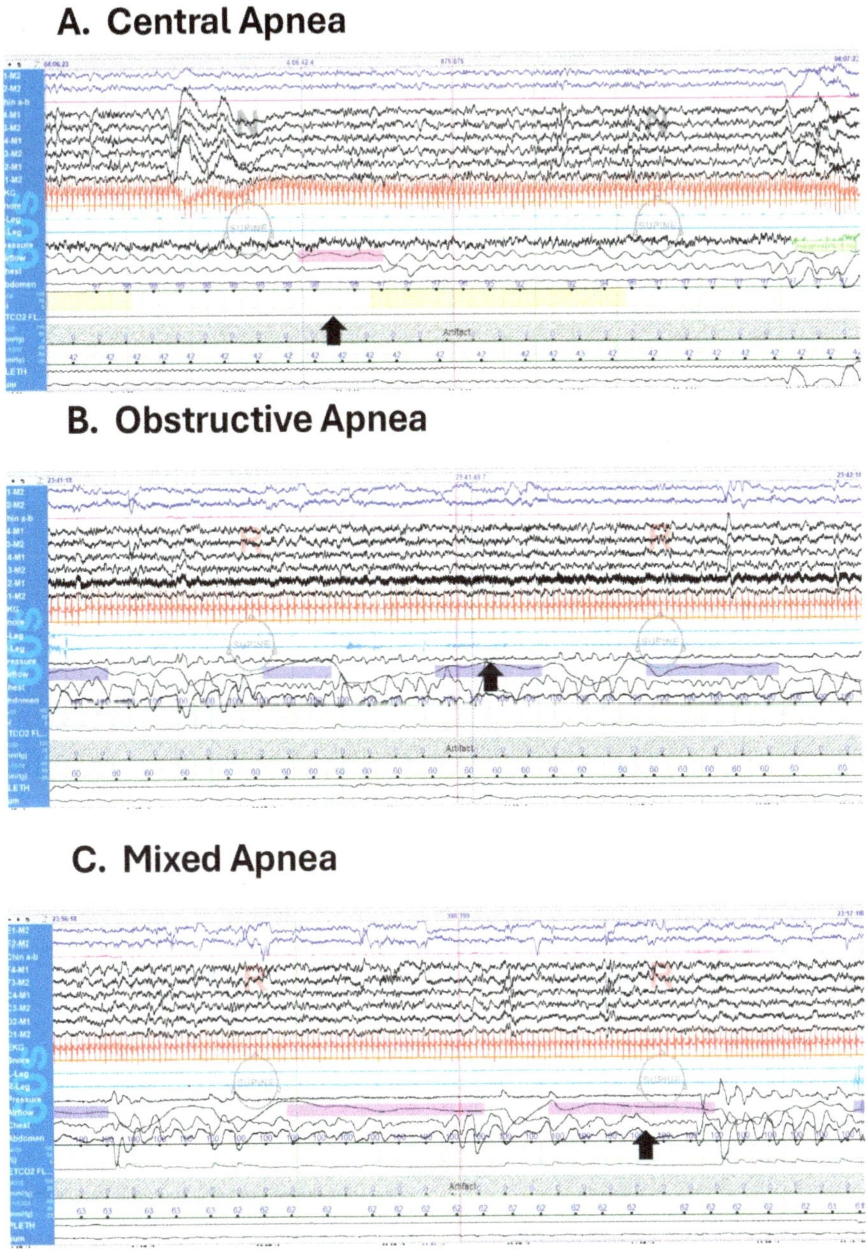

Figure 1. Examples representing 3 types of apneas of prematurity events. Arrows indicate apneas, and colored highlights in pink, purple, and magenta indicate respective events. (**A**)—central apnea: 10 s central apnea with absence of both airflow by nasal thermistor and chest wall movement by transthoracic impedance followed by drop in SpO2 to 92%. (**B**)—obstructive apnea: 9 s obstructive apnea with absent airflow with cardiac deceleration with continued chest wall movement. (**C**)—mixed apnea: 20 s mixed apnea with initial obstruction with absent airflow with chest wall movement followed by central apnea with cardiac deceleration.

3. Oxidative Stress Cellular Mechanisms

Oxidative stress is usually considered as the presence of "a serious imbalance between the generation of reactive oxygen species (ROS) and antioxidant defenses in favor of ROS, causing excessive oxidative damage to biomolecules and cellular organelles". In a biological cellular context, ROS are generated as a natural by-product of aerobic metabolic pathways. Accordingly, mitochondrial respiration is a substantial source of ROS [46]. In addition to mitochondria, ROS are synthesized by a variety of enzymes such as the NADPH oxidase family, xanthine oxidase, nitric oxide synthase, and peroxisomal constituents [47]. In addition, ROS are generated during exposure to ionizing and UV radiations, as well as by the ingestion and activity of a wide range of drugs and xenobiotics. In the endoplasmic reticulum, oxidative moieties are released during the folding of proteins and the formation of disulphide bonds. ROS are characteristically very reactive chemical molecules particularly in their ability to aggregate electrons to the oxygen molecule [48]. In this context, unstable molecules such as superoxide anion ($\bullet O_2^-$), hydrogen peroxide (H_2O_2), hydroxyl radical (OH^-), and singlet oxygen ($^1O_2^-$) are produced. We should emphasize that ROS are intrinsic to the homeostatic functioning of cells, fulfilling critically important functions [49]. However, when ROS exceed antioxidant capacity, oxidative stress (OS) damage occurs [50].

4. Relationship Between IH and OS

As discussed above, oxidative stress (OS) occurs because of the homeostatic imbalance between ROS production and the efficacy of antioxidant systems in canceling OS, thereby leading to an increased and excessive level of free radicals with subsequent damage to cells and organs [2,51]. A free radical is any atom or molecule that contains one or more unpaired electrons that includes reactive oxygen species (ROS), reactive nitrogen species (RNS), and sulfur-centered radicals collectively named oxidants [2,52]. These molecules are highly unstable, highly reactive, and known to cause direct damage to a variety of cellular structures. Under normal physiological circumstances, only a small percentage of oxygen is incompletely reduced to form ROS, and most ROS are neutralized by endogenous or exogenous antioxidants. At low physiological levels, free radicals can be protective and, for example, play an important role in our immune system. However, increased levels of free radicals are produced in response to stressful metabolic situations and can cause oxidative and nitrosative stresses that directly damage tissues [3]. Infections, hypoxia–reoxygenation, ischemia, inflammation, hyperoxia, and radiation are stressors commonly known to induce oxidative stress. In addition to the direct cellular damage caused by ROS and RNS, they have also been shown to trigger pro-inflammatory and pro-apoptotic pathways that accentuate and propagate the damage [3,53,54].

There are several reasons why newborn infants, especially those born preterm, are at risk of oxidative stress. First, relative oxidative stress is generally high within the first weeks of birth after transitioning from the low-oxygen fetal environment of the uterus (PaO2 20–26 mm Hg) to the rich extrauterine oxygen environment (PaO2 100 mm Hg) [55,56]. Second, preterm infants have insufficient antioxidant defenses either de novo or passively acquired transplacentally until the third trimester, which increases their susceptibility to oxidative stress [56,57]. Third, premature infants have a high energy demand for growth and very high metabolic rates, which in the process of generating ATP (adenosine triphosphate) will lead to oxidative product formation in the mitochondria. Lastly, they have a high concentration of non-protein-bound iron (NPBI) that is available to react with H_2O_2 through the Fenton reaction and produce hydroxyl radicals [2].

Intermittent hypoxia (IH) is one of the most important consequences of AOP and the culprit for oxidative stress [58]. Indeed, it is now well established that a balanced oxygen homeostasis is an essential requirement to maintain the regulation of normal cellular systems. Any alterations in the oxygen levels are accompanied by cellular stress responses which may modify the expression of a large number of regulatory genes and proteins. If hypoxia occurs, i.e., critical oxygen delivery to preserve cellular bioenergetics, this will result in oxidative stress and the formation of hypoxia-inducible factors (HIFs) and reactive oxygen species (ROS) [59]. The cumulative evidence indicating that IH leads to an overall pro-oxidant state is derived from animal models that illustrate that IH increases extracellular superoxide concentration, activates NADPH oxidases, induces hypoxia-inducible factor 1 (HIF-1α), and downregulates both HIF-2α and superoxide dismutase [60–62]. Furthermore, it is known that during hypoxia, ATP, the major cellular source of energy, is depleted. This leads to the formation and accumulation of purine derivatives and hypoxanthine. Upon reoxygenation, oxidases are activated, especially the NADPH oxidase family and xanthine oxidases. These oxidases metabolize purine derivatives and generate a burst of superoxide anion, causing significant oxidative stress and subsequent tissue damage [3]. Beyond the direct action of free radicals on cellular components, oxidative stress has also been shown to cause direct alterations of DNA through the interruption of classical gene expression and DNA regulation of DNA repair, in addition to epigenetic modifications [2]. However, the impact of cycle duration, the severity of oxygen deprivation for each cycle, pattern of re-oxygenation, and overall cumulative duration of the intermittent events will need to be explored in much greater detail to deduce a more specific in-depth understanding of how these elements individually and collectively contribute to OS.

The extent of oxidative stress is difficult to quantify in preterm infants because free radicals are highly reactive and unstable, making them difficult to measure directly. Fortunately, the levels of by-products from the oxidation of lipids, proteins, and DNA can be quantified and have shown promising clinical value [63]. In addition, other biomarkers have been identified that help measure the risk of OS. Those currently applicable to neonatal medicine are summarized in Table 1.

5. Clinical Implications of AOP and OS

5.1. Retinopathy of Prematurity (ROP)

The largest body of evidence implicating OS in neonatal disease revolves around retinopathy of prematurity (ROP). ROP is a vasoproliferative retinal disorder affecting premature infants and is a leading cause of visual impairment and blindness worldwide. Despite robust screening protocols and treatment advancements, at least 50,000 children worldwide suffer from blindness annually due to this condition [64]. Gestational age and birth weight are the two risk factors most highly correlated with ROP, both being inversely proportional to ROP incidence. Hyperoxia has long been accepted as the primary contributor to the pathogenesis of ROP through increased ROS and RNS and the subsequent vascular damage and excessive angiogenic effect they cause. More recently, studies in both animal models and premature infants have demonstrated that intermittent hypoxia is also a major contributing factor by further increasing oxidative stress in the preterm retina [65].

Upon birth, the neonate must transition from a relatively hypoxic intrauterine environment to an oxygen-rich extrauterine environment. This transition has important consequences for the preterm retina, whose development is not yet complete. In utero, retinal angiogenesis begins around the 15th week of gestation, originating from the optic nerve and expanding peripherally. The nasal portion of the retina does not become vascularized until 36 weeks of gestation and the temporal area not until 40 weeks gestation [64]. Thus, the preterm retina must develop in a maladaptive hyperoxic environment for which it was not programmed. Further occurrence of hypoxic–hyperoxic events, as is the common corollary of AOP, would further challenge the normal maturation of the retina and promote oxidative stress damage translating into the emergence of ROP as demonstrated by several experiments in animal models and confirmed by clinical cohort studies [3,66–69].

The development of ROP is frequently described as a two-phase occurrence. The first phase is characterized by cessation of normal retinal vascular development due to the hyperoxic extrauterine environment. This occurs largely due to suppression of hypoxia-inducible factor 1α (HIF1-α), which causes downregulation of the vascular endothelial growth factor (VEGF). Oxidative damage to the existing vasculature has been implicated during this stage. The second phase is comprised of abnormal vascular proliferation stimulated in part by the hypoxia-induced production of VEGF and other growth factors [45]. It is most likely during this second stage that IH contributes to ROP by precipitating retinal neovascularization.

Historically, knowledge surrounding hyperoxia and its deleterious effects on the developing retina was the main focus of ROP prevention. Significant debate exists regarding the ideal target oxygen saturation required to mitigate this disease without worsening other outcomes. Studies have found that preterm infants with a lower target oxyhemoglobin saturation (85–89%) compared to a higher target SpO2 (91–95%) have a lower incidence of ROP but incur higher mortality [70]. Thus, other improved options for prevention are needed.

Recent evidence of the impact of IH on developing ROP provides a potential target for prevention. As mentioned, IH creates a pro-oxidant environment through increases in superoxide concentrations and HIF-1α expression, as well as degradation of HIF-2α and downregulation of superoxide dismutase [71]. In preterm infants, Di Fiori et al. demonstrated an association between chronic IH and severe ROP requiring laser therapy [64]. Similarly, a post hoc analysis of the Canadian Oxygen Trial showed that infants experiencing greater frequency or severity of IH events incurred a higher risk for developing ROP [72]. Furthermore, research regarding the timing and patterns of IH have demonstrated that these may be critical in predicting ROP risk. For example, rats exposed to cycles of IH in a clustered manner (three episodes 10 min apart) vs. equally dispersed cycles (2 h apart) will develop more severe retinopathy [73]. These observations provide valuable insights to guide further research regarding preventative strategies and risk stratification for preterm infants at risk for ROP.

5.2. Neurodevelopmental Impairment (NDI)

The brain is one of the most sensitive organs to oxidative damage due to the large amount of polyunsaturated fatty acids, the high metabolic rate and bioenergetic dependence on utilization of oxygen, and the relative scarcity of antioxidant defense mechanisms [74]. Mitochondria and NADPH oxidase are the predominant sources of ROS in the central nervous system and have been shown to play a prominent role in IH-induced neuronal injury [74].

The preterm brain is at particularly high risk given the critical period of rapid brain development that occurs postnatally. It is known that oligodendrocyte maturation, myelination, axon development, and synapse formation occur mainly during the third trimester of gestation and in the first postnatal year and thus oxidative stress occurring during this time likely contributes to the white matter impairments frequently seen in preterm neonates [75].

Increased frequency of IH has been shown to be associated with neurodevelopmental impairment, such as language, cognitive, and motor delays. This has been demonstrated in many single-center studies, as well as in the large, multicenter Canadian Oxygen Trial that examined a cohort of >1000 infants [71]. However, data from rodent studies suggest that this relationship is more than just an association. Several studies have shown that in neonatal rodents, IH exposures evoke disruption in sleep architecture and cause working memory impairments and hyperactivity behaviors, remarkably resembling attention deficit hyperactivity disorder (ADHD), a condition frequently manifested among children who were born prematurely [76–79]. These behaviors were accompanied by a reduction in extracellular levels of dopamine, increased cerebral expression of caspase-3 reflecting increased apoptosis, and marked increases in oxidative stress, neuronal cell losses, and glial proliferation in the CA1 hippocampal region and cortex that could be mitigated by transgenic ablation of NADPH oxidase and inducible nitric oxide synthase, two well-established enzymes that underlie the major generation of ROS [76,80,81]. Subsequent studies led by Gozal and collaborators led to the development of a neonatal mouse model of IH simulating AOP and demonstrated that IH during postnatal days 2–10 induced hypomyelination in the corpus collosum, striatum, fornix, and cerebellum, as well as alterations in myelin-forming processes [76]. A later study by Darnall et al. also reported decreases in white matter integrity in rat pups exposed to IH similar to that experienced by a premature neonate with AOP [82]. Furthermore, preterm infants have recently demonstrated a significant decrease in EEG (electroencephalogram) amplitude during breathing pauses from 5 to >15 s. It was speculated that frequent disruption to brain activity may impact long-term neurodevelopment [83]. An excellent summary of extant studies on the association between IH and nervous system morbidity further confirms the putative adverse effects of IH in prematurely born infants [84].

There are several biomarkers linking oxidative stress to CNS (central nervous system) dysfunction and to white matter injury as seen in preterm neonates [85]. Additionally, non-protein-bound iron (NPBI) has been described as a reliable early predictive biomarker of neurodevelopmental outcomes [86]. Increased levels of isoprostanes (IsoPs) have been reported in preterm infants who developed white matter injury seen at corrected term gestational age. Increased levels of IsoPs have also been found to be associated with poor neurodevelopmental trajectories at 12 months of age [87]. Lastly, augmented adenosine (Ado) plasma level concentration at day 15 in premature neonates has also been found to be significantly associated with white matter injury using MRI. Ado represents a promoter of oligodendrocyte maturation and has previously been reported to be elevated in response to OS in preterm neonates [2].

5.3. Respiratory Disorders

Premature neonates, particularly those born less than 28 weeks, have significant short- and long-term respiratory pathology including respiratory distress syndrome (RDS) and increased risk for developing bronchopulmonary dysplasia (BPD). Infants born prematurely have a limited surface area for gas exchange as well as a limited surfactant which is necessary to keep the lungs from collapsing during exhalation. Thus, premature neonates frequently require positive pressure respiratory support, both invasive and noninvasive, and supplemental oxygen. These therapies can increase the risk for future respiratory pathology, making it challenging to distinguish the impact of IH on such pathologies. However, evidence exists that IH plays a role in both short- and long-term respiratory morbidity. Furthermore, significant lung development and alveolarization occurs postnatally for these infants albeit with curtailed pulmonary vascular angiogenesis [88,89], making the lungs a prime target for oxidative stress during the early postnatal period, similarly to other rapidly developing organ systems [90–93].

Several studies have demonstrated a relationship between the frequency and duration of IH in very-low-birthweight infants and BPD. Animal models have indicated that IH alone is unlikely to cause the parenchymal changes seen in BPD, but that it likely causes a more severe phenotype [94–96]. Wheezing and asthma-like symptoms are common sequelae for preterm neonates, affecting over 30% [3]. Di Fiori et al. found that early exposure to IH during the first week of life was associated with reported asthma medication use at 2 years of age [97].

Biomarkers of OS have been found in infants with these diseases, further providing evidence on the putative role of IH and subsequent OS in their pathology [98,99]. In preterm infants with RDS, elevated plasma levels of MDAs (methyl-D-aspartate), protein carbonyls, AOPPs, 8-OHdG, and H_2O_2 and an elevated oxidant/antioxidant ratio (calculated as protein carbonyls/(superoxide dismutase + glutathione peroxidase)) were found compared to healthy preterm controls [2]. Another study reported that elevated AOPPs (advanced oxidation protein products) and 8-OHdg on days 0–3 correlated with severity of RDS. For preterm infants who developed BPD, higher urinary levels of 8-OHdg were reported compared to those who did not go on to develop BPD. A higher concentration of 8-OHdg has also been reported in serum and tracheal aspiration samples on the day after birth and at day 28 of life in those infants who developed BPD compared to those who did not [2].

In addition to BPD, prematurity has also been implicated as a risk factor for sleep-disordered breathing later in life [100]. The mechanisms underlying this association are unclear, but it is very possible that bone growth patterns may be affected by IH and lead to maladaptive development of the oropharyngeal space and of the craniofacial structures [101–109].

5.4. Necrotizing Enterocolitis (NEC)

NEC is a gastrointestinal disorder that primarily occurs in preterm neonates and has a complex pathophysiology culminating in an inflammatory response that can lead to irreversible intestinal injury [110,111]. While the pathophysiology of NEC is incompletely understood, it has been suggested that there is an association between biomarkers of intrauterine OS events and NEC [112]. As indicated, OS is difficult to measure in vivo, such that correlational biomarkers are used to evaluate host susceptibility to OS by measuring oxidant-induced changes in proteins, lipids, and DNA [113]. Cord blood concentration and intestinal levels of biomarkers of OS have been demonstrated in infants with NEC, establishing an association between perinatal OS and NEC [112–114]. Preliminary evidence would suggest a role for postnatal IH as a contributing risk factor in NEC pathophysiology [115,116].

6. Therapeutic Approaches

Management strategies for AOP with IH events focus primarily on prevention. Desaturation events are a consequence of immature respiratory control superimposed upon an immature respiratory system [71]. There is no consensus on when to initiate therapy for AOP. Generally, treatment is indicated when episodes are recurrent and are associated with bradycardia or desaturation [58]. Two mainstays of management are optimizing oxygenation and enhancing respiratory control (Figure 2).

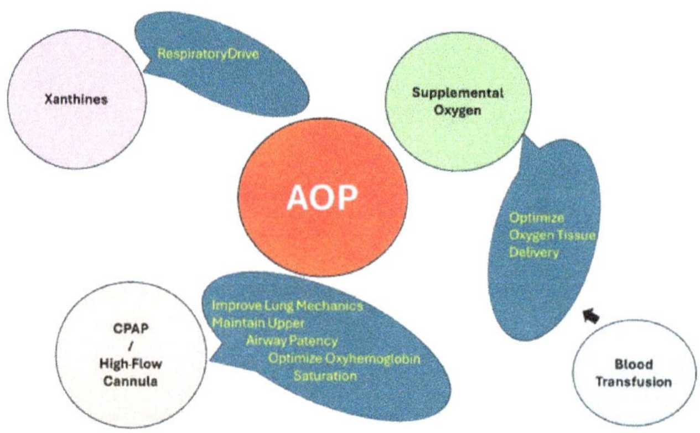

Figure 2. Treatment strategies for apnea of prematurity include a multi-prong approach including xanthines, supplemental oxygen, blood transfusion, and noninvasive ventilation.

6.1. Optimizing Oxygenation

The optimal target oxygen saturation for the preterm infant has not been definitively defined [58,117,118]. Higher oxygen levels may increase survival but increase morbidities, including ROP and BPD. Conversely, lower oxygen levels might increase the likelihood of NEC and death but reduce the risk of problems caused by oxygen toxicity [118]. The benefits of keeping baseline oxygenation in preterm infants at levels of SpO2 90–95% outweigh the risks when restricting the upper limits of SpO2 to <89% [119,120]. Low baseline SpO2 increases the risk of intermittent hypoxemia in the preterm neonate [121], further aggravating the risk of augmenting the maladaptive effects and oxidative stress associated with such occurrence.

The challenges of maintaining SpO2 in a target range with manual titration of inspired oxygen have been described [122]. New devices with automated control of oxygen titration have been described to reduce the incidence of prolonged periods of hypoxemia and hyperoxemia [123]. A recent randomized crossover study, however, demonstrated that automated control with apnea-triggered FiO2 boost resulted in a reduction in post-apnea hypoxemia, but was followed by a greater period of SpO2 overshoot [124].

6.2. Methylxanthine Therapy

Methylxanthines have been the mainstay of treatment for AOP for over 40 years. They act both centrally and peripherally to stimulate respiration and reduce the frequency of apneic episodes. They primarily work through antagonism of adenosine receptors, blocking both excitatory A2A receptors on GABAergic neurons and inhibitory A1 receptors [125–127]. They further improve respiratory dynamics by activating medullary respiratory centers, increasing CO_2 sensitivity, inducing bronchodilation, and enhancing diaphragmatic contraction. Caffeine administration can also modulate inflammation by inhibiting cyclic nucleotide phosphodiesterase and increasing the concentration of cAMP [125,128]. Early studies identified the effectiveness of aminophylline and theophylline in treating AOP, but caffeine has become the preferred treatment given its fewer side effects, wider therapeutic index, and longer half-life that allows for once-daily dosing [129,130].

Despite the widespread use of caffeine, there is variation in clinical practice regarding the optimal timing of initiation, dose, and duration of treatment. In the Caffeine for Apnea of Prematurity (CAP) Trial, the benefits of caffeine were the greatest when therapy was initiated within 3 days after birth [131]. These benefits include reductions in the incidence of BPD, decreased severity of ROP, and improved long-term neurodevelopment [132]. Early use of caffeine has become frequent in the treatment of AOP in the extremely preterm infant and may help dampen this trajectory of increasing frequency of AOP after the first week of life and perhaps limit the magnitude of oxidative stress associated with it [133].

There is also clinical variability in the duration of caffeine treatment. Many clinicians look at corrected gestational age and timing of clinically documented episodes when deciding to discontinue treatment. Therapy is often discontinued around 34–35 weeks post-menstrual age (PMA) at which time the frequency of apneic events typically declines. However, it is well known that nursing documentation of events grossly underestimates the number of cardiorespiratory events [3]. Furthermore, despite a decrease in clinically apparent events after 34 weeks PMA, it has been shown that neonates born <28 weeks still have a significant amount of IH until 43–44 weeks corrected gestational age (CGA). In a multicenter, prospective, randomized trial, Rhein et al. demonstrated that continued caffeine therapy decreased the amount of IH at 35 and 36 weeks CGA by almost 50% [130]. A subsequent study by the same group demonstrated similar findings at 36–38 weeks using a higher dose of caffeine [134]. Another study was conducted to compare the incidence of recurrent AOP after discontinuation of caffeine in a short-course therapy group and a long-course group. There were fewer episodes of hypoxia in the long-course group but an increase in adverse effects of the medication [135]. Because of the immaturity of the metabolic capacity of the hepatic enzyme system during the first weeks of life, most caffeine is excreted unchanged by the kidneys [136]. This observation indicates that newborn infants have a deficiency in their caffeine metabolizing capacity until several months of age when liver enzymes increase to adult levels in their ability to produce demethylated metabolites and caffeine clearance. As a result, the half-life of caffeine decreases from 120 h to 60 h within the first 8 postnatal weeks [137]. A recent review summarized the positive effects when caffeine is administered to the hypoxic rodent, but also the negative effects on the rodent model when caffeine is administered without exogenous oxidative stress. Thus, critical consideration needs to be given to continuing caffeine treatment beyond the recommended corrected gestational age in the preterm infant [138]. Further studies are needed to determine the optimal dosage and ideal time for discontinuation of caffeine therapy.

6.3. Noninvasive Respiratory Support

Nasal continuous positive pressure (CPAP) has been a major therapeutic intervention to treat AOP. It works by splinting open the upper airway to improve patency and increase functional residual lung capacity [139]. More recently, heated humidified high-flow nasal cannula (HHFNC) therapy has been used interchangeably with CPAP [140]. Small studies have found that it delivers similar distending pressures with no differences in the rate of apneic events while providing a more comfortable interface for the neonate [59]. Nasal intermittent positive pressure ventilation (NIPPV) may also be effective in the treatment of AOP and has been compared to CPAP with mixed results [141,142]. However, larger studies are needed to delineate true differences in the effectiveness of these various noninvasive respiratory support strategies in reducing AOP and IH.

In addition to caffeine and noninvasive respiratory support, a variety of other pharmacologic and non-pharmacologic strategies have been proposed in the management of AOP, but because of concerns for side effects, these treatments are not widely employed [58].

6.4. Red Blood Cell Transfusions

Anemia may exacerbate apnea by reducing the oxygen-carrying capacity of blood, decreasing oxygen delivery to the brain [58]. A retrospective study utilizing waveform analysis demonstrated that a short-term lower hematocrit level was associated with an increased frequency of AOP [143]. Another study demonstrated a decrease in IH following packed red blood cell transfusion [144]. A possible side effect of red cell transfusion, however, includes the increased risk of NEC [145] and ROP [146,147]. Lust et al. hypothesized that the increase in adult hemoglobin following transfusion could result in a shift of the oxygen disassociation curve resulting in increased retinal tissue oxygen delivery at a time when it is susceptible to hyperoxia [147].

6.5. Doxapram

Doxapram, a central respiratory stimulant with a brief duration of action, has been used in the management of AOP especially in those neonates resistant to methylxanthine therapy, but because of adverse gastrointestinal and CNS symptoms, it is rarely used today [148]. A recent study demonstrated that in children ages 5–6 years born prior to 32 weeks gestation, doxapram treatment for AOP was not associated with neurodevelopmental disabilities [149]. There is currently a multicenter randomized controlled trial to examine the efficacy and safety of doxapram in the treatment of AOP in preterm infants [150].

6.6. Acid Suppression Medication

Ranitidine and other pharmacologic treatments to reduce gastric acids have been tried, yet the relationship between acid reflux and AOP has not been established [151]. There has been a described relationship between H2 receptor antagonists and necrotizing enterocolitis and late-onset sepsis [152].

6.7. Therapeutic Handling

Prone positioning can stabilize the chest wall and allow for improved ventilation of the lower lobes; however, there is no evidence that it reduces the frequency of AOP [153]. Tactile stimulation including skin-to-skin (kangaroo) care has also been shown to decrease AOP [154,155]. A multicenter neonatal manikin study was undertaken to evaluate tactile stimulation for AOP that demonstrated a large heterogeneity and recommended further prospective studies to provide guidelines for the most appropriate approach [156]. These approaches follow a long set of antecedent trials examining the impact of olfactory, vibratory, kinesthetic, and stochastic resonance methodologies aimed at reducing the probabilistic occurrence of apnea in neonates [157–160].

7. Conclusions

As our understanding of oxidative stress, its damage to cell membranes, and its role in epigenetic modifications evolves, so does our knowledge of its potential impact on all organ systems. This is of particular concern regarding the preterm neonate, who is at a high risk for oxidative damage given the imbalance of pro-oxidative stressors and immature antioxidant systems in combination with the rapid development that occurs postnatally. Apnea of prematurity is ubiquitous in those born <28 weeks gestation, and the intermittent hypoxia that ensues is a large contributor to the oxidative threat experienced by these high-risk neonates. Future studies examining the role of oxidative stress in the overall morbidity catalog of premature infants and the contribution of AOP and its surrogate effector IH are needed.

Author Contributions: All authors participated in the drafting and editing of the manuscript. All authors have read and agreed to the published version of the manuscript.

Funding: D.G. is supported in part by National Institutes of Health grants HL166617 and HL169266.

Acknowledgments: The authors gratefully acknowledge Rakesh Bhattacharjee from UCSD for providing examples of apneic event recordings.

Conflicts of Interest: The authors declare no conflicts of interest.

References

1. World Health Organization. Preterm Birth. Available online: https://www.who.int/news-room/fact-sheets/detail/preterm-birth (accessed on 16 September 2024).
2. Lembo, C.; Buonocore, G.; Perrone, S. Oxidative Stress in Preterm Newborns. *Antioxidants* **2021**, *10*, 1672. [CrossRef] [PubMed]
3. Di Fiore, J.M.; Vento, M. Intermittent hypoxemia and oxidative stress in preterm infants. *Respir. Physiol. Neurobiol.* **2019**, *266*, 121–129. [CrossRef] [PubMed]
4. Pergolizzi, J.V., Jr.; Fort, P.; Miller, T.L.; LeQuang, J.A.; Raffa, R.B. The limited management options for apnoea of prematurity. *J. Clin. Pharm. Ther.* **2022**, *47*, 396–401. [CrossRef] [PubMed]
5. Barrington, K.; Finer, N. The natural history of the appearance of apnea of prematurity. *Pediatr. Res.* **1991**, *29 Pt 1*, 372–375. [CrossRef]
6. Weese-Mayer, D.E.; Di Fiore, J.M.; Lake, D.E.; Hibbs, A.M.; Claure, N.; Qiu, J.; Ambalavanan, N.; Bancalari, E.; Kemp, J.S.; Zimmet, A.M.; et al. Maturation of cardioventilatory physiological trajectories in extremely preterm infants. *Pediatr. Res.* **2024**, *95*, 1060–1069. [CrossRef]
7. Lorch, S.A.; Srinivasan, L.; Escobar, G.J. Epidemiology of apnea and bradycardia resolution in premature infants. *Pediatrics* **2011**, *128*, e366–e373. [CrossRef]
8. Eichenwald, E.C. Committee on Fetus and Newborn, American Academy of Pediatrics. Apnea of Prematurity. *Pediatrics* **2016**, *137*, e20153757. [CrossRef]
9. Finer, N.N.; Higgins, R.; Kattwinkel, J.; Martin, R.J. Summary proceedings from the apnea-of-prematurity group. *Pediatrics* **2006**, *117 Pt 2*, S47–S51. [CrossRef]
10. Martin, R.J.; Wilson, C.G. Apnea of prematurity. *Compr. Physiol.* **2012**, *2*, 2923–2931. [CrossRef]
11. Rigatto, H.; Kwiatkowski, K.A.; Hasan, S.U.; Cates, D.B. The ventilatory response to endogenous CO_2 in preterm infants. *Am. Rev. Respir. Dis.* **1991**, *143*, 101–104. [CrossRef]
12. Lee, D.; Caces, R.; Kwiatkowski, K.; Cates, D.; Rigatto, H. A developmental study on types and frequency distribution of short apneas (3 to 15 seconds) in term and preterm infants. *Pediatr. Res.* **1987**, *22*, 344–349. [CrossRef] [PubMed]
13. Rigatto, H. Respiratory control and apnea in the newborn infant. *Crit. Care Med.* **1977**, *5*, 2–9. [CrossRef] [PubMed]
14. Rigatto, H.; Brady, J.P.; de la Torre Verduzco, R. Chemoreceptor reflexes in preterm infants: II. The effect of gestational and postnatal age on the ventilatory response to inhaled carbon dioxide. *Pediatrics* **1975**, *55*, 614–620. [CrossRef] [PubMed]
15. Rigatto, H.; Brady, J.P.; de la Torre Verduzco, R. Chemoreceptor reflexes in preterm infants: I. The effect of gestational and postnatal age on the ventilatory response to inhalation of 100% and 15% oxygen. *Pediatrics* **1975**, *55*, 604–613. [CrossRef] [PubMed]
16. Khan, A.; Qurashi, M.; Kwiatkowski, K.; Cates, D.; Rigatto, H. Measurement of the CO_2 apneic threshold in newborn infants: Possible relevance for periodic breathing and apnea. *J. Appl. Physiol.* **2005**, *98*, 1171–1176. [CrossRef] [PubMed]
17. Lemke, R.P.; Idiong, N.; Al-Saedi, S.; Kwiatkowski, K.; Cates, D.B.; Rigatto, H. Evidence of a critical period of airway instability during central apneas in preterm infants. *Am. J. Respir. Crit. Care Med.* **1998**, *157*, 470–474. [CrossRef]
18. Di Fiore, J.M.; Martin, R.J.; Gauda, E.B. Apnea of prematurity—Perfect storm. *Respir. Physiol. Neurobiol.* **2013**, *189*, 213–222. [CrossRef]
19. Finer, N.N.; Barrington, K.J.; Hayes, B.J.; Hugh, A. Obstructive, mixed, and central apnea in the neonate: Physiologic correlates. *J. Pediatr.* **1992**, *121*, 943–950. [CrossRef]

20. Poets, C.F.; Quante, M. Rethinking the pathophysiology of cardiorespiratory events in infants born preterm. *J. Pediatr.* **2023**, *262*, 113651. [CrossRef]
21. Al-Matary, A.; Kutbi, I.; Qurashi, M.; Khalil, M.; Alvaro, R.; Kwiatkowski, K.; Cates, D.; Rigatto, H. Increased peripheral chemoreceptor activity may be critical in destabilizing breathing in neonates. *Semin. Perinatol.* **2004**, *28*, 264–272. [CrossRef]
22. Poets, C.F. Apnea of prematurity: What can observational studies tell us about pathophysiology? *Sleep Med.* **2010**, *11*, 701–707. [CrossRef] [PubMed]
23. Di Fiore, J.M.; Poets, C.F.; Gauda, E.; Martin, R.J.; MacFarlane, P. Cardiorespiratory events in preterm infants: Interventions and consequences. *J. Perinatol.* **2016**, *36*, 251–258. [CrossRef] [PubMed]
24. Yackle, K.; Schwarz, L.A.; Kam, K.; Sorokin, J.M.; Huguenard, J.R.; Feldman, J.L.; Luo, L.; Krasnow, M.A. Breathing control center neurons that promote arousal in mice. *Science* **2017**, *355*, 1411–1415. [CrossRef] [PubMed]
25. Cui, Y.; Kam, K.; Sherman, D.; Janczewski, W.A.; Zheng, Y.; Feldman, J.L. Defining preBötzinger Complex Rhythm- and Pattern-Generating Neural Microcircuits In Vivo. *Neuron* **2016**, *91*, 602–614. [CrossRef] [PubMed]
26. Del Negro, C.A.; Kam, K.; Hayes, J.A.; Feldman, J.L. Asymmetric control of inspiratory and expiratory phases by excitability in the respiratory network of neonatal mice in vitro. *J. Physiol.* **2009**, *587 Pt 6*, 1217–1231. [CrossRef]
27. Gourine, A.V.; Dale, N. Brain H^+/CO_2 sensing and control by glial cells. *Glia* **2022**, *70*, 1520–1535. [CrossRef]
28. Dzal, Y.A.; Sprenger, R.J.; Milsom, W.K. Postnatal changes in O_2 and CO_2 sensitivity in rodents. *Respir. Physiol. Neurobiol.* **2020**, *272*, 103313. [CrossRef]
29. Huckstepp, R.T.; Dale, N. Redefining the components of central CO_2 chemosensitivity—Towards a better understanding of mechanism. *J. Physiol.* **2011**, *589 Pt 23*, 5561–5579. [CrossRef]
30. Chernov, M.M.; Erlichman, J.S.; Leiter, J.C. Ionic mechanisms of central CO_2 chemosensitivity. *Respir. Physiol. Neurobiol.* **2010**, *173*, 298–304. [CrossRef]
31. Guyenet, P.G.; Bayliss, D.A.; Stornetta, R.L.; Fortuna, M.G.; Abbott, S.B.; DePuy, S.D. Retrotrapezoid nucleus, respiratory chemosensitivity and breathing automaticity. *Respir. Physiol. Neurobiol.* **2009**, *168*, 59–68. [CrossRef]
32. Gozal, D.; Hathout, G.M.; Kirlew, K.A.; Tang, H.; Woo, M.S.; Zhang, J.; Lufkin, R.B.; Harper, R.M. Localization of putative neural respiratory regions in the human by functional magnetic resonance imaging. *J. Appl. Physiol.* **1994**, *76*, 2076–2083. [CrossRef] [PubMed]
33. Spyer, K.M.; Gourine, A.V. Chemosensory pathways in the brainstem controlling cardiorespiratory activity. *Philos. Trans. R. Soc. B Biol. Sci.* **2009**, *364*, 2603–2610. [CrossRef] [PubMed]
34. Corcoran, A.E.; Hodges, M.R.; Wu, Y.; Wang, W.; Wylie, C.J.; Deneris, E.S.; Richerson, G.B. Medullary serotonin neurons and central CO_2 chemoreception. *Respir. Physiol. Neurobiol.* **2009**, *168*, 49–58. [CrossRef] [PubMed]
35. Putnam, R.W.; Conrad, S.C.; Gdovin, M.J.; Erlichman, J.S.; Leiter, J.C. Neonatal maturation of the hypercapnic ventilatory response and central neural CO_2 chemosensitivity. *Respir. Physiol. Neurobiol.* **2005**, *149*, 165–179. [CrossRef] [PubMed]
36. Blain, G.M.; Smith, C.A.; Henderson, K.S.; Dempsey, J.A. Peripheral chemoreceptors determine the respiratory sensitivity of central chemoreceptors to CO_2. *J. Physiol.* **2010**, *588 Pt 13*, 2455–2471. [CrossRef]
37. Forster, H.V.; Smith, C.A. Contributions of central and peripheral chemoreceptors to the ventilatory response to CO_2/H^+. *J. Appl. Physiol.* **2010**, *108*, 989–994. [CrossRef]
38. Carroll, J.L.; Kim, I. Carotid chemoreceptor "resetting" revisited. *Respir. Physiol. Neurobiol.* **2013**, *185*, 30–43. [CrossRef]
39. Gauda, E.B.; Carroll, J.L.; Donnelly, D.F. Developmental maturation of chemosensitivity to hypoxia of peripheral arterial chemoreceptors—Invited article. *Adv. Exp. Med. Biol.* **2009**, *648 Pt 23*, 243–255. [CrossRef]
40. Kim, I.; Yang, D.; Carroll, J.L.; Donnelly, D.F. Perinatal hyperoxia exposure impairs hypoxia-induced depolarization in rat carotid body glomus cells. *Respir. Physiol. Neurobiol.* **2013**, *188*, 9–14. [CrossRef]
41. Mouradian, G.C., Jr.; Lakshminrusimha, S.; Konduri, G.G. Perinatal Hypoxemia and Oxygen Sensing. *Compr. Physiol.* **2021**, *11*, 1653–1677. [CrossRef]
42. Gauda, E.B.; McLemore, G.L.; Tolosa, J.; Marston-Nelson, J.; Kwak, D. Maturation of peripheral arterial chemoreceptors in relation to neonatal apnoea. *Semin. Neonatol.* **2004**, *9*, 181–194. [CrossRef] [PubMed]
43. Martin, R.J.; DiFiore, J.M.; Jana, L.; Davis, R.L.; Miller, M.J.; Coles, S.K.; Dick, T.E. Persistence of the biphasic ventilatory response to hypoxia in preterm infants. *J. Pediatr.* **1998**, *132*, 960–964. [CrossRef] [PubMed]
44. Alvaro, R.; Alvarez, J.; Kwiatkowski, K.; Cates, D.; Rigatto, H. Small preterm infants (less than or equal to 1500 g) have only a sustained decrease in ventilation in response to hypoxia. *Pediatr. Res.* **1992**, *32*, 403–406. [CrossRef] [PubMed]
45. Martin, R.J.; Wang, K.; Köroğlu, O.; Di Fiore, J.; Kc, P. Intermittent hypoxic episodes in preterm infants: Do they matter? *Neonatology* **2011**, *100*, 303–310. [CrossRef]
46. Mailloux, R.J. An Update on Mitochondrial Reactive Oxygen Species Production. *Antioxidants* **2020**, *9*, 472. [CrossRef]
47. Magnani, F.; Mattevi, A. Structure and mechanisms of ROS generation by NADPH oxidases. *Curr. Opin. Struct. Biol.* **2019**, *59*, 91–97. [CrossRef]
48. Halliwell, B.; Gutteridge, J.M.C. *Free Radicals in Biology and Medicine, 3rd ed*; Oxford Science Publications: Oxford, UK, 1999.
49. Sies, H.; Jones, D.P. Reactive oxygen species (ROS) as pleiotropic physiological signalling agents. *Nat. Rev. Mol. Cell Biol.* **2020**, *21*, 363–383. [CrossRef]
50. Hong, Y.; Boiti, A.; Vallone, D.; Foulkes, N.S. Reactive Oxygen Species Signaling and Oxidative Stress: Transcriptional Regulation and Evolution. *Antioxidants* **2024**, *13*, 312. [CrossRef]

51. Czerska, M.; Mikołajewska, K.; Zieliński, M.; Gromadzińska, J.; Wąsowicz, W. Today's oxidative stress markers. *Med. Pr.* **2015**, *66*, 393–405. [CrossRef]
52. Halliwell, B. Free radicals, antioxidants, and human disease: Curiosity, cause, or consequence? *Lancet* **1994**, *344*, 721–724. [CrossRef]
53. Rahal, A.; Kumar, A.; Singh, V.; Yadav, B.; Tiwari, R.; Chakraborty, S.; Dhama, K. Oxidative stress, prooxidants, and antioxidants: The interplay. *BioMed Res. Int.* **2014**, *2014*, 761264. [CrossRef] [PubMed]
54. Davies, K.J. Oxidative stress, antioxidant defenses, and damage removal, repair, and replacement systems. *IUBMB Life* **2000**, *50*, 279–289. [CrossRef] [PubMed]
55. Saugstad, O.D.; Oei, J.L.; Lakshminrusimha, S.; Vento, M. Oxygen therapy of the newborn from molecular understanding to clinical practice. *Pediatr. Res.* **2019**, *85*, 20–29. [CrossRef] [PubMed]
56. Frank, L.; Sosenko, I.R. Development of lung antioxidant enzyme system in late gestation: Possible implications for the prematurely born infant. *J. Pediatr.* **1987**, *110*, 9–14. [CrossRef]
57. Viña, J.; Vento, M.; García-Sala, F.; Puertes, I.R.; Gascó, E.; Sastre, J.; Asensi, M.; Pallardó, F.V. L-cysteine and glutathione metabolism are impaired in premature infants due to cystathionase deficiency. *Am. J. Clin. Nutr.* **1995**, *61*, 1067–1069. [CrossRef]
58. Erickson, G.; Dobson, N.R.; Hunt, C.E. Immature control of breathing and apnea of prematurity: The known and unknown. *J. Perinatol.* **2021**, *41*, 2111–2123. [CrossRef]
59. Trayhurn, P. Oxygen—A critical, but overlooked, nutrient. *Front. Nutr.* **2019**, *6*, 10. [CrossRef]
60. Fabian, R.H.; Perez-Polo, J.R.; Kent, T.A. Extracellular superoxide concentration increases following cerebral hypoxia but does not affect cerebral blood flow. *Int. J. Dev. Neurosci.* **2004**, *22*, 225–230. [CrossRef]
61. Yuan, G.; Nanduri, J.; Khan, S.; Semenza, G.L.; Prabhakar, N.R. Induction of HIF-1α expression by intermittent hypoxia: Involvement of NADPH oxidase, Ca^{2+} signaling, prolyl hydroxylases, and mTOR. *J. Cell. Physiol.* **2008**, *217*, 674–685. [CrossRef]
62. Nanduri, J.; Wang, N.; Yuan, G.; Khan, S.A.; Souvannakitti, D.; Peng, Y.J.; Kumar, G.K.; Garcia, J.A.; Prabhakar, N.R. Intermittent hypoxia degrades HIF-2α via calpains resulting in oxidative stress: Implications for recurrent apnea-induced morbidities. *Proc. Natl. Acad. Sci. USA* **2009**, *106*, 1199–1204. [CrossRef]
63. Peña-Bautista, C.; Durand, T.; Vigor, C.; Oger, C.; Galano, J.M.; Cháfer-Pericás, C. Non-invasive assessment of oxidative stress in preterm infants. *Free Radic. Biol. Med.* **2019**, *142*, 73–81. [CrossRef] [PubMed]
64. Zhang, L.; Buonfiglio, F.; Fieß, A.; Pfeiffer, N.; Gericke, A. Retinopathy of Prematurity—Targeting Hypoxic and Redox Signaling Pathway. *Antioxidants* **2024**, *13*, 148. [CrossRef] [PubMed]
65. Di Fiore, J.M.; Bloom, J.N.; Orge, F.; Schutt, A.; Schluchter, M.; Cheruvu, V.K., Walsh, M.; Finer, N.; Martin, R.J. A higher incidence of intermittent hypoxemic episodes is associated with severe retinopathy of prematurity. *J. Pediatr.* **2010**, *157*, 69–73. [CrossRef] [PubMed]
66. Beharry, K.D.; Cai, C.L.; Skelton, J.; Siddiqui, F.; D'Agrosa, C.; Calo, J.; Valencia, G.B.; Aranda, J.V. Oxygen-Induced Retinopathy from Recurrent Intermittent Hypoxia Is Not Dependent on Resolution with Room Air or Oxygen, in Neonatal Rats. *Int. J. Mol. Sci.* **2018**, *19*, 1337. [CrossRef]
67. Beharry, K.D.; Cai, C.L.; Ahmad, T.; Guzel, S.; Valencia, G.B.; Aranda, J.V. Impact of Chronic Neonatal Intermittent Hypoxia on Severity of Retinal Damage in a Rat Model of Oxygen-Induced Retinopathy. *J. Nat. Sci.* **2018**, *4*, e488.
68. Beharry, K.D.; Cai, C.L.; Valencia, G.B.; Valencia, A.M.; Lazzaro, D.R.; Bany-Mohammed, F.; Aranda, J.V. Neonatal Intermittent Hypoxia, Reactive Oxygen Species, and Oxygen-Induced Retinopathy. *React. Oxyg. Species* **2017**, *3*, 12–25. [CrossRef]
69. Winners-Mendizabal, O.G.; Orge, F.H.; Di Fiore, J.M.; Martin, R.J.; Kc, P. Hypoxia-hyperoxia paradigms in the development of oxygen-induced retinopathy in a rat pup model. *J. Neonatal-Perinat. Med.* **2014**, *7*, 113–117. [CrossRef]
70. Stone, W.L.; Shah, D.; Hollinger, S.M. Retinopathy of prematurity: An oxidative stress neonatal disease. *Front. Biosci.* **2016**, *21*, 165–177. [CrossRef]
71. Di Fiore, J.M.; MacFarlane, P.M.; Martin, R.J. Intermittent Hypoxemia in Preterm Infants. *Clin. Perinatol.* **2019**, *46*, 553–565. [CrossRef]
72. Poets, C.F.; Roberts, R.S.; Schmidt, B.; Whyte, R.K.; Asztalos, E.V.; Bader, D.; Bairam, A.; Moddemann, D.; Peliowski, A.; Rabi, Y.; et al. Association Between Intermittent Hypoxemia or Bradycardia and Late Death or Disability in Extremely Preterm Infants. *JAMA* **2015**, *314*, 595–603. [CrossRef]
73. Coleman, R.J.; Beharry, K.D.; Brock, R.S.; Abad-Santos, P.; Abad-Santos, M.; Modanlou, H.D. Effects of brief, clustered versus dispersed hypoxic episodes on systemic and ocular growth factors in a rat model of oxygen-induced retinopathy. *Pediatr. Res.* **2008**, *64*, 50–55. [CrossRef] [PubMed]
74. Wang, Y.; Zhang, S.X.; Gozal, D. Reactive oxygen species and the brain in sleep apnea. *Respir. Physiol. Neurobiol.* **2010**, *174*, 307–316. [CrossRef] [PubMed]
75. Cai, J.; Tuong, C.M.; Zhang, Y.; Shields, C.B.; Guo, G.; Fu, H.; Gozal, D. Mouse intermittent hypoxia mimicking apnoea of prematurity: Effects on myelinogenesis and axonal maturation. *J. Pathol.* **2012**, *226*, 495–508. [CrossRef] [PubMed]
76. Gozal, D.; Daniel, J.M.; Dohanich, G.P. Behavioral and anatomical correlates of chronic episodic hypoxia during sleep in the rat. *J. Neurosci.* **2001**, *21*, 2442–2450. [CrossRef] [PubMed]
77. Decker, M.J.; Hue, G.E.; Caudle, W.M.; Miller, G.W.; Keating, G.L.; Rye, D.B. Episodic neonatal hypoxia evokes executive dysfunction and regionally specific alterations in markers of dopamine signaling. *Neuroscience* **2003**, *117*, 417–425. [CrossRef]

78. Decker, M.J.; Jones, K.A.; Solomon, I.G.; Keating, G.L.; Rye, D.B. Reduced extracellular dopamine and increased responsiveness to novelty: Neurochemical and behavioral sequelae of intermittent hypoxia. *Sleep* **2005**, *28*, 169–176. [CrossRef]
79. Ratner, V.; Kishkurno, S.V.; Slinko, S.K.; Sosunov, S.A.; Sosunov, A.A.; Polin, R.A.; Ten, V.S. The contribution of intermittent hypoxemia to late neurological handicap in mice with hyperoxia-induced lung injury. *Neonatology* **2007**, *92*, 50–58. [CrossRef]
80. Li, R.C.; Row, B.W.; Kheirandish, L.; Brittian, K.R.; Gozal, E.; Guo, S.Z.; Sachleben, L.R., Jr.; Gozal, D. Nitric oxide synthase and intermittent hypoxia-induced spatial learning deficits in the rat. *Neurobiol. Dis.* **2004**, *17*, 44–53. [CrossRef]
81. Nair, D.; Dayyat, E.A.; Zhang, S.X.; Wang, Y.; Gozal, D. Intermittent hypoxia-induced cognitive deficits are mediated by NADPH oxidase activity in a murine model of sleep apnea. *PLoS ONE* **2011**, *6*, e19847. [CrossRef]
82. Darnall, R.A.; Chen, X.; Nemani, K.V.; Sirieix, C.M.; Gimi, B.; Knoblach, S.; McEntire, B.L.; Hunt, C.E. Early postnatal exposure to intermittent hypoxia in rodents is proinflammatory, impairs white matter integrity, and alters brain metabolism. *Pediatr. Res.* **2017**, *82*, 164–172. [CrossRef]
83. Zandvoort, C.S.; Dereymaeker, A.; Baxter, L.; Jansen, K.; Naulaers, G.; de Vos, M.; Hartley, C. Apnoea suppresses brain activity in infants. *Imaging Neurosci.* **2024**, *17*, 1–14. [CrossRef]
84. Di Fiore, J.M.; Raffay, T.M. The relationship between intermittent hypoxemia events and neural outcomes in neonates. *Exp. Neurol.* **2021**, *342*, 113753. [CrossRef] [PubMed]
85. Perrone, S.; Grassi, F.; Caporilli, C.; Boscarino, G.; Carbone, G.; Petrolini, C.; Gambini, L.M.; Di Peri, A.; Moretti, S.; Buonocore, G.; et al. Brain Damage in Preterm and Full-Term Neonates: Serum Biomarkers for the Early Diagnosis and Intervention. *Antioxidants* **2023**, *12*, 309. [CrossRef] [PubMed]
86. Buonocore, G.; Perrone, S.; Longini, M.; Paffetti, P.; Vezzosi, P.; Gatti, M.G.; Bracci, R. Non protein bound iron as early predictive marker of neonatal brain damage. *Brain* **2003**, *126*, 1224–1230. [CrossRef]
87. Matthews, M.A.; Aschner, J.L.; Stark, A.R.; Moore, P.E.; Slaughter, J.C.; Steele, S.; Beller, A.; Milne, G.L.; Settles, O.; Chorna, O.; et al. Increasing F2-isoprostanes in the first month after birth predicts poor respiratory and neurodevelopmental outcomes in very preterm infants. *J. Perinatol.* **2016**, *36*, 779–783. [CrossRef]
88. Thébaud, B.; Goss, K.N.; Laughon, M.; Whitsett, J.A.; Abman, S.H.; Steinhorn, R.H.; Aschner, J.L.; Davis, P.G.; McGrath-Morrow, S.A.; Soll, R.F.; et al. Bronchopulmonary dysplasia. *Nat. Rev. Dis. Primers* **2019**, *5*, 78. [CrossRef]
89. Baker, C.D.; Abman, S.H. Impaired pulmonary vascular development in bronchopulmonary dysplasia. *Neonatology* **2015**, *107*, 344–351. [CrossRef]
90. Teng, M.; Wu, T.J.; Jing, X.; Day, B.W.; Pritchard, K.A., Jr.; Naylor, S.; Teng, R.J. Temporal Dynamics of Oxidative Stress and Inflammation in Bronchopulmonary Dysplasia. *Int. J. Mol. Sci.* **2024**, *25*, 10145. [CrossRef]
91. Luo, L.; Li, Y.; Long, Z.; Jiang, F.; Wu, F.; Wang, Q. Exploring research trends and hotspots on oxidative stress and bronchopulmonary dysplasia: Insights from bibliometric and visualized study. *Pediatr. Pulmonol.* **2024**, *ahead of print*. [CrossRef]
92. Marcelino, M.; Cai, C.L.; Wadowski, S.; Aranda, J.V.; Beharry, K.D. Biomarkers of lung alveolarization and microvascular maturation in response to intermittent hypoxia and/or early antioxidant/fish oil supplementation in neonatal rats. *Pediatr. Pulmonol.* **2023**, *58*, 2352–2363. [CrossRef]
93. Kimble, A.; Robbins, M.E.; Perez, M. Pathogenesis of Bronchopulmonary Dysplasia: Role of Oxidative Stress from 'Omics' Studies. *Antioxidants* **2022**, *11*, 2380. [CrossRef] [PubMed]
94. Dylag, A.M.; Mayer, C.A.; Raffay, T.M.; Martin, R.J.; Jafri, A.; MacFarlane, P.M. Long-term effects of recurrent intermittent hypoxia and hyperoxia on respiratory system mechanics in neonatal mice. *Pediatr. Res.* **2017**, *81*, 565–571. [CrossRef] [PubMed]
95. Mankouski, A.; Kantores, C.; Wong, M.J.; Ivanovska, J.; Jain, A.; Benner, E.J.; Mason, S.N.; Tanswell, A.K.; Auten, R.L.; Jankov, R.P. Intermittent hypoxia during recovery from neonatal hyperoxic lung injury causes long-term impairment of alveolar development: A new rat model of BPD. *Am. J. Physiol. Lung Cell Mol. Physiol.* **2017**, *312*, L208–L216. [CrossRef] [PubMed]
96. Ratner, V.; Starkov, A.; Matsiukevich, D.; Polin, R.A.; Ten, V.S. Mitochondrial dysfunction contributes to alveolar developmental arrest in hyperoxia-exposed mice. *Am. J. Respir. Cell Mol. Biol.* **2009**, *40*, 511–518. [CrossRef] [PubMed]
97. Di Fiore, J.M.; Dylag, A.M.; Honomichl, R.D.; Hibbs, A.M.; Martin, R.J.; Tatsuoka, C.; Raffay, T.M. Early inspired oxygen and intermittent hypoxemic events in extremely premature infants are associated with asthma medication use at 2 years of age. *J. Perinatol.* **2019**, *39*, 203–211. [CrossRef]
98. Mandala, V.K.; Bollaboina, S.K.Y.; Changala, B.; Kotha, R., Sr.; Kasula, L. Intermittent Hypoxia in Preterm Neonates and Its Effect on Neonatal Morbidity and Mortality: A Systematic Review. *Cureus* **2023**, *15*, e45561. [CrossRef]
99. Alalaiyan, S.; Shakeeb, D.; Al Hazzani, F.; Binmanee, A. Significance of Intermittent Hypoxic Episodes in Premature Infants Prior to Discharge. *Cureus* **2023**, *15*, e36113. [CrossRef]
100. Rosen, C.L.; Larkin, E.K.; Kirchner, H.L.; Emancipator, J.L.; Bivins, S.F.; Surovec, S.A.; Martin, R.J.; Redline, S. Prevalence and risk factors for sleep-disordered breathing in 8- to 11-year-old children: Association with race and prematurity. *J. Pediatr.* **2003**, *142*, 383–389. [CrossRef]
101. Sadras, I.; Reiter, J.; Fuchs, N.; Erlichman, I.; Gozal, D.; Gileles Hillel, A. Prematurity as a Risk Factor of Sleep-Disordered Breathing in Children Younger than Two Years: A Retrospective Case-Control Study. *J. Clin. Sleep Med.* **2019**, *15*, 1731–1736. [CrossRef]
102. Zalzal, H.G.; Newman, D.; Behzadpour, H.; Nino, G.; Lawlor, C.M. Phenotypical Characterization of Obstructive Sleep Apnea in Premature Infants using Polysomnography. *Laryngoscope* **2024**, *134*, 1933–1938. [CrossRef]

103. Jaleel, Z.; Schaeffer, T.; Trinh, C.; Cohen, M.B.; Levi, J.R. Prematurity: A Prognostic Factor for Increased Severity of Pediatric Obstructive Sleep Apnea. *Laryngoscope* **2021**, *131*, 1909–1914. [CrossRef] [PubMed]
104. Mehta, B.; Waters, K.; Fitzgerald, D.; Badawi, N. Sleep disordered breathing (SDB) in neonates and implications for its long-term impact. *Paediatr. Respir. Rev.* **2020**, *34*, 3–8. [CrossRef] [PubMed]
105. Huang, Y.S.; Hsu, J.F.; Paiva, T.; Chin, W.C.; Chen, I.C.; Guilleminault, C. Sleep-disordered breathing, craniofacial development, and neurodevelopment in premature infants: A 2-year follow-up study. *Sleep Med.* **2019**, *60*, 20–25. [CrossRef] [PubMed]
106. Ortiz, L.E.; McGrath-Morrow, S.A.; Sterni, L.M.; Collaco, J.M. Sleep disordered breathing in bronchopulmonary dysplasia. *Pediatr. Pulmonol.* **2017**, *52*, 1583–1591. [CrossRef] [PubMed]
107. Tapia, I.E.; Shults, J.; Doyle, L.W.; Nixon, G.M.; Cielo, C.M.; Traylor, J.; Marcus, C.L.; Caffeine for Apnea of Prematurity—Sleep Study Group. Perinatal Risk Factors Associated with the Obstructive Sleep Apnea Syndrome in School-Aged Children Born Preterm. *Sleep* **2016**, *39*, 737–742. [CrossRef]
108. Raynes-Greenow, C.H.; Hadfield, R.M.; Cistulli, P.A.; Bowen, J.; Allen, H.; Roberts, C.L. Sleep apnea in early childhood associated with preterm birth but not small for gestational age: A population-based record linkage study. *Sleep* **2012**, *35*, 1475–1480. [CrossRef]
109. Sharma, P.B.; Baroody, F.; Gozal, D.; Lester, L.A. Obstructive sleep apnea in the formerly preterm infant: An overlooked diagnosis. *Front. Neurol.* **2011**, *2*, 73. [CrossRef]
110. Singh, D.K.; Miller, C.M.; Orgel, K.A.; Dave, M.; Mackay, S.; Good, M. Necrotizing enterocolitis: Bench to bedside approaches and advancing our understanding of disease pathogenesis. *Front. Pediatr.* **2023**, *10*, 1107404. [CrossRef]
111. Hodzic, Z.; Bolock, A.M.; Good, M. The Role of Mucosal Immunity in the Pathogenesis of Necrotizing Enterocolitis. *Front. Pediatr.* **2017**, *5*, 40. [CrossRef]
112. Perrone, S.; Tataranno, M.L.; Negro, S.; Cornacchione, S.; Longini, M.; Proietti, F.; Soubasi, V.; Benders, M.J.; Van Bel, F.; Buonocore, G. May oxidative stress biomarkers in cord blood predict the occurrence of necrotizing enterocolitis in preterm infants? *J. Matern. Fetal Neonatal Med.* **2012**, *25* (Suppl. S1), 128–131. [CrossRef]
113. Perrone, S.; Tataranno, M.L.; Santacroce, A.; Negro, S.; Buonocore, G. The role of oxidative stress on necrotizing enterocolitis in very low birth weight infants. *Curr. Pediatr. Rev.* **2014**, *10*, 202–207. [CrossRef] [PubMed]
114. Ozdemir, R.; Yurttutan, S.; Sari, F.N.; Oncel, M.Y.; Erdeve, O.; Unverdi, H.G.; Uysal, B.; Dilmen, U. All-trans-retinoic acid attenuates intestinal injury in a neonatal rat model of necrotizing enterocolitis. *Neonatology* **2013**, *104*, 22–27. [CrossRef] [PubMed]
115. Bodkin, D.; Cai, C.L.; Manlapaz-Mann, A.; Mustafa, G.; Aranda, J.V.; Beharry, K.D. Neonatal intermittent hypoxia, fish oil, and/or antioxidant supplementation on gut microbiota in neonatal rats. *Pediatr. Res.* **2022**, *92*, 109–117. [CrossRef] [PubMed]
116. Mustafa, G.; Cai, C.L.; Bodkin, D.; Aranda, J.V.; Beharry, K.D. Antioxidants and/or fish oil reduce intermittent hypoxia-induced inflammation in the neonatal rat terminal ileum. *Prostaglandins Other Lipid Mediat.* **2021**, *155*, 106565. [CrossRef] [PubMed]
117. Cummings, J.J.; Polin, R.A.; AAP Committee on Fetus and Newborn. Oxygen targeting in extremely low birth weight infants. *Pediatrics* **2016**, *138*, e20161576. [CrossRef]
118. Askie, L.M.; Darlow, B.A.; Finer, N.; Schmidt, B.; Stenson, B.; Tarnow-Mordi, W.; Davis, P.G.; Carlo, W.A.; Brocklehurst, P.; Davies, L.C.; et al. Association Between Oxygen Saturation Targeting and Death or Disability in Extremely Preterm Infants in the Neonatal Oxygenation Prospective Meta-analysis Collaboration. *JAMA* **2018**, *319*, 2190–2201. [CrossRef]
119. Manja, V.; Saugstad, O.D.; Lakshminrusimha, S. Oxygen Saturation Targets in Preterm Infants and Outcomes at 18–24 Months: A Systematic Review. *Pediatrics* **2017**, *139*, e20161609. [CrossRef]
120. Stenson, B.J. Oxygen Saturation Targets for Extremely Preterm Infants after the NeOProM Trials. *Neonatology* **2016**, *109*, 352–358. [CrossRef]
121. Di Fiore, J.M.; Walsh, M.; Wrage, L.; Rich, W.; Finer, N.; Carlo, W.A.; Martin, R.J.; SUPPORT Study Group of Eunice Kennedy Shriver National Institute of Child Health and Human Development Neonatal Research Network. Low oxygen saturation target range is associated with increased incidence of intermittent hypoxemia. *J. Pediatr.* **2012**, *161*, 1047–1052.e1. [CrossRef]
122. Hagadorn, J.I.; Furey, A.M.; Nghiem, T.-H.; Schmid, C.H.; Phelps, D.C.; Pillers, D.-R.M.; Cole, C.H.; AvVIOx Study Group. Achieved versus intended pulse oximeter saturation infants born less than 28 weeks' gestation: The Aviox study. *Pediatrics* **2006**, *118*, 1574–1582. [CrossRef]
123. Dargaville, P.A.; Marshall, A.P.; Ladlow, O.J.; Bannink, C.; Jayakar, R.; Eastwood-Sutherland, C.; Lim, K.; Ali, S.K.M.; Gale, T.J. Automated control of oxygen titration in preterm infants on non-invasive respiratory support. *Arch. Dis. Child. Fetal Neonatal Ed.* **2022**, *107*, 39–44. [CrossRef] [PubMed]
124. Marshall, A.; Ladlow, O.J.; Bannink, C.; Lim, K.; Ali, S.K.M.; Gale, T.J.; Dargaville, P.A. Apnea-triggered increase in fraction of inspired oxygen in preterm infants: A randomized crossover study. *Arch. Dis. Child. Fetal Neonatal Ed.* **2024**, *109*, 81–86. [CrossRef] [PubMed]
125. Dobson, N.R.; Hunt, C.E. Caffeine: An evidence-based success story in VLBW pharmacotherapy. *Pediatr. Res.* **2018**, *84*, 333–340. [CrossRef] [PubMed]
126. Mayer, C.A.; Haxhiu, M.A.; Martin, R.J.; Wilson, C.G. Adenosine A2A receptors mediate GABAergic inhibition of respiration in immature rats. *J. Appl. Physiol.* **2006**, *100*, 91–97. [CrossRef]
127. Wilson, C.G.; Martin, R.J.; Jaber, M.; Abu-Shaweesh, J.; Jafri, A.; Haxhiu, M.A.; Zaidi, S. Adenosine A2A receptors interact with GABAergic pathways to modulate respiration in neonatal piglets. *Respir. Physiol. Neurobiol.* **2004**, *141*, 201–211. [CrossRef]
128. Bancalari, E. Caffeine for apnea of prematurity. *N. Eng. J. Med.* **2006**, *354*, 2179–2181. [CrossRef]

129. Schmidt, B.; Roberts, R.S.; Davis, P.; Doyle, L.W.; Barrington, K.J.; Ohlsson, A.; Solimano, A.; Tin, W.; Caffeine for Apnea of Prematurity Trial Group. Caffeine therapy for apnea of prematurity. *N. Engl. J. Med.* **2006**, *354*, 2112–2121. [CrossRef]
130. Rhein, L.M.; Dobson, N.R.; Darnall, R.A.; Corwin, M.J.; Heeren, T.C.; Poets, C.F.; McEntire, B.L.; Hunt, C.E.; Caffeine Pilot Study Group. Effects of caffeine on intermittent hypoxia in infants born prematurely: A randomized clinical trial. *JAMA Pediatr.* **2014**, *168*, 250–257. [CrossRef]
131. Davis, P.G.; Schmidt, B.; Roberts, R.S.; Doyle, L.W.; Asztalos, E.; Haslam, R.; Sinha, S.; Tin, W.; Caffeine for Apnea of Prematurity Trial Group. Caffeine for Apnea of Prematurity trial: Benefits may vary in subgroups. *J. Pediatr.* **2010**, *156*, 382–387.e3. [CrossRef]
132. Schmidt, B.; Roberts, R.S.; Davis, P.; Doyle, L.W.; Barrington, K.J.; Ohlsson, A.; Solimano, A.; Tin, W.; Caffeine for Apnea of Prematurity Trial Group. Long-term effects of caffeine therapy for apnea of prematurity. *N. Engl. J. Med.* **2007**, *357*, 1893–1902. [CrossRef]
133. Abu Jawdeh, E.G.; O'Riordan, M.; Limrungsikul, A.; Bandyopadhyay, A.; Argus, B.M.; Nakad, P.E.; Supapannachart, S.; Yunis, K.A.; Davis, P.G.; Martin, R.J. Methylxanthine use for apnea of prematurity among an international cohort of neonatologists. *J. Neonatal-Perinat. Med.* **2013**, *6*, 251–256. [CrossRef] [PubMed]
134. Dobson, N.R.; Rhein, L.M.; Darnall, R.A.; Corwin, M.J.; Heeren, T.C.; Eichenwald, E.; James, L.P.; McEntire, B.L.; Hunt, C.E.; Caffeine Study Group. Caffeine decreases intermittent hypoxia in preterm infants nearing term-equivalent age. *J. Perinatol.* **2017**, *37*, 1135–1140. [CrossRef] [PubMed]
135. Pradhap, K.; Solaiappan, M.; Ramya, S.; Muthukumaran, N. Effect of short course caffeine on recurrence of apnea of prematurity in preterm infants less than 32 weeks: A randomized controlled trial. *J. Neonatol.* **2024**, *38*, 375–381. [CrossRef]
136. Aldridge, A.; Aranda, J.V.; Neims, A.H. Caffeine metabolism in the newborn. *Clin. Pharmacol. Ther.* **1979**, *25*, 447–453. [CrossRef] [PubMed]
137. Koch, G.; Datta, A.N.; Jost, K.; Schulzke, S.M.; van den Anker, J.; Pfister, M. Caffeine citrate dosing adjustments to assure stable caffeine concentrations in preterm neonates. *J. Pediatr.* **2017**, *191*, 50–56.e1. [CrossRef] [PubMed]
138. Endesfelder, S. Caffeine: The story beyond oxygen-induced lung and brain injury in neonatal animal models—A narrative review. *Antioxidants* **2024**, *13*, 1076. [CrossRef]
139. Miller, M.J.; Carlo, W.A.; Martin, R.J. Continuous positive airway pressure selectively reduces obstructive apnea in preterm infants. *J. Pediatr.* **1985**, *106*, 91–94. [CrossRef]
140. Sreenan, C.; Lemke, R.P.; Hudson-Mason, A.; Osiovich, H. High-flow nasal cannulae in the management of apnea of prematurity: A comparison with conventional nasal continuous positive airway pressure. *Pediatrics* **2001**, *107*, 1081–1083. [CrossRef]
141. Lemyre, B.; Davis, P.G.; de Paoli, A.G. Nasal intermittent positive pressure ventilation (NIPPV) versus nasal continuous positive airway pressure (NCPAP) for apnea of prematurity. *Cochrane Database Syst. Rev.* **2002**, *1*, CD002272. [CrossRef]
142. Gizzi, C.; Montecchia, F.; Panetta, V.; Castellano, C.; Mariani, C.; Campelli, M.; Papoff, P.; Moretti, C.; Agostino, R. Is synchronised NIPPV more effective than NIPPV and NCPAP in treating apnoea of prematurity (AOP)? A randomised cross-over trial. *Arch. Dis. Child. Fetal Neonatal Ed.* **2015**, *100*, F17–F23. [CrossRef]
143. Zagol, K.; Lake, D.E.; Vergales, B.; Moorman, M.E.; Paget-Brown, A.; Lee, H.; Rusin, C.G.; Delos, J.B.; Clark, M.T.; Moorman, J.R.; et al. Anemia, apnea of prematurity, and blood transfusions. *J. Pediatr.* **2012**, *161*, 417–421.e1. [CrossRef] [PubMed]
144. Kovatis, K.Z.; Di Fiore, J.M.; Martin, R.J.; Abbasi, S.; Chaundhary, A.S.; Hoover, S.; Zhang, Z.; Kirpalani, H. Effect of Blood Transfusions on Intermittent Hypoxic Episodes in a Prospective Study of Very Low Birth Weight Infants. *J. Pediatr.* **2020**, *222*, 65–70. [CrossRef] [PubMed]
145. AlFaleh, K.; Al-Jebreen, A.; Baqays, A.; Al-Hallali, A.; Bedaiwi, K.; Al-Balahi, N.; AlGhamdi, A.; AlKharfi, T.; Alzahem, A. Association of packed red blood cell transfusion and necrotizing enterocolitis in very low birth weight infants. *J. Neonatal-Perinat. Med.* **2014**, *7*, 193–198. [CrossRef] [PubMed]
146. Slidsborg, C.; Jensen, A.; Forman, J.L.; Rasmussen, S.; Bangsgaard, R.; Fledelius, H.C.; Greisen, G.; la Cour, M. Neonatal risk factors for treatment-demanding retinopathy of prematurity: A Danish National Study. *Ophthamology* **2016**, *123*, 796–803. [CrossRef] [PubMed]
147. Lust, C.; Vesoulis, Z.; Jackups, R.; Liao, S.; Rao, R.; Mathur, A.M. Early red cell transfusion is associated with development of severe retinopathy of prematurity. *J. Perinatol.* **2019**, *39*, 393–400. [CrossRef]
148. Vliegenthart, R.J.; Ten Hove, C.H.; Onland, W.; van Kaam, A.H. Doxapram Treatment for Apnea of Prematurity: A Systematic Review. *Neonatology* **2017**, *111*, 162–171. [CrossRef]
149. Treluyer, L.; Zana-Taieb, E.; Jarreau, P.; Benhammou, V.; Kuhn, P.; Letouzey, M.; Marchand-Martin, L.; Onland, W.; Pierrat, V.; Saade, L.; et al. Doxapram for apnea of prematurity and neurodevelopmental outcomes at age 5–6 years. *Arch. Dis. Child. Fetal Neonatal Ed.* **2024**, *109*, F443–F449. [CrossRef]
150. Poppe, J.A.; Flint, R.B.; Smits, A.; Willemsen, S.P.; Storm, K.K.; Nuytemans, D.H.; Onland, W.; Poley, M.J.; de Boode, W.P.; Carkeek, K.; et al. Doxapram versus placebo in preterm newborns: A study protocol for an international double blinded multicentre randomized controlled trial (DOXA-trial). *Trials* **2023**, *24*, 656. [CrossRef]
151. Quitadamo, P.; Giorgio, V.; Zenzeri, L.; Baldassarre, M.; Cresi, F.; Borrelli, O.; Salvatore, S. Apnea in preterm neonates: What's the role of gastroesophageal reflux? A systematic review. *Dig. Liver Dis.* **2020**, *52*, 723–729. [CrossRef]
152. Terrin, G.; Passariello, A.; De Curtis, M.; Manguso, F.; Salvia, G.; Lega, L.; Messina, F.; Paludetto, R.; Canani, R.B. Ranitidine is associated with infections, necrotizing enterocolitis, and fatal outcome in newborns. *Pediatrics* **2012**, *129*, e40–e45. [CrossRef]

153. Ballout, R.A.; Foster, J.P.; Kahale, L.A.; Badr, L. Body positioning for spontaneously breathing preterm infants with apnoea. *Cochrane Database Syst. Rev.* **2017**, *1*, CD004951. [CrossRef] [PubMed]
154. Cramer, S.J.E.; Dekker, J.; Dankelman, J.; Pauws, S.C.; Hooper, S.B.; Te Pas, A.B. Effect of Tactile Stimulation on Termination and Prevention of Apnea of Prematurity: A Systematic Review. *Front. Pediatr.* **2018**, *6*, 45. [CrossRef] [PubMed]
155. Montealegre-Pomar, A.; Bohorquez, A.; Charpak, N. Systematic review and meta-analysis suggest that Kangaroo position protects against apnoea of prematurity. *Acta Paediatr.* **2020**, *109*, 1310–1316. [CrossRef] [PubMed]
156. Ouedraogo, P.; Villani, P.E.; Natalizi, A.; Zagre, N.; Rodriques, P.A.B.; Traore, O.L.; Gatto, D.; Scalmani, E.; Putoto, G.; Cavallin, F.; et al. A multicentre neonatal manikin study showed a large heterogeneity in tactile stimulation for apnoea of prematurity. *Acta Paediatr.* **2024**, *113*, 1519–1523. [CrossRef]
157. Bergeron, S.; Champoux-Ouellet, É.; Samson, N.; Doyon, M.; Geoffroy, M.; Farkouh, A.; Bertelle, V.; Massé, É.; Cloutier, S.; Praud, J.P. Effects of vanilla odor on hypoxia-related periodic breathing in premature newborns: A pilot study. *Arch. Pediatr.* **2024**, *31*, 369–373. [CrossRef] [PubMed]
158. Kesavan, K.; Frank, P.; Cordero, D.M.; Benharash, P.; Harper, R.M. Neuromodulation of Limb Proprioceptive Afferents Decreases Apnea of Prematurity and Accompanying Intermittent Hypoxia and Bradycardia. *PLoS ONE* **2016**, *11*, e0157349. [CrossRef] [PubMed] [PubMed Central]
159. Smith, V.C.; Kelty-Stephen, D.; Qureshi Ahmad, M.; Mao, W.; Cakert, K.; Osborne, J.; Paydarfar, D. Stochastic Resonance Effects on Apnea, Bradycardia, and Oxygenation: A Randomized Controlled Trial. *Pediatrics* **2015**, *136*, e1561–e1568. [CrossRef] [PubMed] [PubMed Central]
160. Henderson-Smart, D.J.; Osborn, D.A. Kinesthetic stimulation for preventing apnea in preterm infants. *Cochrane Database Syst. Rev.* **2002**, *2*, CD000373. [CrossRef] [PubMed]

MDPI AG
Grosspeteranlage 5
4052 Basel
Switzerland
Tel.: +41 61 683 77 34

Antioxidants Editorial Office
E-mail: antioxidants@mdpi.com
www.mdpi.com/journal/antioxidants

www.ingramcontent.com/pod-product-compliance
Lightning Source LLC
Chambersburg PA
CBHW040746250926
51850CB00007BB/1425